Construction Company Management

Construction Company Management will give readers a detailed understanding of the critical aspects of managing a successful construction company in a dynamic and complex construction business environment characterised by intense competition, supply chain disruptions and rapid changes in technology, regulations, client preferences and market conditions.

The book will introduce readers to different dimensions of construction company management. The topics covered reflect current business practices in the construction industry, including company strategy and business models, stakeholder management, contract management, resource management, risk management, knowledge management, company finance, digital innovation, organisational resilience and the regulatory environment. The book also includes much-needed discussions on ethics, integrity and professional standards and diversity, equity and inclusion in construction companies. It explores the opportunities and challenges relevant to construction company management in global contexts with the help of case studies from different regions of the world.

Providing a concise book on this essential subject, *Construction Company Management* serves both students and those educators who teach it in their built environment courses. Practitioners will find the theory-informed company management practices discussed in the book valuable and useful in their practical contexts.

Abid Hasan, PhD, FHEA, is a lecturer and Higher Degree by Research (HDR) coordinator at the School of Architecture and Built Environment, Deakin University. Dr Hasan holds a PhD in construction management from the University of South Australia (UniSA), a master of technology degree in construction technology and management from the Indian Institute of Technology Delhi (IIT-Delhi) and a bachelor of technology (Hons) in civil engineering from Aligarh Muslim University (AMU). He started his career with India's largest construction company, Larsen and Toubro Ltd. Then he worked at Hindustan Petroleum Corporation Limited (HPCL), a Fortune 500 and Forbes 2000 public sector company, for over four years before moving to Australia. Dr Hasan teaches both undergraduate and postgraduate construction management courses at Deakin and is a Fellow of Advance HE (UK). He has published more than 50 journal papers, book chapters and conference papers. He was the lead editor for the special issue on 'health and well-being' in the *Construction Innovation* journal. He regularly reviews manuscripts for leading construction management journals and serves as an organising and technical committee member on several conference committees. He has also served on the academic board of Deakin University and the education committee of the National Association of Women in Construction (NAWIC), Victoria chapter. He has co-authored another book titled *Occupational Diseases in the Construction Industry*.

Asheem Shrestha, PhD, is a senior lecturer and a course coordinator at the School of Architecture and Built Environment, Deakin University, where he teaches professional practise and construction company management. He was trained as a civil engineer and has 20 years of experience in construction practice, teaching and research. He earned his PhD from the University of Melbourne for examining risk allocation in public private partnerships, which also won him the CIOB award for Excellence in Building Research. His current research interest is in procurement of large infrastructure projects, construction economics and sustainable built environment. He has published more than 50 journal papers, book chapters and conference papers on various areas of construction management, and he currently reviews for top construction and engineering journals.

Kumar Neeraj Jha, PhD is a professor of civil engineering at the Indian Institute of Technology Delhi. He started his career with Larsen and Toubro Ltd. and was instrumental in the successful completion of many construction projects. He has published more than 150 papers in a number of international and national journals and conference proceedings. He has supervised 20 PhD students and more than 100 MTech students. His four books, *Construction Project Management, Formwork and Scaffold Engineering, Construction Safety Management,* and *Determinants of Construction Project Success in India,* are widely accepted as textbooks in different universities. He teaches courses in construction engineering and management. He regularly conducts a number of training programs for practitioners and academics and has been involved in several consultancy and sponsored research projects of national importance. He is a reviewer of all major journals in the area of construction engineering and management and is the editor-in-chief of the *Journal of Safety Engineering.*

Construction Company Management

**Abid Hasan, Asheem Shrestha and
Kumar Neeraj Jha**

Taylor & Francis Group

LONDON AND NEW YORK

First published 2025

by Routledge
4 Park Square, Milton Park, Abingdon, Oxon OX14 4RN

and by Routledge
605 Third Avenue, New York, NY 10158

Routledge is an imprint of the Taylor & Francis Group, an informa business

© 2025 Abid Hasan, Asheem Shrestha and Kumar Neeraj Jha

British Library Cataloguing-in-Publication Data
A catalogue record for this book is available from the British Library

Library of Congress Cataloging-in-Publication Data
Names: Hasan, Abid (PhD), author. | Shrestha, Asheem, author. | Neeraj Jha, Kumar, author.
Title: Construction company management / Abid Hasan, Asheem Shrestha, Kumar
    Neeraj Jha.
Description: New York, NY : Routledge, 2024. | Includes bibliographical references.
Identifiers: LCCN 2024007142 (print) | LCCN 2024007143 (ebook) | ISBN 9781032121123
    (hardback) | ISBN 9781032119564 (paperback) | ISBN 9781003223092 (ebook)
Subjects: LCSH: Construction industry—Management.
Classification: LCC HD9715.A2 H36 2024 (print) | LCC HD9715.A2 (ebook) |
    DDC 624.068—dc23/eng/20240512
LC record available at https://lccn.loc.gov/2024007142
LC ebook record available at https://lccn.loc.gov/2024007143

ISBN: 978-1-032-12112-3 (hbk)
ISBN: 978-1-032-11956-4 (pbk)
ISBN: 978-1-003-22309-2 (ebk)

DOI: 10.1201/9781003223092

Typeset in Times New Roman
by Apex CoVantage, LLC

Contents

Acknowledgements

We extend our heartfelt gratitude to the various individuals and organisations whose unwavering support has been instrumental in the creation of this book. We are also grateful to the researchers and practitioners whose work has advanced our knowledge of construction company management and provided a strong foundation for this book.

We extend our sincere thanks to Purva Gupta, V.P.S. Nihar Nanyam, Satya Priya, Sparsh Johari, Padam Narayan, Santu Kar and Bilal Ayub for helping us find the relevant resources and data crucial for writing some of the book chapters. Their invaluable support has been indispensable in shaping this book. We also acknowledge the contribution of the various agencies and organisations that provided data and case studies.

We are indebted to our respective employers, Deakin University (Australia) and IIT-Delhi, for furnishing us with the requisite resources. Their support has been pivotal in ensuring the successful completion of this book.

Last, but not least, we wish to convey our deepest gratitude to our families for their patience, understanding and unwavering support during the development of this book. Their continued love and support made this work possible.

Preface

Construction companies play a crucial role in our societies. They significantly contribute to the economy, offer employment and provide critical infrastructure and housing essential to our survival and a functioning community. However, construction companies operate in highly regulated and competitive markets with slim profit margins. They are exposed to several external risks in an increasingly uncertain and turbulent business environment defined by economic slowdowns and financial uncertainties, geopolitical threats, disasters, regulatory reforms, technological disruptions and supply chain issues. The resilience of construction organisations is continuously tested. As a result, companies that are unprepared to face various crises and disruptions and bounce back or bounce forward cease to exist. The construction industry has one of the highest rates of insolvencies in many countries.

Ensuring that a construction company properly functions and remains profitable is of critical importance to various stakeholders. Failed businesses in the construction industry have significant social and economic costs. They not only affect developers and investors but also severely affect other supply chain partners due to the pyramidical structure of interconnected contractual relationships, as each supply chain member relies on others to stay in business. Moreover, unfinished projects affect the financial condition, health and well-being of affected customers and slow the development of the entire community. Finally, company failure affects the livelihood of workers and owners and puts enormous pressure on their physical and mental health.

Therefore, developing a robust understanding of different facets of construction company management is imperative for students, researchers, practitioners and policymakers associated with the built environment. The book provides theoretical and practical knowledge of construction company management. It draws from a large body of research work and industry practice from different countries, making it essential reading for all involved or planning to get involved in managing a medium or large construction company. It aims to equip the reader with the fundamental knowledge necessary to understand the main dimensions of construction company management.

Abid Hasan
Asheem Shrestha
Kumar Neeraj Jha

1 Introduction

Abid Hasan, Asheem Shrestha and Kumar Neeraj Jha

1.1 Construction

Construction is an ancient human activity and can be traced back to the Stone Age or much before that, depending on the definition of what constitutes a construction activity. Construction activities became much more organised after the agricultural revolution when humans started to stay in one place and thus needed permanent shelters. Buildings and structures constructed several hundred years ago still exist in different countries. The long and continued history of construction and its continuous growth underscore its importance and relevance in the economy and society.

There is no one definition of what constitutes construction. In the literature, construction has been defined in different ways. The broader definition of construction identifies it as an economic activity that involves the entire construction process, from producing raw and manufactured building materials and components and providing professional services such as design and project management to executing the physical work on construction sites (Gruneberg, 1997). Meanwhile, the narrow definition of construction only considers the physical work carried out on the production site, such as new work, renovation, repair or extension of buildings, structures and other heavy constructions (Pheng & Hou, 2019). Between these two extremes, construction can be considered as a process including various activities from conception through design and procurement to execution (Pheng & Hou, 2019), which is also the view adopted in this book.

1.2 The Construction Industry

The construction industry is a major employer and significant contributor to a nation's gross domestic product (GDP) worldwide. It employs about 7% of the world's working-age population. Every year, more than $11 trillion is spent on construction-related goods and services, equivalent to 11% of the global GDP of $100 trillion, according to the World Bank (2022) data. Moreover, the global construction output is expected to grow by 42% or $4.5 trillion between 2020 and 2030 to reach $15.2 trillion or 13.5% of global GDP, with almost half of it attributed to growth in the Asia-Pacific region (Oxford Economics, 2021). The construction industry also greatly contributes to several other sectors that produce materials, equipment and services used in its production process. Therefore, in the modern era, the construction industry can be considered one of the most critical industries in the world. It affects the economy, the environment and society as a whole. It touches everyone's daily lives and serves almost all other industries that need social, transportation and utility infrastructure (World Economic Forum, 2016).

DOI: 10.1201/9781003223092-1

However, on the other hand, the construction industry exerts several negative impacts, especially on the environment. It contributes significantly to resource depletion, energy consumption, waste generation and greenhouse gas emission. It is known for project delays, cost overruns, low productivity and wasting a large sum of resources, including public money and natural resources. Moreover, it has a poor image and reputation for a male-dominated work culture; lack of diversity and inclusiveness; unsafe and unhealthy work conditions causing a large number of injuries, fatalities and mental health disorders; unethical practices leading to quality issues and corruption; lack of innovation; and slow uptake of technology.

1.3 Construction Companies

Construction companies operate in a business environment challenged by low profit margins, high fragmentation, multiple stakeholders, complex and dispersed supply chains and a highly regulated sector, exposing them to several risks over which they have little or no control (World Economic Forum, 2016). Low barriers to entry in the construction industry create a saturated market with intense competition (Tripathi et al., 2021). Moreover, many construction companies are grappling with high inflation, high interest rates, volatility in material prices, shortage of skilled labour and increasing labour costs impacting their day-to-day operations and overall business, especially in the post-COVID-19 era.

As a result, the risks of failure in the construction business are quite high, as reflected in data on insolvencies in the construction industry in many countries (Coggins et al., 2016). The data from the Australian construction industry show that over 25% of the largest 200 residential builders recorded an operating loss in the year 2021–2022. Moreover, construction company insolvencies accounted for close to 30% of all company insolvencies (Reserve Bank of Australia, 2022). Similar upward trends in the number of construction companies going bankrupt can be seen in recent years in other countries, such as the United States and the United Kingdom. For instance, the construction industry in England and Wales experienced the highest number of insolvencies (3,949, 19% of cases with industry captured) in the 12 months ending Q3 2022 (The Insolvency Service, 2022). UK government data further show that the construction industry usually has the highest quarterly number of insolvencies of any industrial grouping (The Insolvency Service, 2022).

There is also considerable regulatory and societal pressure on companies operating in the construction industry to improve their business decisions, practices and processes to address various challenges and provide better value to stakeholders. The stakeholders demand several improvements, such as the adoption of sustainable construction materials and practices, waste reduction, better project outcomes, a safe and healthy work environment, higher productivity and diverse and inclusive workplaces. Rapid technological changes are also forcing construction companies to change their work practices regularly to stay competitive. However, the industry is known for its slow uptake of technology and innovation and is struggling to improve its productivity.

How well construction companies respond to various existing and emerging threats or challenges and meet stakeholders' expectations would determine their future and shape the role and contribution of the industry and its place in the modern world. Therefore, developing knowledge of contemporary issues and various aspects of successful construction company management is crucial for students, researchers, practitioners and policymakers associated with the built environment.

1.4 Construction Company Classification

The construction industry is characterised by its extensive range of projects and the variety of companies participating in its operation. It encompasses all companies or organisations professionally engaged in the construction process, such as those providing consultancy and managerial services and those carrying out execution work on construction sites (Colean & Newcomb, 1952; Ofori, 1990).

Construction companies primarily construct buildings, rail and road infrastructure, transmission lines, oil and gas pipelines, etc. In addition to new construction, many construction companies carry out demolition, repairs and renovations and install utilities. Moreover, construction companies vary from individual contractors handling small, simple works to large corporations managing multi-billion-dollar sophisticated developments. Small businesses, such as tradespeople, provide services directly to consumers or work for other building and construction businesses and the government. In comparison, large businesses are engaged in large construction and infrastructure projects.

Some construction companies specialise in one type of work (e.g., concreting or demolition) or undertake projects in a particular subsector, while others have a more diversified portfolio of projects. Therefore, by type of product, the construction industry consists of companies specialising in residential buildings, industrial and commercial buildings and infrastructure or civil engineering works (Pheng & Hou, 2019). Additionally, construction companies may have a localised presence, with all business activities concentrated in a particular region or state, or have projects at different locations spread over the country or in several countries. Geographical location is an essential consideration for construction companies deciding their reach and presence as they need to consider the availability and transportation costs of project resources to the construction site location. As a result, most construction companies are usually market-located (Gruneberg, 1997).

In short, the construction industry has a diverse mix of companies in terms of size, type, nature of business, scale of operations and geographical reach and presence. The structure of the construction industry can be quite different among countries depending on various social and economic factors and government policies. A simple construction company classification based on size, roles and ownership structure is presented here.

1.4.1 Based on Size

Construction companies vary in size, each with distinct operational characteristics. Typically, companies can be categorised as small, medium or large based on the number of employees or the annual turnover. For example, companies with staff under 20 in Australia are generally considered small, 20–199 are medium-sized, and those with over 200 staff are considered large. In the United States, specific definitions of small and medium-sized enterprises (SMEs) are based on the industry they operate in. SMEs are typically defined as businesses that have fewer than 500 employees.

Staff headcount and financial ceilings determine enterprise categories in the European Union nations. SMEs comprise enterprises that employ fewer than 250 persons and have an annual turnover not exceeding EUR 50 million and/or an annual balance sheet total not exceeding EUR 43 million. Within the SME category, a small enterprise employs fewer than 50 persons and has an annual turnover and/or annual balance sheet total of less than EUR 10 million.

A microenterprise employs fewer than 10 persons and has an annual turnover and/or annual balance sheet total that does not exceed EUR 2 million (European Commission, 2007).

The classification range may differ in other countries as no universally accepted definition of SMEs or large companies exists. The classification of a company's size can also be influenced by its role in the industry and the nature of its services. For instance, a construction contractor with 20 employees might be classified as a small company. In contrast, an architectural firm with the same number of staff could be considered medium-sized or, in some contexts, even large.

The typical construction businesses in many countries are very small in size. For example, more than 50% of construction businesses in Australia do not have any employees at all. Small businesses employing fewer than 20 people account for the overwhelming majority (98.5%) of the Australian construction industry (Master Builders Australia, 2020). Generally speaking, small construction companies are local businesses that typically specialise in a particular trade but exhibit lower levels of managerial capacity and resources compared to medium-sized and large firms. They usually operate under a single manager who oversees various departments, and there are no separate departments for managing human resources (HR), legal and occupational health and safety (OHS) duties or finances. In many of these small companies, the managing director not only leads the company but also holds the majority shareholder position, indicating a close connection between ownership and daily management. Some small companies, such as environmental or engineering consultancies, may provide highly specialised and technical work or services.

Medium-sized companies in the construction sector typically experience an increased volume of turnover compared to smaller companies, reflecting their growth and market presence. Their organisational structure is more segmented into functional departments. Larger companies in the construction industry have more staff and annual revenue than small and medium-sized firms. They are often led by a managing director who oversees the overall strategic direction. Their structure is characterised by decentralised divisions, such as engineering, housing, international operations and regional sectors, enabling specialised focus within each area. Large companies possess considerable resources and a broad range of skills and aim to grow their market share through various strategies, including organic growth and acquisitions.

1.4.2 *Based on Ownership*

Construction companies may also be classified based on the company's ownership structure. Sole proprietorships are the simplest form of business where the company is owned and operated by a single individual. In this structure, the owner has complete control over the operations and bears unlimited personal liability. This type of company is typically limited in its ability to raise funds. It is most suitable for small-scale projects or specialised trades such as plumbing, electrical work or carpentry.

Partnerships refers to businesses owned by two or more individuals or firms that share profits and liabilities. They offer shared decision-making and responsibility, bringing together a broader skill set compared to sole proprietorships. However, like sole proprietorships, partners are personally liable for business debts. Partnerships are commonly seen in professional services within the construction industry, like design or engineering consultancies.

Public companies are corporations whose shares are traded publicly on stock exchanges. They have access to significant capital through public investment and are subject to stringent regulatory requirements. The shareholders are the owners of the company and appoint a board of directors to run its operations. These companies are generally large and are best suited for large-scale projects due to their ability to raise funds and greater resources to deliver larger and more complex projects.

Private companies are ones that do not sell their shares to the public. These companies are privately owned by an individual or a limited number of shareholders. They can either be medium- or large-sized companies. Private companies bear the full liability for their investment; however, the owners maintain operational authority and have fewer obligations to disclose their financial and operational information to the public.

Limited liability companies (LLCs) are companies that are established to take advantage of the benefits of limited liability. In LLCs, the owners' liability is limited to the money invested, which means that their personal assets are protected from company debt or liabilities. From a taxation standpoint, LLCs are often more favourable for owners. There is also the option to form limited liability partnerships, which blend the features of partnerships with the benefits of limited liability, providing an alternative structure for owners seeking both collaboration and protection.

1.4.3 Based on the Role

Each type of company plays a distinct role in the construction industry. There are four principal groups of companies that operate in the construction industry (Flanagan, 2022). There are (a) contractors and subcontractors, (b) consultants and designers, (c) manufacturers and suppliers and (d) ancillary service providers.

Contractors, or primary contractors, are the party with the biggest responsibility in the construction process. They can range from smaller companies undertaking simple tasks to large corporations undertaking complex projects. They may have different ownership structures. Sole proprietorships and smaller firms generally take on smaller projects or work as subcontractors or trades, frequently on behalf of larger primary contractors, to provide specialised works in specific areas of construction. The primary contractor is generally a medium-to-large enterprise. They are the focus of this book, and the term *construction company* refers to contractor organisations in this book.

Consultants and designers offer a range of professional services in the construction industry, including architectural design, civil engineering, mechanical and plumbing systems, etc., as well as specialised services in cost estimation and project management. Traditionally, these firms were smaller and were limited to providing only one specific type of service; however, now, many of these firms have grown into large multinational companies offering a range of services globally. They also represent the interests of clients or owners in many construction projects and act on their behalf to ensure the successful delivery of the project.

Manufacturers and suppliers include companies that are not directly involved in projects but are an integral part of the construction industry. They are responsible for producing and delivering a wide range of products used in construction. The quality, innovation and technological advancements in the manufacturing process directly impact the efficiency, sustainability and safety in construction projects.

Finally, companies that offer ancillary services play a diverse and vital role in supporting the construction industry. These companies provide a wide array of services, including logistics, legal advice, information technology support, financial services and security. Their expertise is not confined solely to the construction sector, as many of these firms also cater to a variety of other industries.

1.5 Construction Company Management – Points of Departure

Construction companies operate very differently and face different opportunities and constraints compared to companies from many other industries (e.g., manufacturing). Some of the differences are discussed below.

1. It generally takes significant time and resources to produce the final product in the construction industry. A house may take a few months to a year to build, whereas a large project such as an airport, highway or refinery may stretch over several years, consuming resources worth billions of dollars. As a result, construction companies, suppliers and clients are exposed to severe financial and other risks over the lengthy construction period.
2. The manufacturing cycle, or the time required to produce the product from raw materials, is significantly longer in the construction industry. While organisations in the manufacturing sector can produce millions and billions of their products every year, it is challenging or nearly impossible for construction companies, even large ones, to execute thousands of construction projects simultaneously. For instance, the Volkswagen group's global vehicle production in the fiscal year 2022 was 8.72 million. Similarly, more than 200 million units of iPhones were sold in 2022. In contrast, most construction companies only engage in delivering a few projects at a given time, and their business depends on those projects. Therefore, construction products are not commonly produced in bulk (except for the prefabricated elements, panels, and structures).
3. In many sectors (e.g., manufacturing), companies determine the price of their products based on market conditions and other considerations and know the exact cost of production. In contrast, the client determines the price of the final product in the construction industry before the project starts (Seymour, 2019). To win the tender, construction companies must commit to a cost (e.g., lump sum contracts) without knowing with absolute certainty how much the final cost would be. Also, they must adjust their profit margins to be competitive. As a result, many construction companies struggle to obtain a reasonable return or profit on their investment.
4. Entry into and exit from the construction industry is easier than in manufacturing industries and often unrestricted. Any person with know-how or construction skills can enter the industry. Also, the need for capital requirement is low. For example, a person with carpentry, plumbing or electrical skills and relevant qualifications can become a sole trader and start a company. Similarly, it is easier for construction companies to diversify and undertake projects in different sectors if they can acquire relevant resources and hire skilled staff.
5. Construction companies work with different stakeholders in different projects, and the construction site (i.e., production space) is often shared with many other teams and companies with different cultures, policies and strategies, work ethics and technological maturity. As a result, conflicts are common, and it is challenging to implement changes or improvements if other parties are not on board. In contrast, companies in other sectors have better control over their workers and workspaces if the production occurs in a factory environment.
6. The labour force many companies employ in construction projects is temporary or on a hire basis. In contrast, manufacturing companies often work with regular or permanently employed workers. As a result, they experience better labour productivity by investing in the training, development and welfare of their employed workers. On the other hand, construction companies are reluctant to invest resources in training temporary workers due to the transient nature of work and their relationship with the company.

7. The final product in the construction industry needs to be built at the point of its utilisation, although offsite manufacturing has the potential to change it altogether in the future. Consequently, locational or local conditions (e.g., bad weather) and the availability of local resources (e.g., local labour force, local subcontractors and suppliers, etc.) influence the construction process in ways typically not observed in a controlled factory production environment. Also, it restricts the ability of a company to take projects anywhere or everywhere, whereas a manufacturing company can produce products in one location and sell them in many.

8. The client's needs drive the growth in demand in the construction industry, and construction companies can do very little to influence them through marketing and advertisements. In contrast, manufacturing companies can grow more quickly by driving demand through persuasive marketing. People may purchase several products they do not require, but the same cannot be said about construction projects, as they are based on the client's careful need and benefit considerations. Each construction product is only constructed after the client decides to procure it (Pheng & Hou, 2019). Therefore, location, number and type of projects or future work supply are determined by the clients, with very little influence from construction companies. Construction companies locate or operate where there is demand and adjust themselves to follow market trends (Seymour, 2019).

9. Local construction companies often have a competitive advantage as many clients prefer them because they are cheaper and have more expertise in the local conditions. Moreover, they have a better network with the local stakeholders, such as material and labour suppliers and regulatory authorities. As a result, companies do not experience high business threat levels in their local markets. In comparison, international companies face several geopolitical and unknown risks. However, clients will most likely prefer other companies if the local company lacks the necessary resources or expertise.

10. Due to the limited capacity of construction companies to deliver several projects, many companies can co-exist in markets with sufficient demand. However, low entry barriers often result in the presence of a large number of companies in a given market competing for new projects. For instance, it is not uncommon to find several companies offering residential construction services in one region. Intense competition means lower profits on projects, which has severe consequences for the long-term sustainability of construction companies.

11. Construction products are highly durable and last several years, even for generations (e.g., highways, bridges, airports, etc.). Many people build a house only once in their lifetime but may purchase the same electronic, electrical, or mechanical products every few years due to shorter product life. As a result, the work comes from either new clients or the same clients investing in new projects. Construction companies must continuously look for new clients and areas for business continuity and growth.

The points discussed above distinguish construction companies from companies operating in other industries, such as manufacturing. The unpredictable production environment, large number of risks, complex stakeholder relationships, unique characteristics of the product, as well as variable demand and supply and the dispersed and fragmented supply chain require a different perspective on the management of construction companies, often missing from literature and books on company management written for companies operating in other sectors. To this end, the present book aims to provide context-specific knowledge on various aspects of construction company management.

1.6 Structure of the Book

The book consists of 13 chapters, including this introduction chapter. These chapters cover a wide range of topics, from basic definitions and fundamental concepts to more advanced knowledge. Practical examples from construction companies operating in different countries are used throughout to illustrate the various dimensions of construction company management.

Chapter 2. Ethics, Integrity and Professional Standards

Chapter 2 focuses on the topic of ethics in construction, discussing the critical and often overlooked aspects of the industry concerning ethics, integrity and professional standards. This chapter unpacks the vital role of ethics in construction, an industry frequently perceived as highly unethical. Unethical practices in construction have detrimental effects on both the economy and society. Thus, it becomes imperative for construction companies to actively participate in reshaping this perception through ethical conduct, which, as research suggests, also leads to better business prospects.

Starting with an overview of ethics as a philosophical framework, the chapter moves on to discuss ethics in professional and business contexts. It addresses why a good understanding of ethical principles is essential for industry professionals, not just as a theoretical concept but as a practical tool for navigating complex business environments. In its second part, the chapter sheds light on the underlying reasons for unethical behaviours in the industry and, using examples of practical scenarios, discusses various ethical challenges construction companies and professionals face. By highlighting the impacts of these issues, the chapter stresses the need for a coordinated effort to resolve them, showing how they affect not only society but also the industry's reputation and performance.

The final section offers a roadmap for fostering ethical behaviour within construction companies. It emphasises the importance of leadership culture in setting ethical standards, the role of ethics education in building awareness and the implementation of codes and practices to guide decision-making. This part of the chapter is particularly important as it provides actionable strategies for companies to integrate ethical considerations into their core operations and external dealings. It delves into the ways leadership can influence and inspire ethical conduct throughout the organisation, where ethical values are exemplified at the highest levels. It also discusses the significance of continuous ethics education, highlighting how training and awareness programs can instil a deeper understanding of ethical issues among employees. Furthermore, it examines the practical application of ethical codes and practises, illustrating how these tools can be effectively used to navigate ethical dilemmas and reinforce a robust ethical culture.

Chapter 3. Stakeholder Management

Chapter 3 considers a crucial aspect of construction company management; that is, stakeholder management. In the dynamic and increasingly complex social, cultural, political and economic environments, construction companies must be able to manage their stakeholders and work with them to achieve their organisational objectives. Construction companies rely on many actors and supply chain partners, such as project clients, suppliers, subcontractors, regulatory bodies, communities and employees, to deliver construction projects and achieve their business goals. Often, these stakeholders have varying and sometimes competing interests, expectations, desires and needs vis-à-vis the company. Businesses that ignore stakeholders and their needs struggle to survive in the long run and eventually fail. Successful companies invest considerable resources in understanding their stakeholders' needs and expectations, both existing and emerging, through regular communication and engagement. Moreover, they closely work with them to create value for all stakeholders, not just shareholders. Therefore, understanding and responding to stakeholders' needs is essential to managing a successful construction company.

The chapter first introduces the concept of stakeholders to the readers, emphasising the difference between shareholders and stakeholders and why construction companies should focus on stakeholders, not just shareholders. Next, it discusses the stakeholder identification and analysis process. It provides knowledge on stakeholder identification, analysis and engagement processes and communication, drawing on research and practical examples. Mistakes or errors at any stage of stakeholder management can be costly and detrimental to the organisation. Therefore, the chapter argues for the need to carefully and thoroughly undertake various stakeholder management processes, dedicating sufficient organisational resources and involving relevant individuals and teams. The chapter then offers practical examples and insights into underlying principles of how construction companies can take a multi-stakeholder value creation approach. Construction companies should not prioritise the needs of a particular stakeholder at the expense of others, as such an approach can affect the business's sustainability due to its reliance on a complex and interconnected network of stakeholders. Finally, the chapter discusses the importance of building a solid corporate reputation and identity in stakeholder management. It also provides a list of questions to be asked while developing a stakeholder management plan to guide the process.

Chapter 4. Bidding and Contracts

Chapter 4 presents a detailed overview of the bidding process and the various types of construction contracts in the construction industry from the perspective of construction companies. For construction companies, bidding marks the critical stage where they first engage with a project. This is then followed by contract negotiation and finalisation before the construction work begins. These pre-construction stages are crucial for construction companies, where bidding strategies and decisions, along with the right type of contract, determine the project's financial viability and operational success. A well-informed bidding strategy is key to securing profitable projects, while poor bidding can lead to significant losses. Similarly, the risks of contractual failures can lead to not only project failures but also far-reaching impacts on the financial health and reputation of construction companies. Industry analyses consistently show a high correlation between project success and well-managed bidding and contract processes.

This chapter is structured to provide a comprehensive overview of these critical stages in the construction process. It begins by examining the bidding process in detail, exploring the rationale behind why companies choose to bid on certain projects and not others. It provides insights into the strategic considerations involved, including project feasibility, potential profitability and alignment with the organisational capabilities and goals. The chapter then discusses various types of contracts utilised in the construction industry. This section offers an overview of different contract models and their suitability for different types of projects. Understanding these contracts is essential for construction firms to ensure that they enter into agreements that best suit the project's nature, scope and risk profile. Finally, the chapter focuses on the critical aspect of risk allocation in contract negotiation. This part explores how risks are allocated between the contracting parties to ensure that responsibilities are clearly defined and managed throughout the project. This chapter offers valuable insights, equipping readers with the knowledge needed to understand these complex processes.

Chapter 5. Strategic Management

Chapter 5 covers the topic of strategic management for construction companies. While strategic management is a well-established concept in general management literature, and many larger and more successful companies exhibit a proficient understanding and application of these principles, its exploration within the construction industry remains relatively underdeveloped compared to other sectors. The construction industry presents its own unique set of challenges, making the application of strategic management both distinct and crucial. Many construction companies fail every year due to a lack of strategic planning. A thorough understanding of strategic management in this context is essential, as it extends the focus from mere short-term project management to a broader, long-term vision for the company's growth and sustainability. This chapter aims to bridge this gap, offering insights and strategies specifically applicable to the nuanced environment of the construction industry.

The chapter presents the fundamental strategic management theories and discusses how they apply to construction companies. The chapter emphasises current trends and developments within the construction industry, highlighting how companies can adapt and maintain their competitiveness in an increasingly challenging and competitive market. Drawing from management literature, some of the core strategies and strategic tools are explained in the construction context. Readers will gain a comprehensive understanding of the application of the fundamental principles of strategic management.

Chapter 6. Business Models

Chapter 6 discusses business models from the perspective of construction companies. Business models are a fundamental component of any successful enterprise, and they have become particularly relevant in today's dynamic and evolving business landscape. In recent times, business models have gained a lot of attention within the construction industry. Yet, there is often a lack of clarity among industry professionals regarding their definition and application. Business models act as functional tools that help companies realise their strategic objectives. They are particularly useful for leveraging new market opportunities and mitigating the effects of disruptive technologies.

This chapter highlights the pressing need for construction companies to adapt and redefine their business models in response to the rapidly changing market. It begins by addressing common misunderstandings surrounding the concept of business models. It provides a clear, well-rounded definition, drawing from various economic and management theories. This foundational understanding is crucial for understanding the unique application of business models within the construction context, setting the stage for a deeper analysis.

The next part of the chapter then discusses the practical application of business models in construction companies. By examining a variety of international examples, the chapter offers a window into the diverse approaches and practices across different geographical landscapes and markets. This global perspective helps us understand how business models can be tailored and executed effectively in various settings. The focus of the chapter is also on business model innovation, particularly in two critical areas: technological innovation and sustainability. This section emphasises the importance of both aspects in reshaping the construction industry's future. Through an in-depth examination of current research and developments, the chapter sheds light on their impact and potential. The final part of this chapter is marked by three case studies from successful companies in different countries. These real-life examples illustrate the effective integration of business models in the construction industry and provide practical insights and lessons for construction companies aspiring to adapt and excel in this rapidly evolving industry.

Chapter 7. Resource Management

Chapter 7 discusses resource management issues and strategies from an organisational perspective. *Resource management* is the process of planning, organising, directing and controlling the flow of resources within an organisation for use according to the defined goal or output. Resource management has specific objectives to be fulfilled to contribute to the fulfilment of the business objectives.

Proper resource management ensures that processes and techniques are in place to provide all the necessary resources for construction companies to complete a project or meet business objectives. Since one of the primary objectives of a construction company is to earn a profit, resource management also aims for cost reduction with various strategies like optimisation, negotiation, process improvement, etc., as resources constitute a major part of the total cost. The chapter provides information on how construction companies can efficiently use resources such as materials, plant and equipment and HR and maintain agility and the ability to pivot faster during supply chain disruptions. It also discusses various HR issues and practices relevant to modern enterprise management.

Chapter 8. Financial Management

Chapter 8 is on financial management, which addresses a critical aspect of construction company management; that is, management of financial resources. This chapter focuses on the role of financial management in the survival and success of construction companies. Financial management is crucial for construction companies as it ensures effective allocation and monitoring of financial resources, which is critical for managing the unique risks and complexities of construction projects. It also enables firms to make informed decisions, balancing costs and revenues to maintain profitability and ensure long-term viability.

The chapter is methodically structured to cover several key areas of construction finance. The complexities inherent in the industry's payment structures, funding mechanisms and financial risks are discussed, and the chapter argues that many construction companies face failure due to financial mismanagement and a lack of sound financial planning. It begins by exploring the role of financial management in construction, detailing how it is integral to both short-term project management and long-term strategic planning. This section explains why financial decisions are important and how they impact the overall health and viability of a construction company. Following this, the next part of the chapter focuses on the sources of funding for construction projects. It delves into the complexities of the diverse financing alternatives accessible to organisations. This is followed by a discussion on the accounting function in construction companies, highlighting the role of financial statements. In doing so, the methods and technical processes are explained. A critical component of the chapter is the discussion on financial analysis, with a specific focus on cash flow management and financial performance measurement. The role of financial ratios is discussed, and insights into effective financial planning and monitoring are offered, explaining how these practices can help companies manage their financial health, anticipate challenges and capitalise on opportunities. Finally, using a detailed case study of a failed company, the chapter illustrates how the concepts discussed in the chapter apply in real life, showcasing how poor financial strategies and missteps can lead to the financial collapse of a construction company.

Chapter 9. Risk Management

Chapter 9 examines the risk management process and practices for construction companies. While risk management is a fundamental aspect of managing any business, in the construction industry, the emphasis is typically placed on managing risks at the project level. Construction companies face complex challenges that require a robust framework, as they bear the dual responsibility of managing risks at both the project and enterprise levels. The inherent risks of construction projects, with their myriad of factors and variables, mean that firms must address the cumulative risks from all their ongoing projects, requiring sophisticated and comprehensive risk management approaches. This chapter highlights the often-overlooked area of company-level risk management, highlighting its importance alongside project-specific risk management strategies.

This chapter commences with a clear exposition of what risks entail for construction companies and elaborates on the processes involved in effectively managing these risks. Understanding the nature of these risks and the strategies to mitigate them is crucial for construction companies to navigate the complex environment in which they operate. Following this, the chapter provides a thorough review of the statutory and regulatory norms that govern risk management in the construction sector. This segment is essential for companies to ensure compliance and to understand the legal framework within which they must operate their risk management strategies. The final section of the chapter provides an in-depth case study of one of India's largest construction conglomerates. This case study illustrates the conglomerate's risk management policies and frameworks and demonstrates how these strategies have been pivotal in their success in domestic and international arenas. This real-world example provides valuable insights and lessons for other construction companies looking to enhance their risk management practices and achieve similar success.

Chapter 10. Digital Transformation

Chapter 10 introduces the readers to communication, information management and collaboration issues in the construction industry and how construction companies can use digital technologies to transform their business practices and achieve efficiency. Construction companies are considered less technology savvy and usually lag in technology adoption, but that needs to change given the enormous opportunities and benefits affordable and accessible technologies can provide. Construction teams and organisations must collaborate, communicate effectively and access and share accurate information on time despite working in a heterogeneous, highly fragmented and project-oriented business environment.

Digital transformation of construction companies can remove information and communication management inefficiencies and help them achieve productivity gains and cost reductions in an information-intensive construction business environment. The chapter discusses a few examples of digital technologies in the context of the construction industry. These include mobile technology; cloud computing; building information modelling; augmented, virtual and mixed realities; digital twins; blockchain; unmanned aerial vehicles; etc. It argues that digital transformation in construction companies would involve not only adopting digital technology but also coordinated changes in the existing processes to integrate them into all business areas and bring fundamental changes to how organisations operate and deliver value to different stakeholders. The chapter discusses technology use, conditions of usage and users and how integrating technology, people and processes is guided by the constraints imposed by the broader contexts of project, organisation, stakeholders and regulatory environments. An in-depth understanding of the critical elements of digital transformation from the perspectives of technology, people, processes and the broader context is necessary for successful digital transformation of construction companies. Finally, the chapter reflects on digital maturity and some of the tools used to measure the digital maturity of companies and highlights data privacy challenges, cybersecurity concerns and legal considerations that come with digital transformation.

Chapter 11. Diversity, Equity and Inclusion

Chapter 11 discusses a critical issue that has remained a persistent problem in the construction industry worldwide; that is, gender inequality. Lack of diversity, equity and inclusion (DEI) has severe negative implications for construction companies grappling with skills shortages and productivity challenges worldwide. Unfortunately, despite several advancements made by other sectors to achieve DEI, the construction industry remains a male-dominated workplace where women are treated differently because of their gender. The benefits of a diverse and inclusive workplace are well known, but construction companies have failed to harness them. The low participation of women and horizontal and vertical segregation in their employment in different construction roles and occupations must be addressed. Industry data show the overrepresentation of women in low-paid administrative jobs and underrepresentation in leadership, management and site-based jobs in most construction companies worldwide.

The chapter helps readers understand the drivers of the lack of DEI and the systemic causes. It discusses the issues of biased recruitment and promotion policies, poor work culture and rigid employment practices, discrimination and harassment, gender stereotypes, gender pay gap and OHS concerns that discourage women from joining or staying in the industry and building a successful career. The chapter also discusses several actions and initiatives at the policy, organisational and workplace levels taken by governments and organisations in different countries to achieve DEI in the construction industry. It advocates that mainstreaming gender perspectives and attaining the goal of DEI would require both top-down and bottom-up approaches – government policies and legislation, organisational strategy, policies, procedures and processes, procurement strategy, recruitment, career growth support and retention strategies, employment and pay conditions, decision-making processes, engagement and dialogue, etc. A sustainable change in DEI would require intervention and actions at different operational, functional and policy levels.

Chapter 12. Knowledge Management

Chapter 12 examines the concepts of knowledge and knowledge management and highlights the significance of proper knowledge management for project-based construction companies. It focuses on both explicit and tacit knowledge. Operating in an information-rich environment, construction companies rely heavily on the knowledge of employees, supply chain partners and other stakeholders to successfully complete various project activities. If not managed properly, the knowledge that resides in employees' minds and developed through their years of experience performing various activities and interacting with other stakeholders can be lost if they leave the organisation. Therefore, knowledge is an important organisational asset and must be managed effectively by construction companies to create and maintain a competitive advantage.

The chapter discusses four knowledge management processes (creation, storage and retrieval, transfer and application) in the context of construction companies and elaborates on them with the help of lessons learned as an example of knowledge management practice widely adopted by construction companies worldwide. It also discusses the role of employees, processes and technology in knowledge management. In particular, it argues that a knowledge-based culture supported by top management, technology and knowledge governance is required to improve knowledge management in construction companies. The chapter also introduces readers to benefits, challenges and best practices concerning knowledge management in construction companies.

Chapter 13. Organisational Resilience

Chapter 13 focuses on organisational resilience. Resilience is critical for the survival of construction companies that operate in a dynamic business environment and are exposed to both slow or abrupt short- and long-term shocks resulting from changes in the regulatory environment, market dynamics, market competition, disruptive technologies and human-caused and natural disasters or pandemics like COVID-19. Construction companies need to build resilience to be able to adapt and change their strategy quickly in a constantly changing business environment.

The community relies on the effective functioning of the construction industry to provide essential services, employment and economic growth. Resilient construction companies are vital for resilient communities. Since construction is one of the largest industries and employers in many countries and construction companies provide critical infrastructure to communities, the lack of resilience in construction companies and business failures leave unfinished jobs and unpaid workers, suppliers and subcontractors, severely affecting the communities they serve. The insolvencies in the construction industry have broader implications for the sector, society and the economy. The chapter discusses several current issues that threaten the resilience level of construction companies. These include financial vulnerabilities, supply chain disruptions, skill shortages, poor OHS, unsustainable practices and technological disruptions. It also provides guidance on how construction companies can develop resilience and address some of the concerns raised in the chapter, drawing on broader literature and examples of practices from construction and other sectors.

References

Coggins, J., Teng, B., & Rameezdeen, R. (2016). Construction insolvency in Australia: Reining in the beast. *Construction Economics and Building, 16*(3), 38–56.

Colean, M. L., & Newcomb, R. (1952). *Stabilizing construction: The record and the potential* (1st ed.). McGraw-Hill.

European Commission. (2007). *Micro-, small- and medium-sized enterprises: Definition and scope* (Document 32003H0361). https://eur-lex.europa.eu/legal-content/EN/LSU/?uri=CELEX:32003H0361

Flanagan, R. (2022). The race to the future for the construction sector. In R. Best & J. Meikle (Eds.), *Describing construction* (pp. 276–293). Routledge.

Gruneberg, S. L. (1997). *Construction economics: An introduction*. Macmillan.

The Insolvency Service. (2022). *National statistics: Commentary – Company insolvency statistics July to September 2022*. https://www.gov.uk/government/statistics/company-insolvency-statistics-july-to-september-2022/commentary-company-insolvency-statistics-july-to-september-2022

Master Builders Australia. (2020). *Facts & stats on how building supports the economy*. https://masterbuilders.com.au/facts-stats-on-how-building-supports-the-economy/

Ofori, G. (1990). *The construction industry: Aspects of its economics and management*. Singapore University Press.

Oxford Economics. (2021). *Future of construction: A global forecast for construction to 2030*. https://resources.oxfordeconomics.com/hubfs/Future%20of%20Construction_Full%20Report_FINAL.pdf

Pheng, L. S., & Hou, L. S. (2019). *Construction quality and the economy: A study at the firm level*. Springer.

Reserve Bank of Australia. (2022). *Box C: Financial stress and contagion risks in the residential construction industry*. https://www.rba.gov.au/publications/fsr/2022/oct/box-c-financial-stress-and-contagion-risks-in-the-residential-construction-industry.html

Seymour, H. (2019). *The multinational construction industry*. Routledge.

Tripathi, K. K., Hasan, A., & Neeraj Jha, K. (2021). Evaluating performance of construction organizations using fuzzy preference relation technique. *International Journal of Construction Management, 21*(12), 1287–1300.

The World Bank (2022). *GDP (current US$)*. https://data.worldbank.org/indicator/NY.GDP.MKTP.CD

World Economic Forum. (2016). Shaping the future of construction: A breakthrough in mindset and technology. https://www3.weforum.org/docs/WEF_Shaping_the_Future_of_Construction_full_report__.pdf

2 Ethics, Integrity and Professional Standards

Asheem Shrestha

2.1 Introduction

The construction industry is an integral part of the global economy. It significantly impacts every aspect of our lives. It also attracts strong interest from politicians, policymakers, economists, unions, media, non-governmental organisations and activists, end-users and the public. Unlike other industries, construction 'products' are resource-intensive, custom-made and location-specific, with many different people, teams and organisations, directly and indirectly, responsible for their planning, design, delivery, management and maintenance. Consequently, construction companies work with a broad spectrum of stakeholders while operating in a highly competitive, low-profit and high-pressure business environment.

The construction process entails a multitude of contracts with uneven power dynamics where large sums of money are exchanged. These characteristics present fertile ground for ethical issues to emerge and foster a climate where moral and institutional norms may not be consistently upheld. Research conducted in Australia, the United Kingdom, South Africa, Hong Kong, Nigeria, Pakistan and elsewhere demonstrates that the sector is beset by problems of unethical practices, misconduct and non-compliance. As a result, there is a great deal of emphasis on ethics in the construction industry to eliminate corrupt business practices.

Ethics is vital for construction for several reasons. For example, construction projects involve inherent occupational health and safety (OHS) risks. Adhering to ethical practices can ensure that OHS measures and standards are followed. Similarly, ethical practices are closely associated with better quality and sustainable products. Additionally, ethical policies and systems promote healthy competition, allow for an equitable sharing of risks and rewards, improve transparency and facilitate the proper utilisation of resources. Therefore, ethics play a significant role in the success of any company. Companies that emphasise ethical behaviour are more likely to have better productivity.

Moreover, good ethical practices can preserve the public's confidence in the industry. Since construction significantly impacts communities, adopting ethical practices and high ethical standards can demonstrate construction companies' commitment to community welfare. Therefore, ethical companies can build trust with stakeholders, enhance their reputation and grow their business while utilising resources more sustainably and reducing their carbon footprint. However, despite this knowledge, construction companies have fallen behind other sectors when prioritising ethics.

In this chapter, we focus on the challenges and opportunities for construction companies to foster ethics both internally and in their external dealings. The chapter first introduces the concept of ethics in general and in relation to businesses and professionals. Next, the nature of the construction industry is discussed, followed by an explanation of the common ethical issues

DOI: 10.1201/9781003223092-2

prevalent in the industry. The chapter then discusses how ethics come into play for construction companies and why they place a low value on ethics. Finally, it explains what may be done to promote ethics in construction companies by looking at approaches to leadership and culture, ethical systems and codes and ethics education and training.

2.2 What is Ethics?

Ethics deals with the philosophy of morality. Throughout the course of human history, civilisations and societies have produced intellectuals, philosophers, leaders, strategists and scientists who have advanced a variety of theories and propositions to help us understand the nature of morality and the principles we ought to live by.

In general, and especially in the Western school of thought, the concept of ethics comprises two fundamental categories: (a) metaphysical ethics and (b) rational ethics. *Metaphysical* or *religious ethics* are viewed as unassailable and permanent, requiring neither human input nor consensus, while *rational ethics* embrace a scientific worldview and are founded on societal consensus. *Utilitarianism* is a well-known notion of rational ethics that emphasises providing the greatest possible benefit to the maximum possible number of people. The principle of utilitarianism underpins the great majority of traditional and modern approaches to economics and business ethics.

Additionally, numerous theoretical approaches to ethics exist, each with a different perspective or focus. However, Starrett et al. (2017) classified ethical theories under three broad categories: consequence-based, principle-based and character-based (Figure 2.1). *Consequence-based ethics* focuses on developing a rationale for evaluating 'pleasure' and 'pain' for actions to produce the desired consequences. Utilitarianism falls into the category of consequential ethics.

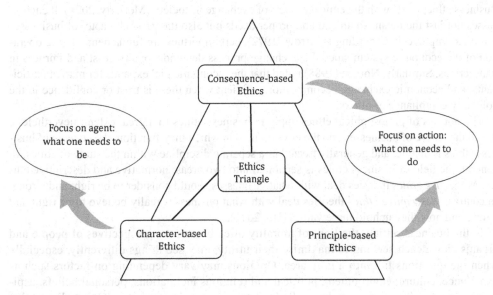

Figure 2.1 The ethical triangle.

Source: Adapted from Starrett et al. (2017)

Consequential ethics, however, has been challenged because it overlooks the possibility of achieving a positive outcome through negative actions. *Principle-based ethics* (also referred to as *deontology*), on the other hand, holds that morality consists of a set of rules and duties that should be adhered to, regardless of the consequences. But unlike these two theories, where the focus is on actions and the qualities associated with the deed, *character-based ethics* concerns virtue and focuses mainly on the qualities of the person performing the act.

Ethics is complex, and theories and ideas are frequently contested, and picking the right approach could be challenging. In many situations, common law (or religious law) provides (imposed) guidance on doing what is 'right'. Laws are formal and externally enforced. Ethics, on the other hand, are often informal and may often come through conscious reflection. In many instances, the ethical reflections of individuals can drive societal morals, ultimately resulting in new laws being written. Sometimes, existing laws are revised or discarded entirely, but as long as they remain in force, they must be observed, regardless of what one thinks of them. However, not all laws are morally justifiable, even when they may have been founded on the basis of metaethics or normative ethics. Moreover, laws may not always provide full coverage for ethical behaviour. For example, laws cannot always tell you what virtues to have as a business or what ethical principles to follow. For this reason, understanding how individuals, communities and organisations can become moral citizens and engage in ethical behaviour is more important.

2.3 Business and Professional Ethics

When discussing ethics in construction, it is crucial to first focus on two specific areas. The first is business ethics, and the second is professional ethics. Each may be seen as a distinct subfield of ethics and has significant relevance for the construction industry.

2.3.1 Business Ethics

Business ethics deals with the ethical aspects of exchange in society (Moriarty, 2021). It encompasses not just the means to an end and proper deeds but also the good character of businesses and their employees. According to Arrow (1972), certain virtues are fundamental to the operation of the economic system since the exchange process demands loyalty, trust and fairness in transactions. Similarly, Noreen (1988) noted that the 'economic pie' expands for market participants, and economic exchange becomes more efficient when there is trust or confidence in the voluntary compliance of others.

The notions of philosophical ethics apply to business ethics to a great extent; nevertheless, due to the nature of businesses and the environment in which they function, the concept of business ethics is distinct and generally treated as a separate discipline within the study of ethics. In general, the field of business ethics is separated into two areas: normative and descriptive ethics. *Normative ethics* theories deal with what businesses should consider to be right and wrong. In contrast, *descriptive ethics* theories deal with what businesses really believe to be right and wrong and how they uphold these values (Ho, 2011).

In the business setting, the lines of morality often blur as the perspectives of people and organisations, each seeking to maximise their utility, may see things differently, especially when the questions fall into a grey area. Opinions may vary depending on factors such as experience, culture, environment, political and religious inclinations, personal beliefs, aspirations, etc. So, *how can ethics be applied to businesses and transactions?* It is believed that companies need to observe the rule of law. Laws are essential and often based on careful consideration of the related ethical aspects. However, laws do not provide comprehensive ethical coverage in most business and professional settings. Speaking specifically about construction, Fellows et al. (2004) claimed that regulations focus mainly on the means but, even then, are not always reliable in serving their purpose. For example, construction contracts place importance on actions and, to an extent, successfully maintain ethical standards. However, many argue that contracts and regulations are insufficient to foster good behaviour. The significant use of contractual agreements in business settings, particularly in the Western world, creates a far greater emphasis on legal rights and obligations, which comes at the expense of relational duties (Fellows et al., 2004). According to Noreen (1988), contracts only cover a small portion of what is needed for a successful transaction. Therefore, effective transactions require more than just legally binding contracts. Much of the trust and understanding that are the essence of business dealings are in the unwritten and unseen elements of agreements.

Ethics is important in business, but why and to what extent? To address this, Moriarty (2021) asked: *Does being ethical really pay for businesses, or, in other words, does it lead to (financial) success?* Research has, in fact, found that being an ethical business does pay off. Related to the 'pain and pleasure' concept from consequential ethics, it is evident that doing things ethically improves reputation and doing things unethically has adverse consequences for companies. In many cases, the consequences can be significant if the outcome emerging from unethical conduct impacts a large section of society or the environment. Examples include the mishandling of dangerous chemicals that led to the recent Beirut explosion, the collapse of the Rana Plaza in Dhaka because of unsafe practices, the Grenfell Tower fire in London caused by substandard cladding or the environmental catastrophe caused by the Gulf of Mexico oil spill due to inadequate standards. In all these instances, the people and organisations involved in the unethical actions leading to the event paid a heavy price, at least in financial and reputational terms.

Unethical behaviour may often go unnoticed or unpunished by the legal system. But even so, businesses should go beyond minimum legal standards and focus on right and wrong in every aspect of their business decisions. There is strong evidence of the correlation between good behaviour and (financial) performance of companies (Kotter & Heskett, 1992). Being of good character builds a company's reputation, and being socially responsible or doing right by the customers and society does pay off and lead to better business prospects in the long term (Miller et al., 2020).

Business ethics also deals with the internal environment of an organisation, which is related to the virtues and behaviours of the management personnel and the employees. Though seen as a 'collective', the company's and its employees' ethical standards can often be at odds. In many instances, individuals in business organisations who tend to believe that they have higher ethical standards than other employees and the company itself face moral dilemmas (Fellows et al., 2004). For example, when management takes action that goes against an employee's moral compass, the employee may decide not to comply.

In all, the importance of business ethics is growing. The availability and accessibility of information are promoting better behaviour from businesses. Consumers and shareholders are now better informed about corporate actions and behaviour, and, at the same time, companies know more about their customers and shareholders and their needs and priorities. Moreover, there also appears to be a growing understanding that businesses exist to serve society and address communal and societal needs rather than only serving their shareholders (Mason, 2009). Consequently, they must create lasting relationships with their partners and society by transitioning from a low trust/low ethics base to a high trust/high ethics base (Wood et al., 2008).

2.3.2 Professional Ethics

No profession can flourish without a commitment to ethics (Brien, 1998). Professional ethics pertains mainly to those with specialised training, knowledge and credentials in their line of work, and it is distinct from philosophical or business ethics. The notions of general ethics may not always apply when dealing with dilemmas of right and wrong in any given profession. Thus, ethical norms for the profession should not be drawn only from philosophy but also from the knowledge within the profession (Starrett et al., 2017). Professional ethics also addresses the morality and actions of the professionals in their field of work and the community in which they operate. These responsibilities include competence, accountability and a commitment to serve their community (Bowen, Pearl, & Akintoye, 2007).

Ethical norms for most professions have become increasingly significant as organisations become more morally conscious (Hamzah et al., 2010). Various professional associations for different professions function as gatekeepers to admit only qualified members and act as facilitators to guarantee that members adhere to the best ethical practices. These associations are aware that unethical behaviour on the part of some may damage public trust in the profession as a whole. So, most have their own set of ethical norms, guidelines and codes of conduct for their members. According to Lynch and Kline (2000), these guidelines should provide systematic guidance for moral reasoning rather than inculcating a certain set of ideals. These guidelines become particularly helpful when sorting out matters that fall into a grey area. The core premise is that while facing an ethical dilemma, one should prioritise one's professional code of conduct over other interests.

A professional should be aware of both the laws and the ethical principles that govern them and utilise both to guide their professional activity. However, professional codes may conflict with one another or legal obligations. For instance, Starrett et al. (2017) discussed a situation where a legal obligation to maintain confidentiality may conflict with an ethical obligation to speak out. Similarly, many professionals in the modern business environment may be required to take on various organisational responsibilities, handling both technical and managerial tasks. For instance, a construction manager previously trained as an engineer may confront a situation where their professional engineer's code of conduct may conflict with their responsibilities as a manager charged with increasing profit at all costs. As a result, they may be asked to prioritise production and schedule demands over safety, presenting a moral dilemma.

2.4 Unethical Practices in the Construction Industry: Underlying Causes

The construction industry in many countries is rife with unethical practices. The underlying causes, however, can be attributed to the very makeup of the sector, how it functions, and its unique attributes. Intense competition, industry fragmentation, temporary relationships, complex contractual relationships, different perspectives, skill shortage and incompetence and a culture that allows opportunistic behaviour are some of the factors that make the construction industry prone to unethical behaviour and are discussed below.

2.4.1 Intense Competition

Construction is a highly competitive industry where contractors compete for razor-thin margins in a saturated market. In the industry, contractor selection is based primarily on who can take on the project at the lowest cost. As a result, contractors often submit the lowest bids possible with very low margins to secure contracts. However, after winning the bid, they may try to recoup the loss by cutting corners in other areas (such as quality and safety) or falsifying information to profit from variations and claims. Moreover, tight and, in many cases, unrealistic project deadlines present the dilemmas of either completing a task ethically, which may result in delays, or doing it unethically by taking shortcuts, leading to lower quality.

2.4.2 Industry Fragmentation

In complex environments where competing perspectives and values are at play and are moulded by unequal power relations, the notions of what is true or untrue, legitimate or illegitimate, right or wrong, are not universally shared (Bontempi et al., 2023). Construction is one such industry; it is complex, dynamic, highly fragmented and involves many stakeholders. It is also fraught with divergent opinions and prone to power abuse. The fragmentation of activities and roles further propagates laterally within the design, construction and operation stages, with several specialists contributing to each step individually. Moreover, many regulatory bodies are directly (and indirectly) involved throughout the process. Individuals, teams and organisations often have conflicting views on values, morals and practices driven primarily by their goals (Tow & Loosemore, 2009). With so many players with varying interests possessing specialised knowledge partaking in the process through a multitude of contractual links between them, there is a high chance that issues of ethics and integrity will emerge.

2.4.3 Temporary Relationships

Temporary partnerships are an additional distinguishing feature of the construction sector. Generally, project partners and teams with various skills, specialisations and backgrounds come together to perform in a high-stress environment but are expected to adhere to professional standards (Alkhatib & Abdou, 2018). But unlike other industries, the team works together only until the (one-off) product is delivered. Team members play their part in the project delivery, and once achieved, they move on to the next project, often with different sets of partners and teams. According to Kliem (2011), this impermanence leads team members to believe that the consequences of their actions do not matter, allowing them to commit ethical violations. Moreover, establishing trust in the relationships with clients or partners in the value chain may not appear as crucial compared to in other industries.

2.4.4 Complex Contractual Relationships

Contracts in the construction industry have many contractual connections between different parties, making relationships more complex and behaviours harder to monitor. Often, lawyers are heavily engaged in the contracting process, focusing more on ensuring legal security than the project's success. Contract negotiations often end up favouring the side with the higher bargaining power. Despite the emergence of new contracting models like collaborative contracts that prioritise the 'best for the project' or 'win–win,' most construction contracts are still of a zero-sum game nature where the more powerful will transfer the high-risk tasks to the less powerful, even when this may not be equitable. When the competition for the contract is high, the client may have more bargaining power, whereas when the competition is low, other parties have greater leverage to negotiate their terms. Additionally, inadequate information sharing and lack of transparency between parties, as generally in construction projects, allows the party that possesses better (private) information to take advantage of the information asymmetry both during contract negotiation as well as during the later stages of the project. For instance, contractors may take advantage of the loopholes or errors in the contract to file claims or seek reimbursements later down the track.

2.4.5 Different Personal and Professional Perspectives

Various individuals and professions involved in the process may adhere to different norms. For example, Bowen, Akintoye et al. (2007) investigated the differing opinions on ethics held by contractors, architects, engineers and quantity surveyors in South Africa and discovered considerable disparities. May et al. (2001) compared the views of contractors and subcontractors on the issue of bid cutting in Australia. They found that the contractors saw the practice as acceptable, while subcontractors viewed it as unethical. On the other hand, Fan et al. (2001) examined the perspectives on the ethical behaviour of quantity surveyors in Hong Kong. They identified a disparity between the perspectives of experienced and inexperienced individuals. All these divergent perspectives and beliefs in a highly fragmented industry can lead to moral dilemmas and may ultimately create disputes and problems related to performance, quality, safety and accountability. Furthermore, new specialisations and professions with distinct perspectives and priorities are emerging due to technological advancements and increased focus on sustainability, further fragmenting an already fragmented industry. Concerns about the environment and social equity are creating complex practical and ethical dilemmas for construction companies and professionals, forcing them to rethink their practices (Hill et al., 2013).

2.4.6 Skill Shortage and Incompetence

Being incompetent or, in other words, not having adequate (technical) knowledge and expertise in one's relevant field is a problem of ethics. For instance, when an unqualified designer produces an unsafe design, or when the authorities give that design the green light, and if the engineers and site managers fail to pick up on it, everyone involved in this scenario would be guilty of being incompetent in fulfilling their duties. With the high demand for construction worldwide, the industry is facing a shortage of skilled professionals, and often, the professionals selected for a position may not possess the skills necessary to do the job effectively. Incompetence among those who sign off on the designs and construction quality standards has been noted as a common industry problem. For instance, in the high-profile case of the Grenfell Tower fires in London, the flammable cladding that was used in the building was approved by the building control inspector despite widespread concerns at the time regarding the danger of using it. Negligence is a major ethical concern, especially when it results in quality and safety issues. For example, Ameh and Odusami (2010) showed that negligence during building design accounts for a very high percentage of building failures in Nigeria. Negligence can occur when one fails (or neglects) to fulfil their responsibilities with due care. Industry professionals can be negligent in identifying and reporting poor designs, materials, workmanship, and safety standards. Indeed, many accidents that occur on construction sites are due to negligence (Othman, 2012).

2.4.7 Industry Culture

Construction industry culture globally is often seen as adversarial, plagued by disputes, power abuse and non-compliance. For instance, the Australian government has convened three separate Royal Commissions to investigate alleged misconduct within the country's building and construction sector; the most recent, in 2015, focused on trade union corruption. Each report highlights the problems of bribery, widespread issues regarding payment security, intimidation and systemic lawlessness. The findings also show that many issues have persisted for decades, demonstrating that attempts to address them have failed thus far. The industry has a culture of underbidding, power imbalance and oppressive behaviour that subsequently contributes to the high rate of insolvencies in the sector (Coggins et al., 2016). The industry culture is reflected in the business practises of companies, where the bigger and more powerful companies are often seen to be aggressive in their dealings with their smaller and less powerful partners, with issues of underpayment, bullying and coercion being all too common. Moreover, the poor culture in the construction industry translates into a low safety record with high rates of injuries and fatalities, long work hours, poor work–life balance, high suicide rates and other mental health issues.

2.5 Common Unethical Practices in the Construction Industry

As discussed in the previous section, the industry environment and culture pave the way for unethical practices within the construction sector. *Corruption* is an umbrella term that covers various unethical behaviours. Typical forms of corruption involve bribery, nepotism, embezzlement, kickbacks, fraud, bid rigging and collusion. Numerous studies and reports discuss the various forms of corruption and how they impact project performance and quality. Corruption can occur at every stage of a project. During project initiation, for example, the self-interests of developers and government officials may play a role in shaping what gets built and where it gets built. Similarly, during the design phase, the developers may pay bribes to local officials to obtain permits for a non-compliant design. During the tendering phase, bribery, bid rigging, collusion, cover pricing, bid shopping and fraud can become prevalent. Similarly, most unethical practices during construction stem from the desire to profit from reduced costs despite substantial effects on the project's performance, quality, safety and environment.

2.5.1 Bribery

Bribery was defined by Johnson (1991, p. 327) as the 'offering of some good, service, or money to an appropriate person for the purpose of securing a privileged and favourable consideration (or purchase) of one's product or corporate project'. It manifests in several forms: cash inducements, gifts and favours, trips and holidays, privileges and entertainment. Bribery is more common in developing countries due to various factors such as opaque systems, the low pace of industrialisation, high rates of poverty among the citizens, inadequate pay for government employees, immunity for public servants and inadequate procurement procedures (Ameh & Odusami, 2010). Alutu (2007), for example, found that it is a common practice in Nigeria for government officers to negotiate their percentage share of the contract. However, developed countries are also not immune. Nearly half of those polled in a Chartered Institute of Building (2013) survey considered corruption widespread in the United Kingdom, with more than a third reporting being offered a bribe at least once. Similarly, the 2002 Royal Commission report on the Dutch construction industry found many instances of corruption where government officials colluded with bidders (van den Heuvel, 2005).

The issue may, however, not be limited to government officials receiving bribes and kickbacks from the main contractors or developers. It may also occur down the contractual chain, where consultants or contractor-side procurement managers may also solicit kickbacks when offering jobs to subcontractors and suppliers. The ex post stages are also associated with high levels of bureaucracy, which include obtaining numerous permits and licenses, each of which presents an opportunity to give or receive bribes in return to expedite the process or even just to approve them. Moreover, concealed construction work (e.g., foundations, plumbing and electrical work, etc.) makes it difficult to see what is underneath, allowing overstating material quantities or hiding poorly executed works in exchange for a bribe (Ameh & Odusami, 2010). Bribery often goes undetected due to the industry's fragmented environment, complex contracts and networks and the high sums of money being exchanged in numerous transactions.

2.5.2 Collusion

One of the most significant issues of ethical misconduct during tendering is collusion. Collusion, or bid rigging, happens when the bidding companies get together to influence the tender outcome and the award price. *Cover pricing* is one of the common collusion practices where all bidders, except the one predetermined, submit a very high or unreasonable low tender price to ensure that the predetermined bidder is awarded the contract. Through this approach, one of the bidders can secure the contract at a favourable price, undermining the competitive process. Moreover, colluding companies take turns to be the winning bidder, an illegal practice referred to as *bid rotation*. Sometimes, bidders may collude to inflate their prices to compensate unsuccessful tenderers. There have been numerous cases of collusion reported in the construction industry globally. For example, in 2008, the Office of Fair Trade launched an investigation in the United Kingdom that imposed fines totalling £129.2 million on 103 construction companies for engaging in cover pricing between 2000 and 2006. Several companies that had won the tenders had faced no genuine competition, resulting in the client unknowingly paying a higher price for the contract. Furthermore, the Office of Fair Trade found many instances where the successful bidder compensated unsuccessful bidders.

2.5.3 Bid Cutting

Bid cutting is another unethical behaviour utilised by the client (or their representative) during the bidding process using coercive means to get a lower price from a bidder after receiving the bids. There are two forms of bid cutting: bid peddling and bid shopping. In *bid peddling*, the client discloses the prices of one bidder to the other, expecting that they will be persuaded into lowering their price. In *bid shopping*, the client pits bidders against each other by revealing their rivals' lower prices (which could be either factual or fictitious) and asking them to cut their prices to win the contract. These practices can occur in the primary contracts where the client (which may be the government) negotiates with the main contractors. However, they are more common in secondary contracts where the main contractor, when awarding a job to the subcontractors or suppliers, tries to profit from the saved costs by adopting this unethical practice (Degn & Miller, 2003; Gregory & Travers, 2010). Bid cutting undermines the competitive process and, more important, contributes to an adversarial environment and a culture of distrust (Degn & Miller, 2003). Furthermore, reduced prices and profit margins may often lead to low levels of quality, safety and project performance.

2.5.4 Fraud

Fraud, according to a report by Ernst and Young (2010), accounts for over 10% of construction expenditures globally. According to the fraud triangle theory developed by Cressey (1953), three factors lead to a fraudulent act: pressure, opportunity and rationalisation. *Pressure* is the impetus to commit fraud, *opportunity* is the setting that allows fraud to occur and *rationalisation* is the attitude and presence (or the lack) of ethical values. As discussed earlier, the construction industry provides an ideal setting for all three factors to co-exist. The intense demands to deliver projects on schedule and budget while making a profit in a highly competitive market provide that impetus, particularly when the preventive systems and measures in place are ineffective. Moreover, the problems are exacerbated by the industry's fragmentation and the high volume of transactions involving individuals who prioritise profit above ethics.

At the project level, potential problems may start at the bidding stage when companies overstate their capabilities but cannot deliver the promised terms. During the construction stage, fraudulent reporting can occur on the part of the contractors (and subcontractors) when reporting on project performance, submitting claims or requesting reimbursement. For instance, the concealed nature of completed work allows one to overstate material quantities or manipulate material quality. Other common issues may include fabricating payment claims for uncompleted work, using inflated labour rates and tampering with or altering contract documents. According to Kliem (2011), fraudulent reporting is one of the most frequently encountered ethical problems and violations in projects.

At the corporate level, tax fraud is reported as a widespread phenomenon within construction businesses globally. An investigation by the Australian Taxation Office (ATO) in 2021 revealed that the construction and building sector was at the top of the list for the most tax evasion (ATO, 2022). Similarly, the issue of payroll fraud is prevalent in the construction industry. Payroll fraud occurs when employees are paid in cash outside the official accounting system or when 'sham' contracting is used to disguise an employment relationship as that of a client and independent contractor. In any case, these practices by the companies allow them to evade their legal responsibilities of paying workers for overtime rates and providing other work entitlements such as insurance, leave payments, compensations and redundancy payments. The scale of the problem was highlighted in the work of Ormiston et al. (2020), which found that in 2017 alone between 75,906 and 125,855 construction workers in the state of New York were either misclassified as independent contractors or were working off the books.

Phoenix activity is another common fraud often reported in the construction industry. *Phoenix activity* is when a company is liquidated but soon after reopens as a new company operating in the same line of business. While this is not always considered illegal or fraudulent, particularly when the liquidation is genuine or occurs as a legitimate business structuring, there are many instances of illegal phoenix activity where companies have an opportunistic agenda to file for bankruptcy to avoid paying their debts. Fraudulent phoenix activity can significantly impact the industry, causing creditors and trades to lose their money and employees to miss out on their wages and entitlements. Smaller subcontractors and suppliers, for instance, might be severely impacted if they do not receive payments. ATO reported that in 2023, the estimated annual cost of illegal phoenix activities in Australia was between AUD 2.85 and 5.13 billion (ATO, 2023).

2.6 Ethical Issues in International Construction

Many construction companies venture abroad to expand their businesses. At the same time, host nations seek them for their specialised knowledge and superior technology or to increase the number of service providers to meet construction demands. For companies, working in a global market adds more layers of complexity and risks. The most obvious challenge is operating in environments with different cultures, regulations and standards. Culture matters because it is related to how an organisation operates in a particular environment. Safety culture, labour relations, working hours, environmental standards and other practices may differ from one culture to another, posing several moral difficulties. It is also possible that power dynamics and interactions with authorities will vary greatly.

Companies that are used to high levels of standardisation have increased challenges while working in an environment where the regulations are less comprehensive and contractual commitments are not prioritised. Some international businesses operating in certain developing countries may relax their quality control requirements regarding environmental, safety and construction standards while being aware that these standards are not good enough (Mirsky & Schaufelberger, 2014). The culture of giving gifts, which may be a widespread practice in many parts of the world, can create a dilemma for international companies and professionals. It may raise the question: *Should I partake in this culture of gift-giving and fit in (and also be in the good books of the individuals who can help my business grow), or should I keep to my moral convictions and not offer gifts (and also risk my chances of being looked at favourably by those people)?*

In some cases, it has been reported that companies actively seek opportunities to pay bribes to win contracts. Much has been written on corruption in international construction, with Hall (1999) noting the susceptibility of developing nations to bribery from companies from developed countries. More than 3,000 company executives were polled from organisations in 28 of the world's leading economies for Transparency International's 2011 Bribe Payers Index Survey. According to the report, public work contracts and construction are significantly more prone to bribery, and all 28 countries had at least one company involved in bribery when operating overseas. While international regulations around bribery and fraud in international projects exist, the problem is still widespread.

Barthorpe (2010) suggested that the relationship between businesses and the society in which they operate is becoming increasingly complex as international trade has brought new ethical challenges for companies. Their supply networks and service outsourcing have extended beyond national boundaries, which may present the problem of ensuring that their domestic social responsibility norms are equally upheld in the entire supply chain and their international operations.

2.7 Consequences of Unethical Practices in the Construction Industry

Construction professionals' poor ethical standards may translate into poor construction safety and quality (Hamzah et al., 2010). For instance, using substandard designs, either intentionally to save costs or unintentionally due to negligence and incompetence, could have severe implications for the project's OHS and quality standards. In extreme cases, this may lead to tragic events resulting in the loss of many lives and substantial financial damages. Similarly, substandard materials and work not meeting engineering standards could result in significant loss of lives and properties in the event of natural calamities and structural collapses (e.g., the collapse of several school buildings in the 2008 Wenchuan earthquake in China).

In the United Kingdom, Oswald et al. (2020) identified many cases of OHS risks being amplified due to cheaper and lower-quality equipment, machinery and temporary structures. Moreover, companies may attempt to save costs by not providing adequate OHS training (or not offering it at all), though often mandated by law. Migrant workers employed at lower rates are at the most risk of being subjected to unsafe working conditions and inadequate training (Shepherd et al., 2021). Accidents and injuries have direct and indirect cost implications due to compensation claims, higher insurance rates and lost productivity.

Corruption also has broader implications; it affects economic growth and development and wastes public money and resources, with people with low incomes being the most vulnerable to its impacts. Transparency International (2011) reported that construction is the most corrupt industry globally. Although most forms of corruption are illegal and, generally, there are severe legal consequences of corrupt practices, their high prevalence in the construction sector shows that the systems and mechanisms to detect corruption are inadequate. This may be partly because there is a moral disparity between what the law or code says and what is generally accepted in the industry.

2.8 Ethics for Construction Companies – Setting an Agenda for Fostering Ethical Behaviours

Building a long-lasting reputation in the construction industry is challenging for any business, but it is more difficult for those who lack a shared ethical compass among their employees and their customers (Alkhatib & Abdou, 2018). Building that common ethical ground for construction companies is essential for navigating the ethical questions they face in the industry. There has been a growing interest in construction ethics, particularly in the last few decades, due to the public's increased awareness and diminishing confidence in the industry's ethical practices (Bowen, Pearl, & Akintoye, 2007). With the advent of information-sharing and communication technologies that make corporate transactions and interactions more transparent, the public is becoming more informed and is demanding that construction businesses function responsibly. Companies proactively seek to be more ethical as a strategy due to ethical consumerism and social awareness (Kercher, 2007).

Moreover, with public opinion shifting towards sustainability, companies need to reciprocate and rise to the challenges to be more environment-friendly to compete. For instance, a poll by KMPG (2013) found that, in response to stakeholder pressure, 95% of CEOs questioned saw sustainability and ethics as critical strategic issues. More important, companies also realise that good ethics lead to good business. All these factors have prompted construction companies to place a greater emphasis on ethics and their codes of conduct. However, while construction businesses are generally making efforts to be more socially responsible in all areas, the attitudes and conduct of many in the industry remain far from ideal.

One of the important challenges that modern enterprises face today is striking a balance between their financial and social obligations. Research examining ethical conduct among different project participants has found contractors to be the most unethical (Ameh & Odusami, 2010; Vee & Skitmore, 2003). Given the slim profit margins for contractors, it is not surprising that many of them use unethical means to win contracts to survive as a business and, if successful, find opportunities to cut corners during construction to make a profit. Social and environmental obligations are hardly the priority, and unethical behaviour is fair game as long as the contract does not directly prohibit it or if it cannot be discovered.

So, how can ethics be improved in the construction industry, and what can construction companies do to foster ethical behaviour in their organisation and among their employees? Considering the widespread prevalence of unethical practices in the construction industry, some may argue that stricter laws and regulations are required. For some time, consequence-based ethics has been the dominant approach in construction practice. Programs are developed to assure compliance by focusing on detecting and penalising violators. However, Gyoo Kang et al. (2014) argued that relying solely on this approach has led to a failure to address the underlying problems of unethical behaviour. Given the persistence of unethical issues in the sector, we can say that regulations alone are ineffective in encouraging moral conduct. So, approaches drawing from principle-based ethics and character-based ethics may be necessary.

People are most likely to behave ethically under three conditions: (a) they are informed by a respected leader about what is expected of them, (b) they work in an environment that values ethical conduct and (c) they are shown how others might benefit from their moral actions (Stout, 2010). According to Ford and Richardson (1994), ethical behaviour is most likely to thrive when company values, leadership influence, processes and systems, codes of ethics, ethics training and incentives work together to promote a healthy ethical environment.

2.8.1 The Role of Leadership in Cultivating an Ethical Culture

According to Bowen, Akintoye et al. (2007), ethical failures in construction companies often stem from a pervasive lack of value placed on ethics by the company's culture and a lack of leadership support for ethical behaviour. Ethics is often overlooked by its leaders and is treated as something required only because the management feels pressured to show shareholders and the government that they are taking steps to ensure ethical behaviour (Kliem, 2011). This mindset must be dispelled if an organisation intends to foster an ethical culture. And, unless all members of an organisation – leaders, managers and employees – embrace the idea that upholding ethical standards benefits the company and its employees, creating and sustaining an ethical culture may not be possible.

The popular belief is that a company's unethical behaviour stems from senior executives lacking professional ethics and morals, with greed being the most prominent cause (Alkhatib & Abdou, 2018). Therefore, interventions to encourage ethical behaviour start with leadership. According to Bass and Steidlmeier (1999), there are three pillars upon which the ethics of leadership stand: (a) the leader's moral character; (b) the moral plausibility of the ideals behind the leader's vision, articulation and agenda; and (c) the morality of the processes of ethical choice and action that leaders and followers engage in and seek collectively. So, the leader must be of moral character and should set up ethical values and norms to be shared by subordinates.

To promote an ethical culture, leaders should also ensure fair treatment of workers, which has been identified as a significant component in shaping the ethical disposition of employees (Tow & Loosemore, 2009). *Fairness* refers to using fair procedures for recruiting, performance evaluations, promotions and compensation. While these measures would be implemented because of government regulations, a culture with strong values would prioritise these issues regardless of legal obligation. Moreover, the leader's ideals should be conveyed clearly to the subordinates, and the leader should be able to demonstrate the links between the organisational strategies, its decisions and its values. Ethical boundaries that are explicit and attainable must be set not just for the employees but also for the leaders themselves.

Over the centuries, scholars have tried to define what makes a good leader. Studies on business ethics have focused mainly on two types of leadership approaches: (a) transactional leadership, where leaders motivate employees through rewards and punishments, and (b) transformational leadership, where the leaders lead through influence, inspirational motivation, intellectual stimulation and individualised consideration (Bass & Steidlmeier, 1999). However, recent studies examining leadership in the construction industry emphasise transformational leadership. Jung et al. (2014) found that leaders are more successful in driving ethical behaviour when they rely on influence and persuasion rather than authority and directives.

Moreover, leaders' behaviours around matters they prioritise and pay attention to influence subordinates' behaviours. The importance of leaders helping their teams appreciate the value of ethics in their work is vital. So, to inspire trust and confidence, leaders should make their values consistent with their actions. According to Jennings (2006), leaders who exemplify what is expected of others through their actions are more likely to inspire their subordinates to engage in ethical conduct. The employees' commitment to the business and sense of belonging are positively correlated with the ethical ideals espoused by the organisation's leadership and culture (Ferrell & Fraedrich, 2016).

A leader's capacity to identify the character in the selection of new employees or business partners is also critical to the success of the culture-building process. An important aspect of a leader's ability to infuse their principles into the organisation's culture is hiring appropriate individuals who share their values and can effectively execute their vision (Tow & Loosemore, 2009). The leader must also ensure that all new employees are fully qualified for their positions. Similarly, the leader's capacity to choose external collaborators reinforces this culture. Wherever possible, leaders should collaborate with clients and business partners who share their values and those who prioritise ethics. According to research by Wood et al. (2008), this results in a trust-based partnership that pushes both parties to raise their ethical standards and enhance their overall ethical performance in business.

2.8.2 Ethical Decision Making

Ethical dilemmas often arise when construction companies and practitioners make decisions, as ethical decision-making in a business context involves a sequence of complex judgements. In most cases, this may concern whether to prioritise profits above fulfilling their duties as responsible corporate citizens. For construction companies, this may be related to how they operate on projects, manage client expectations and demands, deal with stakeholders and contractual partners and treat their employees. Every unethical behaviour or action, including those discussed in the previous sections, results from a decision taken either at an individual level (by an employee) or at the company level (as a collective decision) to produce an outcome.

In today's rapidly evolving business climate, the outcomes are increasingly difficult to predict, and there is never complete knowledge of whether the outcome would benefit everyone. The inherent uncertainties in predicting the outcome allow ethical concerns to enter into the decision-making process (Zarkada-Fraser, 1998). Particularly, those who are risk averse with a low tolerance for ambiguity would likely be concerned regarding the uncertainties and could take (unethical) decisions solely considering their own utility.

Much effort has been invested in studying ethical decision-making, and several models have been developed in the business ethics literature. Most of these models have taken a descriptive approach to focus on the influencing factors guiding the decision-making process (Ho, 2011). Overall, most models identify similar factors, such as the importance of the decision maker's values and experiences, as well as internal and external influences. The general idea is that a better ethical climate leads to ethical decisions. Yet, depending on the context, these factors may exert varying degrees of influence and function in unison or isolation. An example is the model by Bommer et al. (1987), which illustrates how several factors influence a decision that is either ethical or unethical (Figure 2.2).

Normative models on how ethical decisions should be made are less common. As normative principles and standards differ from business to business, creating a uniform model that applies in every context is challenging. However, those that exist draw from the ethical principles of utilitarianism, justice and virtue ethics. In a rare study looking specifically at construction companies, Gyoo Kang et al. (2014) highlighted that the existing frameworks in the business ethics context may not always apply to construction, mainly because of the unique nature of the construction industry. They presented a normative model for construction companies using decision trees to navigate ethical dilemmas at three levels: the individual, project and corporate levels. For each level, the model draws from virtue ethics, which relates to the character of the decision maker; deontology, which concerns the principles and obligations (irrespective of the result); and consequentialism, which is concerned with the outcome of the decision, with the right outcome being something that maximises the total utility for the greater number of people

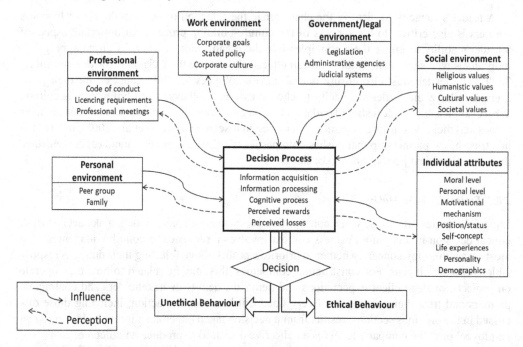

Figure 2.2 Ethical decision-making model.

Source: Adapted from Bommer et al. (1987)

(utilitarianism). Overall, the key message from most published literature is clear: businesses need normative frameworks to help them make critical decisions concerning ethics. Considering the ethical challenges in the construction industry, companies would greatly benefit from this to navigate ethical dilemmas.

2.8.3 Setting Up Systems for Ethical Behaviour

Developing an ethical culture demands the commitment of the leaders and all employees. So, businesses should make ethical behaviour the norm at all levels of the organisation. Systems and frameworks put in place to encourage ethical behaviour can have a positive impact on the organisation's ethical culture. According to Booth and Schulz (2004), the organisational environment has a greater influence on employees' ability to act ethically than the employees themselves. So, regardless of workers' ethical beliefs, an environment with mechanisms to encourage ethical behaviour results in a general propensity to act in the best interests of their organisations and a lessened tendency to act opportunistically.

According to Kliem (2011), ethical behaviour cannot be practised without courage. In many professional settings, there can be much pressure to act within traditional norms and customs, even when the practices are unethical. This restricts those who want to do the right thing for fear of being penalised. In the 2018 Ethics at Work survey conducted by the Ethics Centre Australia and the Institute of Business Ethics, it was reported that even when employees are aware of misconduct at the workplace, 35% do not speak up for fear that it would jeopardise their jobs, while many feel pressured to compromise on ethical behaviour (Dondé & Somasundaram, 2018). Therefore, systems should be in place to foster an atmosphere where team members feel safe and are encouraged to voice ethical concerns. Organisations with a dedicated compliance and ethics hotline provide a confidential channel for workers to report ethical violations. Some workers may feel more comfortable reporting issues this way than going straight to management.

Ethical standards are generally low when the top management focuses purely on positive outcomes and where employees are under relentless pressure to attain them. In this kind of culture, unethical practices are often ignored and discounted; even worse, achievements are rewarded even when acquired through unethical means (Jennings, 2006). By recognising the ethical actions of workers, an ethical culture may be fostered. Rewards and incentives are important means for companies to promote ethical behaviour among managers and employees. Many ethical issues may be managed by changing the way employees are compensated. Individuals are more likely to act ethically if they are financially rewarded for doing so and punished for not following the rules. Often, top-level managers are offered bonuses or company shares when the company performs well financially. Similar incentives can be provided for ethical performance. Ethical conduct may also be embedded into the design of ongoing performance reviews for all staff so that it becomes the standard in the organisation. Whether the outcome is positive or negative, rewarding ethical behaviour sends a strong message that leadership appreciates the right thing.

Besides monetary rewards, employees care equally about their contributions and recognition at work. There should also be non-monetary rewards in the form of opportunities for promotions and career advancements, empowerment through delegation of important tasks and invitations to provide input in strategic decision-making. Moreover, simple praise, acknowledgement and validation can motivate employees to continue acting ethically and remind them of their positive impact. With the incentive in line with moral drive, performance may also increase. On the other hand, control and sanctions for unethical conduct can be just as crucial. Indeed, this constitutes the foundation of many laws. According to Gurley et al. (2007), the effectiveness of punishments may depend on the chances of getting caught and the severity of the consequences. Therefore, to deter workers from engaging in unethical conduct, there must be effective methods to monitor unethical behaviour and appropriate sanctions for noncompliance. Punishments could be in the form of negative feedback, criticism, reprimand or disciplinary action for more serious breaches.

2.8.4 Corporate Social Responsibility

One of the means to incentivize companies to act ethically and in line with community standards is the adoption of corporate social responsibility (CSR) standards. CSR is often defined as a set of practises that include some or all of the following: fair treatment of workers, community and environmental assistance, respect for human rights and ethical dealings with rivals, suppliers and consumers (Barthorpe, 2010). It gained traction in policymaking in the 1950s as an analytical tool for monitoring the ethical practices of multinational companies. Many countries' corporate social laws mandate that these companies act in socially responsible ways, with human rights, minority relations and environmental protection at the centre of these regulations. When companies voluntarily go above and beyond these mandates, it sets them apart from their competitors (Miller et al., 2020). CSR ratings are often produced for companies, which enables stakeholders to gauge a company's reputation in terms of its CSR.

CSR in the construction industry started to gain momentum as construction companies began to evaluate the impact of their operations on the environment and society (Barthorpe, 2010). It is gradually being accepted as a means to improve organisational competitiveness and overall performance and build an excellent corporate reputation. As a result, the behaviour of companies seems to have improved in some cases. However, the adoption of CSR in construction is falling far behind other sectors (Loosemore & Lim, 2018). Nguyen (2024) found that the uptake of CSR has been slow for various reasons, the most important being the lack of strategic vision, where construction companies still have the traditional mindset of prioritising short-term profit, and their inability to see the value of CSR. Also, there are issues related to the lack of understanding of CSR on the part of the other stakeholders, the lack of means to measure CSR and the lack of financial resources for the companies to make the changes to their practices to focus on CSR.

For many companies, corporate responsibilities are adopted more as a compliance exercise than a genuine concern for the community and environment, and according to Loosemore and Lim (2018), all the improved connections with communities that construction companies manage to develop through this exercise mainly go to waste. Not having a genuine concern for communities may also be a conscious choice, partly due to the nature of projects where the relationships between the company and the communities are temporary. While the company may have to tick the compliance boxes, they may not feel the need to spend their already scarce resources to go above and beyond to build a high level of trust with the communities, as they will be moving away to serve another community in their next project. Additionally, there have been issues with inaccurate CSR reporting. For example, Bontempi et al. (2023) discussed the case of WeBuild, an award-winning Italian construction company, to illustrate the mismatch between the company's narrative of its CSR and its actual conduct in 38 global hydropower projects.

With the growing focus on the ethical behaviour of construction companies, attention is generally placed on larger companies, which may have to answer to their shareholders (and the government) on their conduct through statements in their annual reports and the disclosures of their CSR efforts. Corporate responsibility legislation has emerged in several nations to drive better behaviour of companies. For example, the Modern Slavery Act has been introduced in the United Kingdom and Australia, where corporations must ensure the fair treatment of workers throughout the entire supply chain. However, construction, a highly fragmented industry, employs small and medium-sized enterprises (SMEs) for most of the work. Many SMEs may lack the resources and drive to engage in CSR. Moreover, they may not have to justify their actions at the same level as large enterprises. SMEs are thus seen as having a relatively low priority for CSR initiatives (Loosemore & Lim, 2018).

2.8.5 Ethical Codes

Ethical codes establish overarching standards and recommendations for how one ought to act. They represent an ideal of what people, professions or organisations can be and should strive towards. Many different codes of practice apply to construction, ranging from international standards that apply globally to codes drafted by government agencies at the federal and local levels that apply to a specific region. Professional bodies and institutions in most countries also formulate their own codes that apply to particular professions. While some codes focus on the technical aspects of construction, providing guidelines for quality, performance and safety, others focus specifically on ethics-related issues.

While there is no legal requirement for businesses to adopt a code of ethics, having a code helps them define who they are and what they stand for in the eyes of their customers and other stakeholders. Large companies often tend to have formal codes or written documents outlining the moral standards that all employees must uphold. They generally cover both the compliance-based code of ethics and the values-based code of ethics. Compliance-based codes are centred around the rules and standards in construction practice and provide set guidelines and consequences for any violations. On the other hand, a values-based code of ethics emphasises the organisation's core values and specifies standards of responsible behaviour in relation to society and the environment. It relies more on self-regulation and encouragement than coercion.

Ethical codes are based on several fundamental principles. For example, in the United Kingdom, the Society of Construction Law's 2003 report highlighted that organisations and practitioners should adhere to seven principles: honesty, fairness, fair reward, reliability, integrity, objectivity and accountability in all business dealings (see Mason, 2009). These principles were intended to be relevant to the work of all construction industry professionals and address issues related to corruption, inappropriate behaviour towards others and other unethical concerns during project procurement and delivery (see Table 2.1). In most cases, ethical codes in any industry or organisation are based on these same basic principles with varying degrees of breadth and depth.

For construction companies, having a formal code is effective in managing ethical behaviour (Tow & Loosemore, 2009). Codes provide an important means through which the morals and values of the leaders permeate all levels of the company. Codes also bring clarity to the rules for behaviour, especially when they are well defined and effectively communicated to the employees. Adams et al. (2001) found that even when employees do not recall the specific language of the code, just knowing that such a code exists can positively affect their perceptions of ethical behaviour in the workplace. Yet, for ethical codes to be fully effective, it is also necessary to have control and monitoring procedures in place to see whether they have been observed.

When designing ethical codes for construction companies, aligning them with the professional codes followed by the organisation's employees is necessary. Thus, professionals will not have to choose between business norms and professional codes. Codes should be designed to help employees navigate ethical dilemmas and to provide legitimacy for one's ethical action, even when the result is inferior to what may be attained by unethical means. Even when normative decision-making frameworks may not exist within the company, well-defined and attainable

Table 2.1 Principles for ethical codes of conduct

Principles	Implications
Honesty	Acting with honesty and avoiding actions that might lead others to be misled in any way
Fairness	Avoiding any benefit that might be a result of unjust treatment of others
Fair reward	Avoiding acts that might cause someone else to get less than they deserve for their efforts
Reliability	Maintaining current knowledge and skills and delivering services exclusively within your area of expertise
Integrity	Considering the public's interests, especially those who will use or acquire an interest in the project in the future
Objectivity	Identifying possible conflicts of interest and disclosing them to anyone who could be affected by them
Accountability	Providing information and warning people of issues you are aware of that might negatively impact them. Providing warning within sufficient time to allow the taking of effective action to avoid detriment

Source: Mason (2009)

codes of ethics give norms and directions for operating in both routine and complex situations. Kliem (2011) described a code of ethics as a guidepost for project management professionals when operating under challenging situations. For instance, many companies have zero tolerance for the exchange of gifts. When construction companies have a good set of ethical codes that focus not only on compliance but also on company values, it can help managers and employees make ethical decisions.

2.8.6 Ethics Training and Education

The focus on driving ethical behaviour in construction companies has largely been on the application of consequence-based ethics (incentives, motivation and sanctions) and principle-based ethics (effective processes, systems and codes). While these are effective, there are some who argue that the underlying ethical issues cannot be fully addressed without incorporating character-based ethics. Character-based ethics focuses on virtue and contends that moral actions are the result of introspective reflection. Creating awareness about the benefits of acting ethically through education and training is an effective way to encourage people to adopt more moral perspectives.

Effective application of ethical codes or raising ethics consciousness among employees requires a structured ethics training programme. Drawing from this, Tow and Loosemore (2009) suggested that raising ethical standards for construction companies can only come from formal training to communicate and establish their ethical codes among employees. However, there is a lack of ethics education and training in the construction industry. The ethics education construction professionals receive during their tertiary and professional education is often inadequate (Tow and Loosemore, 2009). Education for construction professionals often focuses on improving their technical and performance skills, with less attention paid to methods for instilling and creating ethical ideals and behaviour (Mohamad et al., 2015).

More recently, academic institutions are taking steps to incorporate ethics education within their curricula. Similarly, professional associations now offer their members a range of ethics training programmes. For newly recruited employees, ethics training could begin early through induction programmes incorporating both compliance- and value-based ethics. Ethics and leadership training programmes, continuing professional development programmes, seminars and courses might be offered regularly to current staff members, which can aid in expanding knowledge and skills and developing personal qualities. Many studies have reported that training programmes emphasising the practical aspects rather than the philosophical side are more effective in encouraging ethical behaviour (Booth & Schulz, 2004; Ford & Richardson, 1994). Workers who have been in the industry for some time may go through phases of moral growth, during which their values and perspectives may shift (Gyoo Kang et al., 2014). Therefore, ethics education may help employees cultivate new ideals aligned with organisational values and societal needs.

2.9 Conclusion

The construction industry is beset with ethical problems. These problems arise mainly due to the industry's characteristics and culture. Changing the industry's fundamental characteristics is difficult, but the culture could change, and construction companies can play an essential role in driving that change. Construction companies and their leaders must realise they have a duty not just to their shareholders but also to the community they serve and the broader society if they are to play a vital role in effecting change. They must also recognise that being ethical is good for business. Once these are recognised, the leaders should consider how they can foster a culture of ethics within their organisation and in their interactions with their partners and stakeholders. Where possible, companies may participate in CSR initiatives that are genuine and relevant to the communities they serve.

In addition, companies could establish normative decision frameworks that directly translate into their ethical codes. This can bring clarity to the company's values and provide employees with guidance not only in their routine operations but also when faced with difficult ethical decisions. Incentive programmes that are generally seen as effective means to motivate ethical behaviour should be supported by sound systems that monitor and reward ethical behaviour. Lastly, ethics education and training programmes should be implemented to create awareness of why doing the right thing as individuals and as an organisation is essential and how it can positively impact the industry-wide culture and society.

References

Adams, J. S., Tashchian, A., & Shore, T. H. (2001). Codes of ethics as signals for ethical behavior. *Journal of Business Ethics, 29*, 199–211.

Alkhatib, O. J., & Abdou, A. (2018). An ethical (descriptive) framework for judgment of actions and decisions in the construction industry and engineering – Part I. *Science and Engineering Ethics, 24*(2), 585–606.

Alutu, O. (2007). Unethical practices in Nigerian construction industry: Prospective engineers' viewpoint. *Journal of Professional Issues in Engineering Education and Practice, 133*(2), 84–88.

Ameh, O., & Odusami, K. (2010). Professionals' ambivalence toward ethics in the Nigerian construction industry. *Journal of Professional Issues in Engineering Education and Practice, 136*(1), 9–16.

Arrow, K. J. (1972). Gifts and exchanges. *Philosophy & Public Affairs, 1*, 343–362.

Australian Taxation Office. (2022). *Tipped off: ATO reveals most dobbed-in industries.* https://www.ato.gov.au/Media-centre/Media-releases/Tipped-off--ATO-reveals-most-dobbed-in-industries/

Australian Taxation Office. (2023). *Illegal phoenix activity.* https://www.ato.gov.au/about-ato/tax-avoidance/the-fight-against-tax-crime/our-focus/illegal-phoenix-activity#ato-Aboutillegalphoenixactivity

Barthorpe, S. (2010). Implementing corporate social responsibility in the UK construction industry. *Property Management, 28*(1), 4–17.

Bass, B. M., & Steidlmeier, P. (1999). Ethics, character, and authentic transformational leadership behavior. *The Leadership Quarterly, 10*(2), 181–217.

Bommer, M., Gratto, C., Gravander, J., & Tuttle, M. (1987). A behavioral model of ethical and unethical decision making. *Journal of Business Ethics, 6*, 265–280.

Bontempi, A., Del Bene, D., & Di Felice, L. J. (2023). Counter-reporting sustainability from the bottom up: The case of the construction company WeBuild and dam-related conflicts. *Journal of Business Ethics, 182*(1), 7–32.

Booth, P., & Schulz, A. K.-D. (2004). The impact of an ethical environment on managers' project evaluation judgments under agency problem conditions. *Accounting, Organizations and Society, 29*(5–6), 473–488.

Bowen, P., Akintoye, A., Pearl, R., & Edwards, P. J. (2007). Ethical behaviour in the South African construction industry. *Construction Management and Economics, 25*(6), 631–648.

Bowen, P., Pearl, R., & Akintoye, A. (2007). Professional ethics in the South African construction industry. *Building Research & Information, 35*(2), 189–205.

Brien, A. (1998). Professional ethics and the culture of trust. *Journal of Business Ethics, 17*, 391–409.

The Chartered Institute of Building. (2013). *A report exploring corruption in the UK construction industry.* https://www.ciob.org/media/212/download

Coggins, J., Teng, B., & Rameezdeen, R. (2016). Construction insolvency in Australia: Reining in the beast. *Construction Economics and Building, 16*(3), 38–56.

Cressey, D. R. (1953). *Other people's money; a study of the social psychology of embezzlement.* Free Press.

Degn, E., & Miller, K. R. (2003). Bid shopping. *Journal of Construction Education, 8*(1), 47–55.

Dondé, G., & Somasundaram, K. (2018). *Ethics at Work 2018 survey of employees – Australia.* Institute of Business Ethics.

Ernst & Young. (2010). *Driving ethical growth – New markets, new challenges. 11th Global Fraud Survey.* EYGM Limited.

Fan, L., Ho, C., & Ng, V. (2001). A study of quantity surveyors' ethical behaviour. *Construction Management and Economics, 19*(1), 19–36.

Fellows, R., Liu, A., & Storey, C. (2004). Ethics in construction project briefing. *Science and Engineering Ethics, 10*(2), 289–301.

Ferrell, O., & Fraedrich, J. (2016). *Business ethics: Ethical decision making & cases.* Cengage Learning.

Ford, R. C., & Richardson, W. D. (1994). Ethical decision making: A review of the empirical literature. *Journal of Business Ethics, 13*, 205–221.

Gregory, D. W., & Travers, E. B. (2010). Ethical challenges of bid shopping. *The Construction Lawyer, 30*, 29.

Gurley, K., Wood, P., & Nijhawan, I. (2007). The effect of punishment on ethical behavior when personal gain is involved. *Journal of Legal, Ethical and Regulatory Issues, 10*(1), 91.

Gyoo Kang, B., Edum-Fotwe, F., Price, A., & Thorpe, T. (2014). The application of causality to construction business ethics. *Social Responsibility Journal, 10*(3), 550–568.

Hall, D. (1999). Privatisation, multinationals, and corruption. *Development in Practice, 9*(5), 539–556.

Hamzah, A.-R., Chen, W., & Xiang, W. Y. (2010). How professional ethics impact construction quality: Perception and evidence in a fast developing economy. *Scientific Research and Essays, 5*(23), 3742–3749.

Hill, S., Lorenz, D., Dent, P., & Lützkendorf, T. (2013). Professionalism and ethics in a changing economy. *Building Research & Information, 41*(1), 8–27.

Ho, C. M. F. (2011). Ethics management for the construction industry. *Engineering, Construction and Architectural Management, 18*(5), 516–537.

Jennings, M. M. (2006). *The seven signs of ethical collapse: How to spot moral meltdowns in companies . . . before it's too late.* St. Martin's.

Johnson, D. G. (1991). *Ethical issues in engineering.* Prentice Hall.

Jung, Y., Jeong, M. G., & Mills, T. (2014). Identifying the preferred leadership style for managerial position of construction management. *International Journal of Construction Engineering and Management, 3*(2), 47–56.

Kercher, K. (2007). Corporate social responsibility: Impact of globalisation and international business. *Corporate Governance eJournal, 1*(1). https://doi.org/10.53300/001c.6906

Kliem, R. L. (2011). *Ethics and project management.* Taylor & Francis.

KMPG. (2013). *The KPMG survey of corporate responsibility reporting 2013.* https://assets.kpmg.com/content/dam/kpmg/pdf/2013/12/corporate-responsibility-reporting-survey-2013.pdf

Kotter, J. P., & Heskett, J. L. (1992). *Corporate culture and performance.* Free Press.

Loosemore, M., & Lim, B. T. H. (2018). Mapping corporate social responsibility strategies in the construction and engineering industry. *Construction Management and Economics, 36*(2), 67–82.

Lynch, W. T., & Kline, R. (2000). Engineering practice and engineering ethics. *Science, Technology, & Human Values, 25*(2), 195–225.

Mason, J. (2009). Ethics in the construction industry: The prospects for a single professional code. *International Journal of Law in the Built Environment, 1*(3), 194–204.

May, D., Wilson, O., & Skitmore, M. (2001). Bid cutting: An empirical study of practice in south-east Queensland. *Engineering, Construction and Architectural Management, 8*(4), 250–256.

Miller, S. R., Eden, L., & Li, D. (2020). CSR reputation and firm performance: A dynamic approach. *Journal of Business Ethics, 163*(3), 619–636.

Mirsky, R., & Schaufelberger, J. (2014). *Professional ethics for the construction industry.* Routledge.

Mohamad, N., Rahman, H. A., Usman, I., & Tawil, N. M. (2015). Ethics education and training for construction professionals in Malaysia. *Asian Social Science, 11*(4), 55.

Moriarty, J. (2021). *Business ethics: A contemporary introduction.* Taylor & Francis.

Nguyen, M. V. (2024). Barriers to corporate social responsibility performance in construction organizations. *Engineering, Construction and Architectural Management, 31*(4), 1473–1496.

Noreen, E. (1988). The economics of ethics: A new perspective on agency theory. *Accounting, Organizations and Society, 13*(4), 359–369.

Ormiston, R., Belman, D., & Erlich, M. (2020). *Payroll fraud in New York's construction industry: Estimating its prevalence, severity and economic costs.* Institute for Construction Research and Training.

Oswald, D., Ahiaga-Dagbui, D. D., Sherratt, F., & Smith, S. D. (2020). An industry structured for unsafety? An exploration of the cost–safety conundrum in construction project delivery. *Safety Science, 122*, 104535.

Othman, A. A. E. (2012). A study of the causes and effects of contractors' non-compliance with the health and safety regulations in the South African construction industry. *Architectural Engineering and Design Management, 8*(3), 180–191.

Shepherd, R., Lorente, L., Vignoli, M., Nielsen, K., & Peiró, J. M. (2021). Challenges influencing the safety of migrant workers in the construction industry: A qualitative study in Italy, Spain, and the UK. *Safety Science, 142*, 105388.

Starrett, S., Lara, A., & Bertha, C. (2017). *Engineering ethics: Real world case studies* (1st ed.). American Society of Civil Engineers.

Stout, L. (2010). *Cultivating conscience*. Princeton University Press.

Tow, D., & Loosemore, M. (2009). Corporate ethics in the construction and engineering industry. *Journal of Legal Affairs and Dispute Resolution in Engineering and Construction, 1*(3), 122–129.

Transparency International. (2011). *Bribe Payers Index 2011*. https://www.transparency.org/en/publications/bribe-payers-index-2011

van den Heuvel, G. (2005). The parliamentary enquiry on fraud in the Dutch construction industry collusion as concept between corruption and state-corporate crime. *Crime, Law and Social Change, 44*(2), 133–151.

Vee, C., & Skitmore, M. (2003). Professional ethics in the construction industry. *Engineering, Construction and Architectural Management, 10*(2), 117–127.

Wood, G., McDermott, P., & Swan, W. (2008). The ethical benefits of trust-based partnering: The example of the construction industry. *Business Ethics: A European Review, 11*(1), 4–13.

Zarkada-Fraser, A. (1998). *Tendering ethics: A study of collusive tendering from a marketing perspective*. Queensland University of Technology.

3 Stakeholder Management

Abid Hasan

3.1 Introduction

Contemporary construction organisations operate in dynamic and increasingly complex social, cultural, political, and economic environments. Moreover, the construction business environment is continuously going through dramatic changes due to the emergence of new project-based delivery models, decentralised planning and operation, global supply chain partners, growing digitalisation and automation, regulatory reforms and increased societal awareness of the construction industry's impact on communities and the environment. As a result, construction companies need to adapt their business practices and operations to match the contemporary market and societal expectations and demands. To do so effectively, they must understand their stakeholders' needs and expectations, both existing and emerging, and engage and collaborate with them to achieve business goals.

For instance, clients, activist groups and the public increasingly demand a sustainable and carbon-neutral built environment, given the significant carbon footprint and negative impact of construction activities on the environment. In addition, regulatory reforms and revisions in planning and construction codes in recent years have focused more on improving sustainability in the construction industry. These changes in the needs and expectations of stakeholders could have significant implications for construction companies that are not paying attention to the adverse environmental and social impacts of their business activities. It may result in negative media coverage, loss of business opportunities, penalties resulting from non-compliance, strained relationships with stakeholders and damage to their corporate image and reputation. Therefore, understanding and responding to stakeholders' needs is essential to managing a successful construction company.

Organisations that neglect their stakeholders for short-term gains often face costly repercussions in the long run because, unlike some other businesses, construction companies cannot function in silos or an isolated environment dealing with very few stakeholders. They rely on a large number of actors and supply chain partners, such as project clients, suppliers, subcontractors, regulatory bodies, communities and employees, to deliver construction projects and achieve their business goals. Often, these stakeholders have varying and sometimes competing interests, expectations, desires and needs vis-à-vis the company. Similarly, construction companies operating in different regions with business interests in national and overseas construction markets must engage and work with diverse individuals, groups, organisations and government bodies, often from different cultures and with different or conflicting views on various strategic and operational matters. Unfamiliar economic and regulatory environments and stakeholders with different social norms and cultural beliefs in different geographical regions and countries could severely affect international projects and operations of multinational construction companies (Aarseth et al., 2014).

DOI: 10.1201/9781003223092-3

Therefore, irrespective of size, location and business activities, construction companies must effectively engage with stakeholders and develop and maintain positive relationships to achieve sustainable growth and increase revenue and profit. They can only function and fulfil their strategy if they adequately manage the expectations and actions of various stakeholders. Proper stakeholder management is critical for successfully managing a construction company and is fundamental to its existence and growth. To sustain and grow in a highly competitive, fragmented, regulated construction industry, stakeholder management should be at the core of a company's short-term and long-term strategy and day-to-day operations.

The chapter first introduces the concept of stakeholders. Next, it discusses the stakeholder identification and analysis process. The subsequent sections of the chapter discuss stakeholder engagement, communication and underlying principles. Finally, the chapter provides insights into the role of corporate reputation in stakeholder management. It also poses vital questions to be considered for developing a stakeholder management plan and delivering it successfully.

3.2 *Stakeholder* Definition

Before we discuss stakeholder management in more detail, it is essential to understand what the term *stakeholder* means in the organisational context. The definition of a stakeholder has been widely contested in the literature. For example, Friedman (1970) argued that a business should primarily focus on the shareholders' interest by generating a positive return on investment and maximising their profit. In his book *Capitalism and Freedom*, Friedman (1962, p. 133) stated that

> there is one and only one social responsibility of business – to use its resources and engage
> in activities designed to increase its profits so long as it stays within the rules of the game,
> which is to say, engages in open and free competition without deception or fraud.

Therefore, value creation for shareholders must be central to the enterprise strategy. To do so, however, it can devote resources fairly for the betterment of the community and its government (Friedman, 1962). Friedman (1962) further reasoned that to call such activities 'social responsibility' would be wrong as the primary intention of an organisation would be to serve self-interest, maximise profits and create better value for its shareholders.

Freeman (1984), on the other hand, argued that the purpose of an organisation is to create value for all its stakeholders, including shareholders. Even if the sole focus of an organisation is to increase shareholder wealth, it cannot successfully do so without paying attention to the needs of different stakeholders because it relies on them to achieve various business objectives. While the conceptualisations of the term stakeholder by Freeman (1984) and Friedman (1970) might appear contradictory at first glance, as widely debated as a stakeholder/shareholder dichotomy in the literature, Freeman et al. (2010) viewed them as compatible concepts. After all, investors and shareholders are one of the most critical stakeholders for a business or organisation. Suppose a construction company cannot provide healthy positive returns to shareholders on their investment. In that case, it will likely struggle to raise capital, support its projects and operations and invest in new initiatives to grow the business. Therefore, organisations must also ensure that they meet their shareholders' expectations by achieving growth and profitability, among other things.

However, *should meeting or exceeding the shareholders' expectations be the sole purpose driving an organisation's social and economic activities?* The answer to this question forms the basis for the fundamental distinction between Friedman's (1970) and Freeman's (1984) conceptualisations of a stakeholder. The former supports the view that the driving principle for a business is maximising profit for shareholders. In contrast, the latter sees profit as a by-product of a successful business when it looks beyond the shareholder profit maximisation approach and focuses on value creation for all stakeholders, including shareholders.

Let us look at some other definitions of stakeholders by different researchers. Mitroff and Mason (1980) defined stakeholders as 'those who depend on the organisation for the realisation of some of their goals, and in turn, the organisation depends on them in some way for the full realisation of its goals'. Clarkson (1995, p. 106) defined stakeholders as

persons or groups that have, or claim, ownership, rights, or interests in a corporation and its activities, past, present, or future. Such claimed rights or interests are the result of transactions with, or actions taken by, the corporation, and may be legal or moral, individual or collective. Stakeholders with similar interests, claims, or rights can be classified as belonging to the same group: employees, shareholders, customers, and so on.

The *PMBOK® Guide* introduced the *perception of being affected* as an essential consideration while defining stakeholders. According to *A Guide to the Project Management Body of Knowledge (PMBOK® Guide) – Seventh Edition*,

Stakeholders can be individuals, groups, or organisations that may affect, be affected by, or perceive themselves to be affected by a decision, activity, or outcome of a portfolio, program, or project. Stakeholders also directly or indirectly influence a project, its performance, or outcome in either a positive or negative way.

(Project Management Institute, 2021, p. 58)

Corporate stakeholders can also be defined as 'any entity that has an interest in the organisation and/or can be affected by the corporation and/or can influence the organisation' (Heikkurinen & Mäkinen, 2018, p. 5).

Considering different viewpoints, a stakeholder could be defined narrowly as 'those groups without whose support the organization would cease to exist' (Stanford Research Institute, 1963, cited in Freeman, 1984, p. 31) or, more broadly, as 'any group or individual who can affect or is affected by the achievement of the organization's objectives' (Freeman, 1984, p. 46). The first definition covers the network of primary stakeholders (e.g., financiers, customers, suppliers, employees and communities) fundamental to running a business. In contrast, the second definition adds secondary stakeholders (e.g., media and special interest groups) that can influence various activities and decisions of a company. Essentially, a construction company and its stakeholders operate in a system bounded by formal (e.g., contract-based arrangements with clients, suppliers, etc.) and informal (e.g., with the public, media, etc.) relationships based on interest and the potential to influence or affect each other.

3.3 Stakeholder Identification

Stakeholder identification is the first step in stakeholder management. The definitions of stakeholders proposed by Freeman et al. (2010) and other researchers provide a foundation for identifying stakeholders for a construction company. However, the answer to the question *Who are the stakeholders of a construction company?* is not always straightforward and is often complex and context dependent. While a standard list of stakeholders would generally include employees, suppliers, clients, customers, partners, competitors, community and various regulatory bodies, developing an exhaustive list of stakeholders requires careful consideration of the social, cultural, legal and business ecosystems surrounding the company and its operations. Given the significant adverse impacts of construction activities on the environment, global warming and climate change, the future generation can also be considered an important stakeholder for construction companies.

Let us discuss a few examples of stakeholders, their interests and how they can affect or are affected by the decisions and actions of a construction company.

Employees of a construction company are among influential stakeholders who can profoundly affect the company in several ways due to their direct relationship with it, involvement in its day-to-day operations and direct or indirect participation in many crucial business decisions. Employees also constitute the most critical resource and help the company build strong relationships with other stakeholders. They represent the company to other stakeholders and are expected to act in its best interests. At the same time, employees can disrupt routine business activities (e.g., by participating in an industrial action). Their unethical conduct could damage the company's reputation and limit business opportunities.

On the other hand, a construction company's business decisions, such as starting a new construction project at a new location, may require employees to relocate, affecting their personal and family lives. Similarly, adopting new work practices and technology may require employees to undergo training and skill development. Likewise, other human resource management decisions, such as salary increases, employment benefits and redundancies, could significantly impact employees. Therefore, employees' financial goals, career ambitions and other personal and professional goals can be directly affected by their organisation's success, failure and actions (Greenwood, 2007).

Project clients, *suppliers* and *subcontractors* can significantly impact a construction company's business activities. As discussed in Chapter 1, unlike manufacturing companies of electronics and many other products, construction companies do not usually first build products (e.g., hospitals, highways, etc.) and then wait for clients to buy them. They rely on clients to initiate or fund new construction projects. As a result, a good business relationship with the client is crucial for acquiring new work in the future to maintain business continuity and growth in the construction industry. On the other hand, delays, poor quality and other issues in construction projects, along with disputes and poor relationships with the client, can severely affect the company in a highly competitive construction market. Likewise, a construction company's poor performance leading to late project completion, cost overruns and substandard construction quality can affect their client's business and reputation.

Construction companies, subcontractors and suppliers depend on each other for successful project delivery and to sustain and grow their businesses. They also have a strong interest in each other's businesses and decisions. For instance, subcontractors or material or equipment suppliers regularly working for the company would be interested in its business growth as it could lead to new opportunities for them in the form of more work and new purchase orders. Likewise, the company would monitor the business performance of its suppliers and subcontractors to ensure continuity of its operations and supply chain.

Governments enjoy a unique status as stakeholders due to their power to regulate the industry and influence the business environment. For example, the unprecedented actions and government directives in the public health interest during the COVID-19 pandemic led to disruptions in the supply chain, skill shortages, price escalations and delays. Government interventions during the pandemic severely affected the functioning of construction companies (Abubakar et al., 2022). The financial losses incurred from the stoppage of construction work, skill and resource shortages, reduced productivity and high inflation forced many construction companies into insolvency. However, at the same time, governments in many countries invested in new construction projects to revive the economy and help construction businesses recover from the pandemic. The government also declared the pandemic a *force majeure* in some countries to protect the contractors from delays and associated liabilities.

The government also acts as a client in public projects. Governments can sometimes also help construction companies facing financial problems avoid bankruptcy if it is in the public interest, especially when a company is involved in delivering public projects and their business failure would affect the public or community. The government can also introduce new acts and regulations to protect the financial interests of construction companies. For example, The Security of Payment Act (known as the SOP Act) in Australia was introduced to ensure that any person who carries out construction work or supplies related goods and services under a construction contract gets paid. It provides a fast and inexpensive process to recover payments due under a construction contract without involving lawyers.

Reforms in construction laws and regulations and changes in the regulatory environment may also force construction companies to change how they operate or do business. For instance, government-led changes such as mandatory building information modelling usage in public projects in many countries (for example, in the United Kingdom and many European nations), quotas for women in different construction roles in public projects (for example, in Victoria, Australia), higher energy performance requirements in buildings, prohibition on the use of certain building materials such as engineered stone (e.g., in Australia), etc., can have severe implications for construction companies that are not prepared to adapt to these regulatory reforms. On the other hand, governments rely on construction companies to build the necessary infrastructure for the communities, contribute to the economy and employment and work ethically for the betterment of society. Poor employment practices and substandard construction quality can lead to adverse health and safety consequences for workers and the general public. Similarly, delays and cost overruns can lead to the wastage of public money and delays in providing essential services.

Media can be an influential stakeholder, especially in this digitalised world. It has the potential to both benefit and harm a construction company, depending on how it reports corporate news. Negative media coverage could quickly change public and stakeholder perceptions towards a company, leading to drops in its share price and market value, loss of business opportunities, interventions from regulatory bodies, changes in leadership and stoppage of projects or business activities. On the other hand, positive media coverage can help a company achieve a competitive advantage (Voinea & van Kranenburg, 2017). At the same time, the media can allow a company to reach out to its stakeholders and gain their trust by rebutting negative news and providing more information on its strategy and action plan. Therefore, the media could shape a firm's legitimacy, reputation, and broader perception of society and other stakeholders of its business activities and ethical behaviour. Nowadays, the increased media coverage due to the speed and reach of digital media platforms can immediately impact a business and its reputation.

The following case study shows how negative media coverage could reduce the confidence of stakeholders in a construction company but, at the same time, how a company can use the wider audience reach of digital media to communicate and engage with stakeholders and regain their trust.

Media speculation about bankruptcy

In May–June of 2022, just weeks after the unexpected death of the CEO and founder of Metricon, Australia's most prominent home builder, there was a widespread rumour in the mainstream media about the company's imminent collapse. The media sources reported that the company faced extreme financial constraints due to rising labour costs, short supply and high demand for materials, fixed price contracts and rising interest rates after the first 2 years of the COVID-19 pandemic. At that time, Metricon had 2,500 directly employed workers and more than 4,000 homes under construction. In addition to private home buyers, the government was among the company's major clients. The media reports created panic among employees, suppliers, subcontractors and private and public clients. Many potential customers started looking elsewhere for a new builder to construct their houses.

In response, the acting CEO of the company held a press conference to deny the news. It was stressed that the company's financial health was not a cause for concern while admitting that it was going through a tough time like any other construction company. The company adopted a series of measures, including a $30 million injection of cash into the company by its shareholders and restructuring. The company executives met with the state government representatives. Many stakeholders, such as suppliers, subcontractors and Master Builders, rallied behind the company to extend their support and publicly deny the rumours using both conventional (e.g., newspaper, press release) and digital media platforms (e.g., social media such as LinkedIn), which helped the company change negative sentiments and re-establish consumer confidence. The company is in business at the time of writing this case study (June 2024).

There can be many other stakeholders relevant to a construction company. Defining its core purpose and what it would mean to different individuals, groups and organisations, and vice versa, could help construction companies identify the relevant stakeholders. For example, the stakeholders for a local small residential building construction company will differ from those of a large multinational construction company with projects and business interests in different regions and countries. Similarly, stakeholders for a construction company that provides affordable public housing would be different from those of a company delivering expensive custom-built homes. Moreover, the stakeholders a construction company will engage with at the project and organisational levels could differ. In any case, construction companies must identify and consider a network of primary and secondary stakeholders in their strategy and business decisions for successful stakeholder management, which is necessary to sustain and grow their business.

Figure 3.1 presents a simplistic view of the stakeholders of a construction company from the perspective of project and organisation with business activities spread over different regions, locally and internationally.

While identifying stakeholders, it is essential to remember that some stakeholders, such as *communities* and *councils*, may appear to be shared across local, inter-state and international

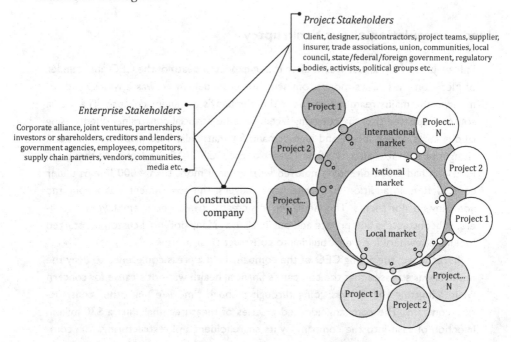

Project Stakeholders

Client, designer, subcontractors, project teams, supplier, insurer, trade associations, union, communities, local council, state/federal/foreign government, regulatory bodies, activists, political groups etc.

Enterprise Stakeholders

Corporate alliance, joint ventures, partnerships, investors or shareholders, creditors and lenders, government agencies, employees, competitors, supply chain partners, vendors, communities, media etc.

Construction company

Project 1

International market

Project... N

Project 2

National market

Project 2

Project... N

Local market

Project 1

Project 1

Project... N

Project 2

Figure 3.1 Typical arrangement of stakeholders of a construction company.

projects. However, the specific organisations and entities could differ. For example, even for a small construction company, there could be many different clients, customers, councils and communities as stakeholders associated with various projects. Stakeholders at the project level could vary depending on the nature, type and location of the project and need to be carefully identified. Moreover, in some projects, such as infrastructure projects spread over hundreds of kilometres, stakeholders like community and local councils could include a cluster of many communities and local councils along the project route, each with different needs and expectations. Government bodies at state and national levels could have very different perspectives on the built environment policies and macroeconomic factors affecting the business activities of a construction company. Similarly, regulatory requirements could vary from one council to another, from one state to another and across countries.

Identifying heterogeneous groups or stakeholders with different characteristics as large amorphous groups like 'the public' or 'society' should be avoided (Freeman et al., 2007). While it may simplify the stakeholder identification process by reducing the number of stakeholders, failure to capture the differences in their expectations and impacts on the business could have severe repercussions for the organisation. For example, a public client could have different expectations than a private client, and the latter, in turn, could be segmented by location, organisation size and end-users, which necessitates the need to segment the stakeholders further once they have been identified at the generic level. Similarly, the company may need to create different engagement opportunities to effectively communicate and consult with culturally and linguistically diverse groups of people.

It is essential to note that stakeholder identification is not a one-time or one-person activity but an ongoing, comprehensive and resource-intensive exercise to ensure that all existing or new stakeholders are identified and accounted for in the stakeholder management planning process. Identifying a comprehensive list of stakeholders requires significant experience, skill and input from multiple people or groups, especially in large or multinational construction companies with many stakeholders spread over different projects, products and regions (Figure 3.1). Therefore, construction companies should seek the opinions of diverse people and teams while developing a list of potential stakeholders.

Some entities may not get involved or react until they are affected by the company's operations and, as a result, may be overlooked in normal circumstances. Rosenhead (2009) discussed a case study of a road project executed by a local council in the United Kingdom that failed to identify the railway authority as a key stakeholder. Since the railway authority had a bridge nearby, they became worried about the potential impact of the construction work on this structure and initiated actions to stop the project. The council lost the legal proceedings and had to pay £0.25 million in legal costs and amend the scheme, adding a delay of more than a year to the project schedule.

Similarly, the role and importance of stakeholders such as the media and special interest groups could be easily underestimated, but they can benefit or harm a construction company in certain situations. The fact that stakeholders who may not exist or are hidden or silent would need to be considered and that the ones who exist could be knowingly or unknowingly omitted makes stakeholder identification a tricky and complex exercise for construction companies. Construction companies and project teams should add all apparent and potential stakeholder entities to the initial list of stakeholders and remove the irrelevant entities later, as the omission of a stakeholder can have significant implications for the company.

3.4 Stakeholder Analysis or Mapping

The next step after identifying stakeholders is to analyse them to understand how a particular stakeholder can affect or be affected by the actions of the organisation. Among several stakeholders identified by a construction company, not all of them will have the same level of interest or involvement in the business activities of the company. Stakeholder analysis helps a company comprehensively understand the relevant stakeholders and their interrelationships, needs and objectives and position them within the decision-making system (Reed et al., 2009). Grouping stakeholders into different categories based on their influence level and mapping out their relationship with the company can be very useful in determining their potential influence on the company (Aly et al., 2019).

There are multiple ways to classify or group stakeholders. For example, stakeholders could be classified as internal and external stakeholders, direct and indirect stakeholders, primary and secondary stakeholders, salient or non-salient groups, contracted and un-contracted stakeholders, normative and derivative stakeholders, social and non-social stakeholders, voluntary or involuntary risk-bearers, human and non-human stakeholders and so on. Harris (2010) further divided primary and secondary stakeholders into different levels. The first level of primary stakeholders included clients, customers, etc., whereas project managers, designers, contractors, etc., formed the second level of primary stakeholders. Stakeholders such as subcontractors, supply chain vendors and insurance organisations were identified as the third level of primary stakeholders. Mitchell et al. (1997) grouped stakeholders into seven different types based on various combinations of attributes of power, legitimacy and urgency – three possessing only one attribute (latent stakeholders), three possessing two attributes (expectant stakeholders) and one possessing all three attributes (definitive stakeholders). In contrast, entities with no power, legitimacy or urgency in relation to the company should not be considered stakeholders.

Table 3.1 shows a few examples of different classification systems of stakeholders.

For stakeholder analysis or mapping exercises, the *power–interest matrix* is widely used. It groups various stakeholders based on their power and interest levels in a 2 × 2 matrix. *Power* refers to a stakeholder's ability to influence business outcomes. In comparison, *interest* refers to the stakeholders' positive interest or negative concerns related to the business activity. The four

Table 3.1 Stakeholder classification systems

Classification system	Definition	Typical examples
Primary and secondary stakeholders (Clarkson, 1995)	A primary stakeholder group is one without whose continuing participation the company cannot survive. There is a high level of interdependence between the company and its primary stakeholder groups.	Shareholders and investors, employees, customers, suppliers, governments, communities, etc.
	Secondary stakeholder groups are defined as those who influence or affect or are influenced or affected by the company, but they are not engaged in transactions with the company and are not essential for its survival. The company is not dependent on secondary stakeholder groups for its survival. Such groups, however, can cause significant damage to a company.	News media, special interest groups, etc.
Normative and derivative stakeholders (Phillips, 2003)	Normative stakeholders are those stakeholders to whom the company has a moral obligation, an obligation of stakeholder fairness, over and above that due other social actors, simply by virtue of their being human.	Shareholders, customers, employees, etc.
	Derivative stakeholders are those groups whose actions and claims must be accounted for by managers due to their potential effects on the company and its normative stakeholders. A group may be considered a derivative stakeholder in a sense, either beneficial or harmful to the company.	News media, activists, etc.
Voluntary and involuntary stakeholders or risk-bearers (Clarkson, 1994)	Voluntary stakeholders bear some form of risk as a result of having invested some form of capital, human or financial, something of value in the company.	Investors, employees, etc.
	Involuntary stakeholders are placed at risk as a result of the company's activities.	Impacted communities and businesses, etc.

categories of stakeholders based on the power–interest matrix are (a) low power–low interest, (b) high power–low interest, (c) low power–high interest and (d) high power–high interest. For instance, decision-makers such as the project sponsor, project manager and project team have high power and interest. In comparison, government and regulatory agencies are often included in the high power–low interest quadrant.

Another stakeholder analysis method is the *threat–cooperation matrix* based on stakeholders' potential for threat and potential for cooperation (Savage et al., 1991). It identifies four groups of stakeholders, recommending different management strategies for each. The ideal group of stakeholders (i.e., supportive stakeholders) shows low threat potential and high cooperation potential. In contrast, the non-supportive stakeholders exhibit high threat potential and low cooperation potential and may cause distress to a company. For instance, staff and managers could be identified as supportive stakeholders, while employee unions could be non-supportive. However, it must also be noted that the same stakeholder could behave differently in harmonious and stressful situations. For example, the same employees could support salary increase or bonus decisions but become hostile if the company is restructuring, resulting in job losses, more work or transfer to a new department. Therefore, the behaviour of a particular stakeholder cannot be assumed to be consistently supportive, threatening or neutral. Moreover, a stakeholder may decide not to exercise their capacity to threaten or cooperate depending on the circumstances or factors unrelated to the company. A variation of the threat–cooperation matrix analysis is the *threat–opportunity matrix* that classifies stakeholders based on the potential for threat and opportunity.

It is important to note that different stakeholder classifications or mapping systems may group the same stakeholder differently depending on the criteria used for their grouping. Moreover, a stakeholder may be included in different groups at different points depending on the context of the business activity. For instance, the local council could be identified as a primary stakeholder in a council-led project (e.g., the development of a public park) because of their involvement as a client. Moreover, they will be in the high power–high interest stakeholder quadrant. However, the same council may be considered a secondary stakeholder in a private project (e.g., a residential dwelling construction) and placed in the high power–low interest stakeholder quadrant. Similarly, the positioning of a stakeholder in a particular quadrant may change with time. For instance, the low interest of not-for-profit environmental and animal rights groups initially in a construction project could convert into high interest as they learn about the project's negative impacts on flora and fauna. Accordingly, their potential for threat could increase from neutral to threatening behaviour. Therefore, the stakeholder behaviour or action may change over time and must be evaluated for different situations or circumstances.

Additionally, a stakeholder sometimes could play multiple roles and, therefore, it could be challenging to identify them as a member of a particular group only. For instance, an employee of a construction company could be a company shareholder, a community member, a member of the worker union and a board member of a not-for-profit environmental protection organisation. Though a company may require employees to declare a potential conflict of interest, in many cases, it cannot prevent them from being a member of another stakeholder group if it is legal and allowed. Similarly, multiple stakeholders could represent the stake of a particular stakeholder (Starik, 1995). For instance, construction workers' interests can be represented by themselves, unions, supervisors, managers, employers and regulatory bodies.

Depending on the situation or issue, stakeholders' interests, capabilities and needs could vary and, thus, their significance (Savage et al., 1991). Therefore, irrespective of the classification system used for mapping stakeholders or determining how stakeholders will be arranged in different groups, it is crucial to understand the nature of their interest in organisational activities and their influence on the firm's decisions and goals. Similarly, attention must be paid to understanding how the company's decisions and actions will affect the stakeholders. Understanding the stakeholders' stakes and expectations is vital to the stakeholder mapping process and developing targeted stakeholder management strategies (Bourne & Weaver, 2010).

Although it is not always possible to accurately predict the actions of various stakeholders and how their decisions and actions will affect the organisation, anticipating stakeholder behaviour based on detailed stakeholder analysis could help construction companies better prepare to handle adverse conditions. Construction companies must make genuine efforts to identify stakeholders, their interests, influence or power, objectives, behaviour and predictability and map their relationships with the company and each other. Predicting how a stakeholder would respond in different situations and planning for various possibilities or best- or worst-case scenarios is essential for successful stakeholder management.

3.5 Stakeholder Engagement

Though resource- and time-consuming, the stakeholder analysis based on detailed diagnosis or mapping of stakeholders could clarify the relative influence and interest that different stakeholders can exercise and their impacts on the company. It helps companies develop initial strategy and identify engagement opportunities and, thus, inform stakeholder management practices. Recognising the potential impacts of the company's actions on stakeholders and vice versa and preparing to manage various risks emerging from those actions are fundamental to effective stakeholder management.

While diagnosing stakeholders along the dimensions of potential for threat and cooperation or power and interest could help identify suitable engagement methods, a construction company should always work towards improving the relationship with stakeholders. Not all stakeholders will always be friendly, and some (e.g., competitors) may have a greater interest in the failure of the business than its success. Still, construction companies must endeavour to engage with all stakeholders and address their concerns productively. On many occasions, construction companies are required to work with their competitors to resolve industry-level issues; thus, developing a healthy business network always pays off.

It may also be noted that not all stakeholders will always be relevant to all decisions taken by a company. For instance, while implementing a new software platform across various departments of a construction company, the concerned stakeholders would be different from that if the company decides to expand its business to an overseas market. Therefore, in some situations, the company would require working closely with its employees, unions and suppliers, while in others it would need to engage with the government, regulatory bodies and other institutional stakeholders. However, in all situations, the stakeholder's capacity, opportunity and willingness to threaten or cooperate with the company must be considered while developing stakeholder engagement strategies (Savage et al., 1991).

Building a positive relationship with potential for high-threat and low-cooperation stakeholders could increase their cooperation level and, thus, reduce the need for a defensive or aggressive strategy. In many instances, increased collaboration could lead to better outcomes for all concerned stakeholders. For example, construction companies could engage with employee unions, suppliers and clients to improve work conditions and occupational health and safety outcomes for construction workers and professionals. Such improvements will likely result in better work productivity, less absenteeism and turnover and better reputation benefitting all these stakeholders.

Kivits and Sawang (2021, p. 10) defined *stakeholder engagement* as

the wide range of tools and practices an organization can use as a mechanism for consent, control, cooperation, accountability, employee involvement and participation, enhancing trust, enhancing fairness and corporate governance by involving stakeholders in its organizational activities.

The definition covers the broad scope of engagement activities with various stakeholders using different engagement tools at different operational levels and points in time. Engagement with stakeholders in a positive manner for mutual benefits, inclusivity and cooperation is essential for the success of organisational activities. Engaging stakeholders, no matter who or what they are, is crucial for a construction company's success.

Regular consultation and interactions with stakeholders allow companies to determine their needs and preferences and work out strategies for managing expectations. They could reduce the potential for threat and better predict stakeholder behaviour and actions if they regularly engage with stakeholders. However, a construction company does not need to engage with each stakeholder with the same intensity. The level of engagement could vary from one-way engagement (i.e., information dissemination) to participative engagement, which involves a two-way relationship and active participation. How a construction company would deal with a stakeholder with high power and interest must differ from its engagement with a low interest–low power stakeholder. For instance, a construction company must work closely with or collaborate with high power–high interest stakeholders while keeping low interest–low power stakeholders informed about the business activities. Similarly, involving stakeholders with low power–high interest could provide insightful opinions and help make better decisions.

Savage et al. (1991) argued that the level of power a stakeholder can exercise is often a function of the organisation's dependence on the stakeholder. When a construction company solely depends on a stakeholder, let us say a particular client or a specific material supplier, it could take away some of the negotiating power of the company. It may also put the company in a vulnerable position as those stakeholders can exercise a high level of power and take advantage of its high dependence. In contrast, stakeholders (e.g., subcontractors) who depend on the construction company will be more willing to cooperate. Therefore, while developing long-term relationships with stakeholders is recommended, identifying alternatives to businesses such as material and equipment suppliers and developing new business opportunities by diversifying the client base could be a vital stakeholder management strategy to reduce dependency. Additionally, determining the predictability of the stakeholder behaviour or potential impacts is essential in deciding the stakeholder engagement plan (Newcombe, 2003). For example, a stakeholder with low predictability could act differently from a stakeholder whose actions are more predictable. Consequently, stakeholders with low predictability would be required to be monitored more closely.

The impact of stakeholder participation on a decision increases as the company moves from informing (least impact) to consulting or involving (less impact) to collaborating (high impact). Though preferred, it is not always possible to satisfy stakeholders even after extensive engagement. Nonetheless, a transparent and comprehensive engagement process could help build trust, consensus and cooperation, and stakeholders will be more willing to compromise and understand mutual concerns and expectations. In most cases, early and ongoing engagement processes provide access to stakeholders' ideas and concerns that construction companies can use to improve business outcomes and create value for other stakeholders.

Kivits and Sawang (2021) recommended six principles to guide the stakeholder engagement process: inclusiveness, reaching out, mutual respect, integrity, affirming diversity and adding value. Heikkurinen and Mäkinen (2018) proposed an integrative perspective on stakeholder engagement. According to this perspective, economic engagement with stakeholders should not be at the expense of democratic rights and structures, the common good and moral and ethical considerations. Stakeholder engagement also promotes values such as representativeness, transparency, accessibility, responsiveness, accountability and sustainability (Franklin, 2020). In many cases, stakeholder engagement could lead to better outcomes for the company due to the sharing of ideas and information. Therefore, construction companies should take it from the perspective of value creation processes for all stakeholders.

Stakeholder engagement plan (Larsen and Toubro)

Larsen and Toubro (L&T), an Indian multinational conglomerate and India's largest technology-driven engineering and construction company, ranks among the world's top engineering and construction companies and has a presence in 31 countries. The company recognises that its success and business sustainability depend on support from its stakeholders and makes it imperative to understand their needs and interests. For instance, it collects customer feedback on various aspects of customer service such as schedule, product quality, safety standards, response to client requirements, communication, courteousness, etc., on a scale of 1 to 10 (10 = *excellent*, 1 = *poor*) on a quarterly basis to be reviewed during the management review meetings and actions are taken to improve performance in the areas rated below an average of 8.0 points.

The stakeholder engagement plan adopted by Larsen and Toubro is shown in Table 3.2.

Table 3.2 A typical stakeholder engagement plan

Stakeholder	Engagement approach
Employees	Regular communication, project updates, town halls, departmental meetings and connect sessions where employees voice their ideas and concerns are heard. Enterprise-wide employee portal called 'L&T Scape' for regular connect
Customers	Continuous ongoing interaction with customers through various channels such as customer meets, information desks, workshops and conferences, exhibitions and trade fairs, advertising campaigns, bulletins and news, one-on-one interactions, periodic reviews, annual reviews, customer satisfaction surveys and feedback forms
Suppliers and contractors	Periodic partner meets, e-tendering and e-procuring and supplier meets. Regular visits to suppliers' and contractors' facilities
Shareholders and investors	Quarterly calls, face-to-face meetings and annual general meetings to provide information and seek their perspectives on the company's performance and strategy
	Investor grievance channels
	Presentation of financial and other relevant business reports
	Regular announcements and filings with the stock exchanges

(*Continued*)

Table 3.2 (Continued)

Stakeholder	Engagement approach
Government	Regular interaction with local governments Member of important industry associations Active role in policy formulation
Communities	Regularly engage with the community through corporate social responsibility initiatives, volunteer activities, a quarterly review of integrated community development projects Continuous engagement with village panchayats and local authorities
Media	Regular press meets and periodic media visits

As shown in Table 3.2, engagement involving employees and managers who regularly participate in the decision-making processes concerning the company's internal matters usually occurs within organisational boundaries. However, external engagement occurs outside the organisational boundaries and requires different engagement strategies from internal engagement. Written and verbal communication, corporate social responsibility (CSR) activities, press releases, surveys and feedback or complaints channels are examples of various means a construction company can use for stakeholder engagement (Table 3.2).

It is important to note that stakeholder engagement is not a single one-off activity. Moreover, it may require considerable resources and support from management and employees. Stakeholder engagement practices should be broken down into a series of actions and tasks that can be implemented, monitored and reviewed periodically to identify the problem areas and solutions. Moreover, the lessons learnt during the process must be utilised to inform the engagement policies and practices and shared across the organisation. In addition to the engagement process, the consequences of engagement must be monitored and reviewed (Greenwood, 2007; Lane & Devin, 2018).

3.6 Stakeholder Communication

Effective and timely communication is vital to any relationship. Likewise, establishing and maintaining a productive relationship with stakeholders requires a comprehensive stakeholder communication strategy. Large construction organisations often have dedicated staff and teams for stakeholder communication and engagement. The team typically includes a communications and stakeholder manager leading communications and engagement activities, a media spokesperson, a social media manager and other staff. They are responsible for communicating with stakeholders and managing public opinions about the company, its operations and various corporate decisions. On the other hand, small companies might overlook the importance of effective stakeholder communication due to a lack of resources, or some companies might intentionally ignore it despite having sufficient resources, which could affect their achievement of business goals and reputation in the long run.

Rather than assuming different stakeholders' positions concerning the company's actions and business decisions, transparent and timely communication could provide better information about their needs and expectations and build mutual trust, paving the path to future collaboration and negotiations with stakeholders. Moreover, in today's digitally connected world, due to the affordability and reach of mobile devices and the internet, even unvalidated news or views could quickly build positive or negative perceptions of the company and cause a loss of trust among stakeholders and shareholders. In such cases, proper and timely communication could help companies maintain the trust of stakeholders.

When developing a communication plan for stakeholder management, it is essential to remember that what works for one stakeholder might not work for others, and a one-size-fits-all approach should not be adopted. The means of communication could include both formal (website, documents and electronic communication) and informal (informal meetings) communication channels. The choice of communication media, style and frequency depends on the stakeholders and how effectively they can be reached and engaged with various communication tools. Since construction companies interact with some stakeholders (e.g., employees, subcontractors) more regularly than others (e.g., local councils), they adopt different strategies for communication with them.

The content or information being shared and the urgency of communication also inform communication choices. Similarly, the company must ensure that communication requirements stipulated in the contract documents and as required under relevant laws and regulations are met to avoid legal and contractual implications. For example, the contractor needs to follow the contract terms and conditions for claiming variation claims, regulatory guidelines for reporting incidents and council or government requirements for stakeholder communication, complaint management and reporting.

For each stakeholder, the company needs to consider the stakeholder characteristics, nature of information, communication format, method and communication frequency in the communication plan and whether it fulfils the organisation's legal, moral and ethical duties. Understanding and incorporating the specific communication needs of culturally and linguistically diverse

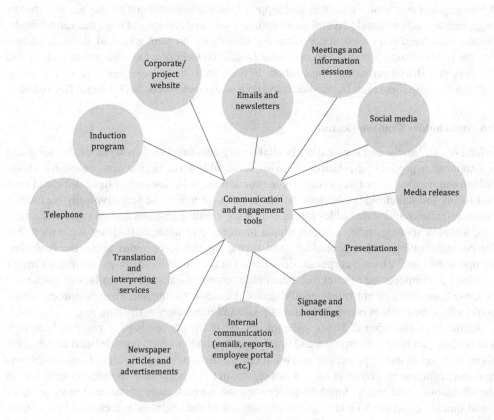

Figure 3.2 Communication tools and channels.

stakeholders in the communication plan is crucial for effective communication, especially in multicultural societies. Therefore, many communication channels and media are used to reach and engage different stakeholders. Wherever possible, the organisation should take the feedback of the concerned stakeholders on the communication plan and try to accommodate their preferences. Figure 3.2 shows some communication tools and channels typically used by construction companies in large projects.

The stakeholder communications plan should allow for two-way communication with stakeholders; therefore, stakeholders must also be able to communicate easily with the company. Information on how stakeholders can contact the company and communicate with its representatives should be available on the corporate website and project sites, along with complaint processes. Additionally, information on how, when and what kind of information will be shared with stakeholders and the mechanisms for feedback and response should be shared with the stakeholders in advance.

The communications plan should aim to not only provide timely and informative communication to stakeholders but also promptly respond to their concerns and queries to build mutual trust and transparency. The plan must demonstrate that the company is committed to minimising any potential impacts of its operations on the community and stakeholders. Moreover, the complaint protocol should ensure that each complaint is received, acknowledged, managed, closed or escalated to the appropriate staff or team within a promised time frame and recorded in the complaint register. The communication plan should also include information on communication during emergencies and unplanned events. The Metricon case study discussed earlier shows that timely and effective communication with stakeholders and public display of support from stakeholders could be crucial in times of crisis.

A typical stakeholder communications plan includes the following components:

1. Company background
2. Communication principles, approach, strategy and objectives
3. Key messages
4. Stakeholder identification and analysis
5. Communication and engagement activities
6. Roles and responsibilities
7. Risk and mitigation strategies
8. Complaints management system
9. Media management
10. Monitoring, reporting and evaluation

Since the stakeholder communication plan is expected to evolve as the conditions and stakeholder mapping change, it is essential to review and monitor the relevance and effectiveness of the existing plan and amend it as required. It should also be ensured that the plan is complied with and any deviations are noted. Irrespective of the organisation's size, accountability and responsibility concerning stakeholder communications must be fixed along with providing necessary support and assistance to individuals or teams managing various communication tasks.

3.7 Stakeholder Management: Value Creation Principle

Even with the best intentions for stakeholders or after following the best stakeholder engagement and communication practices, a construction company may experience the negative influence of the stakeholders. The differences could be simply because different stakeholders have different goals or objectives that do not always align. Therefore, developing and implementing a robust strategy on how the company can work proactively to eliminate or reduce conflicts with stakeholders is critical for its success.

According to Freeman et al. (2010, p. 28), a practical solution is to adopt a stakeholder approach to business that recommends 'creating as much value as possible for stakeholders, without resorting to trade-offs'. Companies may adopt a specific stakeholder or multi-stakeholder approach depending on whether their focus for value creation is on a selected few or multiple stakeholders. However, the underlying principle of value creation for stakeholders remains the same irrespective of the definition of stakeholders or the importance or priority of stakeholders, which could vary depending on business situations. Therefore, an organisation must identify value creation opportunities for its stakeholders.

It is important to reiterate that stakeholders are inherently connected in the process of value creation (Freeman et al., 2010). For example, a construction company cannot grow its business without recognising the needs of other stakeholders, such as clients, employees, suppliers, subcontractors and customers. The sole focus on maximising profit without considering the needs and expectations of stakeholders could damage the company's reputation, result in loss of community support and limit its business opportunities.

Prioritising the needs of a particular stakeholder at the expense of others can affect the performance of a company due to a complex business environment involving an interconnected network of stakeholders. For example, while clients are an important stakeholder group, construction companies cannot create value for clients without concentrating on their relationships with other stakeholders, such as employees, suppliers, regulators and the community. Therefore, a comprehensive view of stakeholder management instead of a narrow focus on one or two stakeholders is paramount for construction companies. Furthermore, the multi-stakeholder value creation thinking could inform the business strategy and actions of the company and help it manage the expectations of diverse stakeholders.

In the current business environment where attention to sustainability has become increasingly important, a construction company with a strong focus on sustainability and ambitions to achieve a carbon-neutral or net zero built environment would share common goals of protecting the environment with suppliers of sustainable materials and technology, the public, activists, investors and governments concerned about the negative impact of the built environment on the environment. The well-aligned interests and shared goals could create new opportunities for the company due to new investments and partnerships, favourable policies, subsidies and positive perceptions among media and communities.

On the other hand, if a construction company does not heed the calls of special interest groups to protect the environment or tackle climate change, complaints from the activists and the public could force the government to tighten the regulatory framework concerning safeguarding the environment, resulting in negative implications for the company and potential loss of business. Environmental degradation will also affect the quality of life of the employees, the community and future generations. Therefore, the company would struggle to achieve its business objectives or receive stakeholders' support if it does not create value for the stakeholders.

However, serving and satisfying stakeholders' expectations could be challenging due to competing social, environmental and business interests. Building consensus and working towards common goals or interests requires regular engagement, communication and negotiation. The leadership team and managers must ensure that the interests, desires and needs of various stakeholders are kept in balance (Freeman et al., 2007). In any case, construction companies must focus on creating and sustaining value for key stakeholders. Doing so will require managing stakeholder relationships thoughtfully and adopting a multi-stakeholder value creation approach (Freeman et al., 2007).

The stakeholder management practices of the construction giant Larsen and Toubro, discussed earlier in the chapter, recognise that value creation for stakeholders is integral to its stakeholder management strategy. It engages in regular and timely dialogue with stakeholders to understand their concerns and act accordingly. It enhances value creation by listening to its stakeholders and understanding what really matters to them. The regular dialogue also helps the company manage risks and opportunities proactively and set clear goals to deliver long-term shared value. The 2020–2021 integrated report of the company mentions that it trained more than 0.9 million vendors, dealers and subcontractors in financial year 2021. It also contributed around $1.5 billion to the exchequer in the same year. Furthermore, it helped organise training and skill development programs for 35,522 rural and urban youth, along with women and differently abled persons from underprivileged communities, to improve their employability.

Many construction companies sometimes adopt a narrow view of value creation thinking that involves investing in community projects as a part of their CSR activities but ignoring the broader issues. For instance, a construction company engaged in a highway construction project might donate money to improve health or education infrastructure along the project route as a part of their CSR program with the intention of earning the goodwill of the people and community support. However, its careless decisions and actions could damage the environment, deliver a poor-quality project or result in the mistreatment of employees and workers if it is not genuinely concerned about all stakeholders. Can we say that the company is fulfilling its social responsibility, though it might be spending a large sum of money and other resources on CSR activities?

Construction companies are a part of the society and community in which they operate. Corporate citizenship implies that they must meet their economic, social, legal and ethical obligations and contribute to the upliftment and well-being of their communities. Socially responsible construction companies understand their responsibilities to stakeholders and recognise the ethical obligation to create social value as a core of their enterprise strategy. Their social responsibility informs their business decisions and is reflected in their actions. Therefore, creating social value is not an ad hoc measure that can be achieved by setting up a separate CSR department or investing a part of the profit in community welfare projects. Protecting the environment by reducing the carbon footprint of construction activities, delivering high-quality projects, abiding by laws and regulations and ensuring the health, safety and well-being of their employees, workers and communities are all actions that contribute towards multi-stakeholder value creation, social responsibility and good corporate citizenship behaviour.

Table 3.3 shows some of the value creation opportunities for construction companies adopting a multi-stakeholder approach to value creation.

Table 3.3 Multi-stakeholder value creation opportunities

Stakeholders	Value creation opportunities
Employees	Remuneration and other benefits
	Health, safety and well-being
	Professional development and career growth opportunities
	Inclusive work culture
	Participatory decision-making process
	Flexible work arrangements

(*Continued*)

Table 3.3 (Continued)

Stakeholders	Value creation opportunities
Shareholders and investors	Increased revenue and market capitalisation
	Solid financial performance
	Consistent credit ratings
	Attractive return on investment
	High dividend
Customers	On-time project completion
	Cost reduction
	Exceeding 'customer expectations'
	Customer engagement at multiple levels
	Long-lasting relationships
	Energy and greenhouse gas emission reductions
Suppliers/contractors	New business opportunities
	Fair competition and transparent procurement
	Knowledge sharing and transfer
	Health, safety and well-being
	Training programs for vendors, dealers and subcontractors
	Extending access to technology
Government	Revenue generation – corporate tax
	Successful public project delivery
	Critical infrastructure development
	Participate in government schemes
	Mandatory and non-mandatory information disclosure
	Following laws, rules and regulations
Communities	Corporate social responsibility programmes
	Community outreach programmes
	Skill and training development programs
	Employment opportunities
	Environment protection
	Critical infrastructure development
Media	Regular and timely updates
	Critical information sharing
	Media briefings and presentations

3.8 Stakeholder Management: The Role of Corporate Reputation

Puncheva (2008) argued that when a stakeholder has never directly interacted with an organisation, they rely on the perceived reputation of the organisation before deciding to engage with the organisation. In such cases, both social legitimacy (institutional actions conforming to social norms and expectations) and pragmatic legitimacy (performance per industry standards) play a critical role in deciding the relationship opportunity and engagement (Puncheva, 2008).

In recent years, we have seen consumers boycotting the products manufactured in countries that do not protect human rights or use child labour. Similarly, clients, financers and building occupants might not be keen to engage or develop relationships with a construction company that is not serious about adopting sustainable building materials and processes, reducing energy consumption and carbon emissions and protecting the health and well-being of their employees, workers and future building users.

Building a solid reputation and identity becomes particularly important for a construction company operating in a competitive market. For instance, volume builders of residential houses compete with several other companies offering similar designs at similar prices, often to such an extent that it becomes confusing for potential clients when choosing a contractor. In such cases, social and pragmatic legitimacies and the organisation's overall reputation could become deciding factors in the decision-making process.

In the past, information about a company was available from limited sources such as corporate websites and media outlets and obtaining more details involved considerable time and cost. In comparison, a significant amount of data can be easily collected nowadays from various online and publicly available sources, although establishing the credibility of information has become more challenging. It is also difficult to predict what sources of information a stakeholder would use while determining an organisation's social and pragmatic legitimacies and reputation. For example, a bad experience of a stakeholder posted as a blog or social media post or a negative media story could harm the organisation's reputation.

Therefore, construction companies must focus on corporate reputation management and build a strong reputation among the community and other stakeholders. Additionally, they need to reinforce their reputation through actions that meet social and industry norms and how they are reported to stakeholders. They also need to decide what activities or aspects of the business will be used to build their reputation. For instance, the answer to whether the company wants to be known for delivering high-quality projects, affordable houses or a strong focus on sustainability relates to the organisational purpose or goal and the customers it intends to serve. Often, the company would need to build a reputation in multiple areas to maintain existing relationships while creating new engagement opportunities with stakeholders. Similarly, it would need to carefully consider the social and pragmatic legitimacies of its current and future stakeholders, which would affect its reputation.

The following case study discusses how a lack of attention to stakeholders' needs and value creation could affect a company's reputation and relationship with other stakeholders.

Carmichael (thermal coal mine) and rail link project (Queensland, Australia)

The multi-billion-dollar Adani's Carmichael (thermal coal mine) and rail link project in Queensland, Australia, attracted strong criticism from many activist groups and media for concerns related to this project's negative environmental and indigenous community impacts. When the #StopAdani campaign and objections to the project and the organisations involved in delivering the project gained momentum in the local and global media, several companies, including international banks and insurance providers, had to reconsider their participation in the project. Consequently, many organisations decided to withdraw their support or involvement in the project. On the other hand, Siemens, a large global firm with a relatively modest contract for the firm's delivery of signalling systems for the coal mine, decided to respect its contractual obligations at the cost of strong backlash and resistance organized by the activists against the firm. For Siemens, withdrawal would have meant a severely damaged reputation and credibility as a reliable supplier and possible negative implications for the organisation's future and thousands of employees worldwide.

However, Siemen's CEO released a formal statement on the project and Siemen's involvement, acknowledging the concerns of the activist groups and community while also clarifying the company's view and decision and assuring that they have secured the right to pull out of the contract if our customer violates the very stringent environmental obligations. The company tried to balance its business interests and different stakeholders' interests but realised it must pay closer attention to activist groups and younger generations on the need to protect the environment to avoid being in a controversial situation like the Carmichael project. Furthermore, learning from these events, Siemens established a Sustainability Committee with external members to identify and escalate projects with environmental concerns. Such proactive measures will likely reduce the instances where Siemens must resort to trade-offs and fail to create value for its stakeholders.

Construction companies must also consider the relationship between different stakeholders and how the action of one stakeholder could trigger unexpected reactions from other stakeholders. For example, the negative media coverage of the Carmichael and rail link project and campaign from activist groups forced many stakeholders to reconsider their involvement. Another stakeholder (Siemens) had to change its approach to appraising projects. The case study shows that a company's reputation is vital to its relationship with stakeholders.

When the corporate reputation is damaged, stakeholders may reconsider their decision to enter a short- or long-term relationship with the company. A poor reputation may result in a change in stakeholders' attitudes, their level of support and business relationships. In contrast, a strong reputation can help a company better engage with stakeholders and seek favourable actions.

3.9 Key Questions

Construction companies must recognise the need for an ongoing strategic stakeholder management process to establish good relationships with current and potential stakeholders in a dynamic and competitive business environment. Companies that invest resources in maintaining a good relationship with stakeholders and show a solid intent to work with stakeholders by adopting a multi-stakeholder value creation approach will perform better and build a strong corporate reputation and image than those focusing on one or two stakeholders only while ignoring others. Table 3.4 shows some of the questions construction companies may consider to inform the development of their stakeholder management strategy and plan. Please note that the list of questions is not exhaustive and only summarises some of the critical issues discussed in the chapter.

Table 3.4 Example questions for successful stakeholder management

S. No	Domain	Key questions
1	Strategy	Is there a corporate strategy to drive the organisation's stakeholder management approach? Is the strategy proactive or reactive in approach? Are there standards, policies and procedures to support stakeholder management? What are the regulatory requirements and moral and ethical duties concerning different stakeholders? Do the enterprise strategy and business ethics acknowledge stakeholder management and value creation as integral components and demonstrate the leadership team's strong commitment? Does the strategy focus on enhancing stakeholder relationships and engagement? Is the strategy based on the values shared across the company?

(Continued)

Table 3.4 (Continued)

S. No	Domain	Key questions
2	Stakeholder identification	Have all entities that could influence or affect or be influenced or affected by the organisation identified? Are entities that may become relevant in the future also identified? Is stakeholder identification a one-off effort or an ongoing process? Have all relevant individuals, teams and departments been consulted to identify stakeholders?
3	Stakeholder analysis	Does stakeholder analysis consider the needs and expectations of stakeholders and what the company needs from them? Are stakeholders grouped based on interests, influence or power, objectives, behaviour and predictability? Are stakeholder relationships with the company and with each other mapped? Do the analysis results inform stakeholder management engagement and communication practices?
4	Stakeholder engagement	Does the engagement plan aim to build strong relationships and trust with stakeholders through regular interactions and engagement? Does it include proactive measures for engagement rather than reactive approaches? Does the organisation have the right skills and enough resources to support a robust engagement process? Are relevant stakeholders consulted while making decisions concerning them? Is the engagement program inclusive? Is the engagement strategy informed by local context and stakeholder characteristics? Does the engagement process aim to satisfy the needs and expectations of different stakeholders rather than constantly trading off one group's interests for another? Does a multi-stakeholder value creation approach underpin the strategy? Is the engagement process tailor-made, regularly monitored and improved instead of adopting a 'one-size-fits-all' approach? Is engagement progress reviewed and used to update the strategy?
5	Stakeholder communication	Does the communication plan set out procedures and mechanisms for the regular distribution of accessible information? Does the communication plan set out procedures and mechanisms for receiving enquiries or feedback and responding to them promptly? Are complaint management policies and procedures well-established? Does the communication plan set out guidelines and means to resolve issues and mediate disputes? Does the communication plan include provisions for communication during emergencies and crises? Does the communication plan identify roles and responsibilities?
6	Reputation and corporate image	Does the company take its image and brand reputation seriously? Does the company understand the importance of reputation in stakeholder management? Does the company consider other organisations' reputations before doing business with them? Does the company actively work towards building a solid corporate image?

3.10 Conclusion

Strong and mutually productive relationships with stakeholders are linked to organisational success. Construction companies depend on primary and secondary stakeholders for their functioning and growth. After analysing and classifying stakeholders into different groups, companies must identify and implement various ways to effectively engage and communicate with stakeholders. It is essential to understand the needs and expectations of various stakeholders and develop a comprehensive plan to satisfy them. In a highly competitive environment, ignoring the needs of stakeholders or failing to satisfy them could make a business irrelevant.

Since many of the operations and projects of a construction company could directly affect the community, it is imperative to develop and implement proactive and open communication channels with community groups to notify them about any changes that may affect the community. The stakeholder engagement and communication plan should aim to establish and maintain open and effective communication with the stakeholders, build their understanding of the company's operations and increase confidence in the process and value engagement.

Not all stakeholders can be expected to benefit the company, and some may present threats that can affect the realisation of organisational goals. Still, construction companies must endeavour to find a way to accommodate all stakeholder interests without trading off one against another for sustainable growth. Focus on maximising the profit and, in the process, disregarding employees' health, safety and well-being; engagement in unethical practices; delivery of subquality products and projects; and degradation of the environment will not help construction companies sustain and grow due to their poor approach to stakeholder management. Therefore, construction companies cannot afford to focus on one or two stakeholders while neglecting others or ignoring their expectations.

Adopting a holistic view of stakeholder management could help companies predict and reduce, if not eliminate, the adverse actions of stakeholders and proactively engage with them, which is crucial for successful construction company management and long-term growth. Developing comprehensive and robust stakeholder engagement strategies, identifying tools for engagement based on salient features of stakeholders and their mappings, effective communication with the stakeholders and maintaining a solid reputation are essential aspects of successful stakeholder management.

References

Aarseth, W., Rolstadås, A., & Andersen, B. (2014). Managing organizational challenges in global projects. *International Journal of Managing Projects in Business, 7*(1), 103–132.

Abubakar, M. E., Hasan, A., & Jha, K. N. (2022). Delays and financial implications of COVID-19 for contractors in irrigation projects. *Journal of Construction Engineering and Management, 148*(9), 05022006.

Aly, A., Moner-Girona, M., Szabó, S., Pedersen, A. B., & Jensen, S. S. (2019). Barriers to large-scale solar power in Tanzania. *Energy for Sustainable Development, 48*, 43–58.

Bourne, L., & Weaver, P. (2010). Mapping stakeholders. In E. Chinyio & P. Olomolaiye (Eds.), *Construction stakeholder management* (pp. 99–120). Wiley-Blackwell.

Clarkson, M. B. E. (1994). A risk based model of stakeholder theory. In *Proceedings of the Second Toronto Conference on Stakeholder Theory*. Faculty of Management, University of Toronto.

Clarkson, M. B. E. (1995). A stakeholder framework for analysing and evaluating corporate social performance. *Academy of Management Review, 20*(1), 92–117.

Franklin, A. L. (2020). *Stakeholder engagement*. Springer.

Freeman, R. E. (1984). *Strategic management: A stakeholder approach*. Pitman.

Freeman, R. E., Harrison, J. S., & Wicks, A. C. (2007). *Managing for stakeholders: Survival, reputation, and success*. Yale University Press.

Freeman, R. E., Harrison, J. S., Wicks, A. C., Parmar, B. L., & de Colle, S. (2010). *Stakeholder theory: The state of the art*. Cambridge University Press.

Friedman, M. (1962). *Capitalism and freedom*. University of Chicago Press.

Friedman, M. (1970, September 13). A Friedman doctrine – The social responsibility of business is to increase its profits. *New York Times Magazine*, 17.

Greenwood, M. (2007). Stakeholder engagement: Beyond the myth of corporate responsibility. *Journal of Business Ethics, 74*(4), 315–327.

Harris, F. (2010). A historical overview of stakeholder management. In E. Chinyio & P. Olomolaiye (Eds.), *Construction stakeholder management* (pp. 41–55). Wiley-Blackwell.

Heikkurinen, P., & Mäkinen, J. (2018). Integrative stakeholder engagement – A review and synthesis of economic, critical, and politico-ethical perspectives. In A. Lindgreen, F. Maon, J. Vanhamme, B. Palacios Florencio, C. Vallaster, & C. Strong (Eds.), *Engaging with stakeholders* (1st ed., pp. 3–18). Routledge.

Kivits, R., & Sawang, S. (2021). *The dynamism of stakeholder engagement: A case study of the aviation industry*. Springer.

Lane, A. B., & Devin, B. (2018). Operationalizing stakeholder engagement in CSR: A process approach. *Corporate Social Responsibility and Environmental Management, 25*(3), 267–280.

Mitchell, R. K., Agle, B. R., & Wood, D. J. (1997). Toward a theory of stakeholder identification and salience: Defining the principle of who and what really counts. *Academy of Management Review, 22*(4), 853–886.

Mitroff, I. I., & Mason, R. O. (1980). A logic for strategic management. *Human Systems Management, 1*(2), 115–126.

Newcombe, R. (2003). From client to project stakeholder: A stakeholder mapping approach. *Construction Management and Economics, 22*(9/10), 762–784.

Phillips, R. (2003). Stakeholder legitimacy. *Business Ethics Quarterly, 13*(1), 25–41.

Project Management Institute. (2021). *A guide to the Project Management Body of Knowledge (PMBOK® Guide) – Seventh edition and the standard for project management.*

Puncheva, P. (2008). The role of corporate reputation in the stakeholder decision-making process. *Business & Society, 47*(3), 272–290.

Reed, M. S., Graves, A., Dandy, N., Posthumus, H., Hubacek, K., Morris, J., Prell, C., Quinn, C. H., & Stringer, L. C. (2009). Who's in and why? A typology of stakeholder analysis methods for natural resource management. *Journal of Environmental Management, 90*(5), 1933–1949.

Rosenhead, R. (2009). Let's save lives. In E. Chinyio & P. Olomolaiye (Eds.), *Construction stakeholder management* (pp. 374–376). John Wiley & Sons.

Savage, G. T., Nix, T. W., Whitehead, C. J., & Blair, J. D. (1991). Strategies for assessing and managing organizational stakeholders. *Academy of Management Executive, 5*(2), 61–75.

Starik, M. (1995). Should trees have managerial standing? Toward stakeholder status for non-human nature. *Journal of Business Ethics, 14*(3), 207–217.

Voinea, C. L., & van Kranenburg, H. (2017). Media influence and firms behaviour: A stakeholder management perspective. *International Business Research, 10*(10), 23–38.

4 Bidding and Contracts

Kumar Neeraj Jha

4.1 Introduction

Construction involves a wide range of activities that include erection, repair and maintenance and demolition. These activities are performed for a variety of structures, including residential houses, commercial buildings and large infrastructure, and involve numerous specialised firms, each handling different aspects of the construction. In this complex system, contracts serve as the backbone for guiding relationships and ensuring clarity of purpose, roles and responsibilities (Murdoch & Hughes, 2002).

All aspects of a construction project, including its scope, timeline, budget and quality standards, are governed by the legal and procedural framework established by the contract. It serve as a binding agreement between parties to address the unique challenges and risks inherent in construction projects. To ensure that construction projects are completed successfully, it is essential to have well-defined contracts that outline expectations, provide legal protection for all parties and foster a cooperative environment to work in.

Bidding is an essential first phase that forms construction contracts. It involves firms submitting detailed proposals with costs and timelines for a project. This process leads to the selection of a company that enters into a contract with the client to deliver the 'product' for an agreed fee. Bidding plays a critical role in the construction process by functioning as the primary means by which construction companies obtain new work.

The objective of this chapter is to present a comprehensive overview of the ex ante stage in the construction process, specifically where construction companies first become involved in projects through the bidding process, followed by contract negotiation and finalisation. This chapter covers three main areas. The first part discusses the bidding process from the perspective of construction companies. Here, the bidding process is explained, and the construction company's rationale for bidding is explored. The second part of this chapter provides an overview of the various types of contracts employed in the construction industry for delivering projects. Finally, the third part focuses specifically on risk allocation and discusses how it is utilised to apportion responsibilities to attain successful construction projects.

4.2 The Bidding Process

Construction bidding is the process through which contractors provide the client with a proposal to construct or manage the construction of a facility (Figure 4.1). In many projects, especially in public-funded projects, price is crucial, and the client generally selects the lowest bid. However, in some projects, qualifications or other factors are also considered equally important as the price.

DOI: 10.1201/9781003223092-4

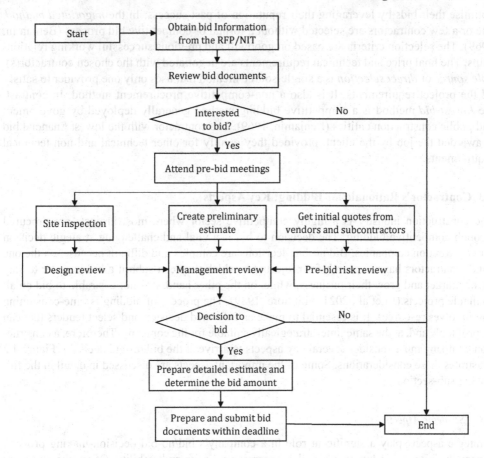

Figure 4.1 Bidding process from the contractor's perspective.

An estimate provides the calculations for the contractor's costs of executing the project, while a bid is a final offer made to the client, considering the profit over the estimate. The process of forming a bid begins with examining the preliminary drawings and determining the quantity of work and materials. Based on these, a highly accurate cost estimate is prepared, and then the bid amount is decided, including the profit margins and risk factors. However, there are a lot of nuances and complexities behind this seemingly simple process.

The process of construction project procurement involves bid solicitation, bid submission and bid selection. First, the client sends out an 'Invitation for Bid' through advertisement, letter or email to notify prospective contractors of the opportunity to bid on a construction project. It involves sharing project specifications, scope drawings and plans, requirements, contract type, delivery method and requests. The process is also known as making a 'request for proposal' or 'notice inviting tender'. Next, interested contractors submit bids within the stipulated deadline. The client reviews bids and decides on a winner. A contract is signed between the parties based on the final pricing and negotiated contract terms to lay the legal groundwork for the project.

The bid selection process could be based on various considerations. For instance, the *best value method* involves evaluating and ranking contractors based on their bids and qualifications and assessing who is the most qualified for the best price. In this method, contractors can optimise their bids by leveraging their reputation of past success. In the *negotiated method*, one or a few contractors are selected without a formal or competitive bid process (Benjamin, 1969). The selection criteria are based on goodwill and previous successful working relationships. The final price and technical requirements are negotiated with the chosen contractor(s). *Sole source* or *direct selection* is a single-source method that uses only one provider to satisfy all the project requirements. It is also a non-competitive procurement method. In contrast, the *lowest bid* method is a competitive bidding method generally deployed by government and public construction entities (Benjamin, 1969). The contractor with the lowest financial bid is awarded the job by the client, provided they qualify for other technical and non-technical requirements.

4.3 Contractor's Rationale for Bidding: Key Aspects

The construction industry is a highly competitive sector where most projects are procured through competitive bidding. The decision to bid is a vital and challenging strategic decision for construction companies. Bid/no-bid decisions are complex and difficult because, on the one hand, contractors have to search for projects for which they can submit a bid in order to stay in the market and grow their business, while, on the other hand, it is not desirable to bid on all available projects (Li et al., 2021; Skitmore, 1989). The process of bidding is time-consuming and involves resources. It is essential to make an informed decision and select tenders that can be profitable and, at the same time, strategically suitable for the company. Therefore, a construction company must consider several key aspects to arrive at the bid/no-bid decision. Figure 4.2 illustrates these considerations. Some of the important aspects are discussed in detail in the following sub-sections.

4.3.1 Strategic Aspects

Strategic aspects play a significant role in a company's bid/no-bid decision-making process, especially in terms of long-term goals concerning revenue and stability. Often, the choice to undertake a project is influenced not just by its profitability but also by key strategic factors. These include alignment with the company's core business goals, expansion into new geographic locations, revenue generation to meet sales targets and technical competency and business development in a specific sector.

Additionally, maintaining or establishing a monopoly in existing markets, entering new markets and leveraging relevant experience and past performance in similar projects can be important considerations. Furthermore, the size of the project is evaluated in relation to the company's overall business, with some companies setting a minimum project value as a threshold for tender consideration. These strategic factors collectively guide a company's project selection to ensure alignment with its broader business objectives.

The profile and relationship with the client are also key considerations when assessing potential projects. From the company's perspective, it is crucial to investigate the client's financial capability and stability. Factors such as long-term relationships and past experiences with the client play a significant role in the decision-making process. In situations involving a new client, several additional aspects need to be taken into account, such as the experiences of other contractors with the client, the ownership structure of the client organisation (such as shareholding patterns, joint ventures or umbrella joint ventures and the management team or key decision makers) and their financial health (revenue; EBITA, or earnings before interest, taxes, and amortization; net worth; debt-to-equity ratio; etc.). Moreover, understanding the sources of project funding or the status of financial closure is essential to evaluating the financial feasibility of the project.

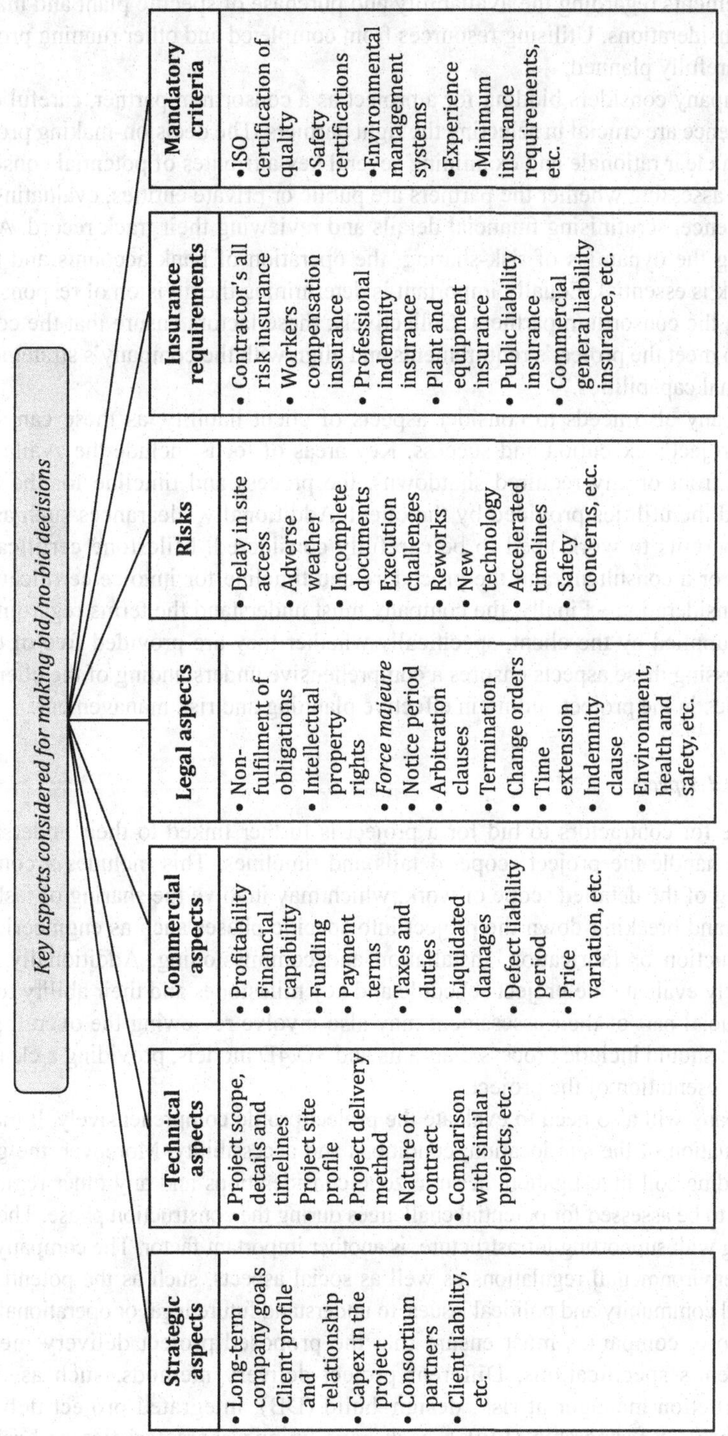

Key aspects considered for making bid/no-bid decisions

Strategic aspects	Technical aspects	Commercial aspects	Legal aspects	Risks	Insurance requirements	Mandatory criteria
• Long-term company goals • Client profile and relationship • Capex in the project • Consortium partners • Client liability, etc.	• Project scope, details and timelines • Project site profile • Project delivery method • Nature of contract • Comparison with similar projects, etc.	• Profitability • Financial capability • Funding • Payment terms • Taxes and duties • Liquidated damages • Defect liability period • Price variation, etc.	• Non-fulfilment of obligations • Intellectual property rights • *Force majeure* • Notice period • Arbitration clauses • Termination • Change orders • Time extension • Indemnity clause • Environment, health and safety, etc.	• Delay in site access • Adverse weather • Incomplete documents • Execution challenges • Reworks • New technology • Accelerated timelines • Safety concerns, etc.	• Contractor's all risk insurance • Workers compensation insurance • Professional indemnity insurance • Plant and equipment insurance • Public liability insurance • Commercial general liability insurance, etc.	• ISO certification of quality • Safety certifications • Environmental management systems • Experience • Minimum insurance requirements, etc.

Figure 4.2 Contractor's rationale for bidding: Key aspects.

The company should also consider the capital expenditure (CAPEX) required for the project. Asset requirements regarding the availability and purchase of specific plant and machinery are important considerations. Utilising resources from completed and other running projects would need to be carefully planned.

If the company considers bidding for a project as a consortium partner, careful deliberation and due diligence are crucial in selecting the right partners. The decision-making process should be based on a clear rationale and examining several key attributes of potential consortium partners, such as assessing whether the partners are public or private entities, evaluating their technical competence, scrutinising financial details and reviewing their track record. Additionally, understanding the dynamics of risk-sharing, the operation of bank accounts and the detailed scope of work is essential. Equally important is determining the division of responsibilities and scope among the consortium partners. Collectively, these factors ensure that the consortium is well suited to meet the project's requirements and align with the company's strategic objectives and operational capabilities.

The company also needs to consider aspects of client liability, as these can significantly impact the project's execution and success. Key areas of focus include the availability of the site or work front or any required shutdowns, the process and timeline for the approval of drawings and the utilities provided by the client. Additionally, clearances such as gate entry passes and permits to work need to be carefully considered. Milestone certification, either by the client or a consultant, and the procedure and timeline for invoice certification are also important considerations. Finally, the company must understand the terms regarding resources or services supplied by the client, specifically whether they are provided free of charge or at a cost. Addressing these aspects ensures a comprehensive understanding of the client's role and responsibilities in the project, aiding in effective planning and risk management.

4.3.2 Technical Aspects

The rationale for contractors to bid for a project is further linked to their understanding and capability to handle the project scope, details and timelines. This includes a comprehensive understanding of the detailed scope of work, which may involve the sharing of tasks with project partners and breaking down the project into specific phases such as engineering, procurement, construction or fabrication, installation and commissioning. Additionally, contractors need to closely evaluate the project schedule and key milestones and their ability to meet these targets. A crucial part of their assessment may also involve reviewing the overall project plan layout, which should include process diagrams and 3D/4D models, providing a clear visual and technical representation of the project.

The company will also need to evaluate the project profile comprehensively. It may include a detailed evaluation of the site location, conditions and accessibility. Moreover, the geographical profile, including soil investigation, seismic zone considerations and any other regional-specific issues, needs to be assessed for potential challenges during the construction phase. The availability of land, along with supporting infrastructure, is another important factor. The company should pay attention to environmental regulations as well as social aspects, such as the potential resistance from the local community and political issues, to understand future legal or operational hindrances.

Furthermore, companies must ensure that the proposed project delivery method aligns with the client's specifications. Different project delivery methods, such as design–bid–build, construction manager at risk, design–build (DB), integrated project delivery (IPD), and public–private partnership (PPPs), each have unique characteristics and requirements. The choice of delivery method impacts how the project is structured and managed, and it influences the contractor's responsibilities, risks and potential rewards. Similarly, the type of contract, whether it is a lump-sum turnkey, item rate or another form, is a significant factor in the bidding decision. It influences the financial planning and risk management aspects of the project. Companies need to evaluate whether they can meet the project requirements within the constraints of the contract type and whether the financial structure aligns with their business model and risk tolerance.

To gauge the feasibility and viability of the project, contractors often compare the technical and commercial aspects of similar projects. This comparison includes examining factors like built-up area, formwork efficiency (measured as sqm/sqm), reinforcement density (kg/sqm) and ducting requirements (MT). Such comparisons provide insights into expected resource requirements and cost efficiencies, enabling contractors to make an informed decision about whether to bid for the project.

Front end engineering design, or FEED, is an engineering design approach adopted prior to detailed engineering, procurement and construction. It is mostly conducted after a conceptual design or feasibility study. It is an important engineering phase that is used to control project expenses and thoroughly plan a project before bid submission. Benchmark studies have shown that FEED constitutes roughly 2% of the project cost, but properly executed FEED projects can reduce up to 30% of costs during design and execution.

4.3.3 Commercial Aspects

Profitability is generally the most important factor behind the rationale for bidding. The financial viability and profitability of the project are assessed based on a detailed cost estimate. The estimate includes an accurate and comprehensive account of labour costs, taxes, insurance, workers' compensation, vacation pay, plant and equipment and any and all additional costs. The company will also assess their financial capability by looking at its current cash flow and other projects to determine whether it can support the necessary workforce, equipment, materials and other resources to commit to the new project. The company must have adequate cash flow to complete the work within the required timeline without jeopardising other obligations.

The payment terms of the contract are important commercial aspects and play an essential role in bidding decisions. The company must ensure that these terms are favourable to its cash flow. It is also important to consider the source and the availability of funding for the project before bidding. The mode of funding must be secure and dependable. The company may also look into other financial aspects, such as the applicability of taxes. If there is a possibility for a tax exemption clause in the contract, the provision for reimbursement of taxes and duties should be considered in case of a change of law or other changes. In the absence of such a clause or provision, the impact of taxes is to be taken into account for bid estimates.

4.3.4 Legal Aspects

The initial contract conditions must be thoroughly reviewed in the early stages of the tendering process. Sometimes, there are terms within the contract that businesses are unprepared for and/ or unable to fulfil; for example, unlimited liability or uncapped liquidated damages. Although there is the opportunity to negotiate specific clauses in pre-bid meetings, some clients may not agree to any modifications to their contract conditions.

Some of the critical contract conditions to be reviewed carefully are as follows:

- Contractual protection in case of non-fulfilment of client obligations resulting in schedule delay, cost escalation, overstay cost and other such circumstances.
- Contract conditions on intellectual property rights and digital data.

- The definition of the *force majeure* clause and all events it covers related to natural calamities, acts of government, strikes, riots, political unrest and weather conditions specific to the project site.
- Reasonable and sufficient notice period.
- Governing law and arbitration clauses: The jurisdiction under which the project falls in case of disputes and the governing law for the project must be well investigated in case of legal recourse. The clauses for alternate dispute resolution methods and arbitration are specified in the document or not.
- Termination arrangements: The company must examine the conditions of termination mentioned in the document and the compensation terms in case the client initiates the termination. Also, whether the contract allows termination by the contractor if required.
- Change orders and an upfront procedure in the document to execute them.
- Liabilities including:

  - Overall contractual liability in the project and the conditions for determination and fixation of the liability.
  - The limitation (upper cap) for liability could vary from 20% of the contract value to 100% of the contract value, and there could be unlimited liability in some cases.
  - The course of action in case the liabilities arise out of subsequent statutory changes in the client's scope.
  - Consequential losses are expressly excluded or included in the document.
  - The clause for bank guarantee is with open-ended tenure or not and whether the liability under the bank guarantee is detrimental to the company's interests.
  - The extent of the company's share in the liability when the liability event is triggered by other consortium members.

- Time extension related conditions include the following:

  - The details of the time extension clause for delays due to the client's default or contractor's default.
  - The apportionment of delay damages in case of delays attributable to both client and contractor.
  - In case of delay on the client's part, the levy of compensation to the contractor for indirect costs is included in the contract or not mentioned.

- Foreign currency policy and exposure in the project.
- Indemnity clauses for events where indemnity is to be provided by the contractor; for example, loss of life or damage to property/environment.
- Environment, health and safety (EHS) requirements, including the following:

  - The safety standards for work requirements such as work at height/hazardous locations/difficult terrain/extreme temperatures.
  - Availability of trained EHS personnel.
  - Any other special EHS requirement.
  - Penalties attached to non-compliance on EHS.
  - Environmental risks (accident/gradual) are covered under insurance coverage by the client or to be taken by the contractor.

4.4 Risks

Before deciding to bid on a project, the company should carefully identify all the potential risks that could arise on the project (such as safety, environmental, contractual, resource and financial) and agree on an acceptable risk level for the company's business at the pre-tendering stage. The company should review the bidding documents, plans and specifications for the project and review the historical data from similar projects. Based on these, it must identify and carefully evaluate all risks related to the tender. On the flip side, the opportunities that open up for the company and its business upon winning the bid must be explored, and how these opportunities might introduce some new risks could also be considered.

Some of the risks to be taken into account while making the bidding decision include the following:

- Delays in receiving full access to the site can delay the whole work. If the compensation in case of delay on the part of the client is not included in the bid document, the cost of idle equipment and labour and other expenses are generally borne by the contractor. Therefore, the risk of incurred costs due to a delay in access should be considered while making a decision.
- Work sites located in areas prone to extreme weather conditions such as heavy rainfall, extreme heat, heavy snowfall or storms may pose execution challenges, delays in completion and damage to supplies and completed work.
- Incomplete construction documents and drawings prolong the project unnecessarily and impact the cost. They may lead to rework, unforeseen changes and coordination issues in the project.
- Site challenges such as utility shifting, unknown site conditions, hazardous waste, high groundwater levels and poor soil-bearing capacity may lead to delays and disputes.
- New technology in the projects is an ever-present risk and opportunity. It could improve productivity and quality, but it could also cause disruptions in the existing processes. The company should pay attention to the risks associated with the use of new technology, such as data privacy and security, and the impact on their work processes if the client mandates its use.
- Rushing a project could lead to unexpected outcomes. A poorly planned construction project with unrealistic timelines could lead to frustration and demotivation and pose health and safety risks for the staff and workers. Also, it could lead to poor-quality construction and increased repairs and reworks.
- An unsafe site or design is a significant risk for the company.
- The location of the project site could have significant implications for the availability of resources and construction costs. If it is a remote site or a place where local suppliers and workers are unavailable, it will incur extra cost to the company to arrange various resources at the work location.

These risks must be categorised in terms of impact and probability of occurrence and prioritised based on the time, money and work each risk will require to manage effectively. A tender risk assessment matrix could be used for tender submissions. The same document can then be transferred to the project team in case the company gets the work. Please see the chapter on risk management for more information on risks, risk assessment and risk management.

4.5 Insurance Requirements

In large and complex construction projects, the potential for loss can be devastating in case of untoward events. The size and nature of construction projects pose a significant necessity for requiring specific insurance to reduce exposure to different types of risks. Therefore, these provisions must be finalised early in the process, preferably at the planning and design stage, so that the contract clearly asserts the insurance requirements and provides the bidding companies with all the provisions and details applicable under the given circumstances. The insurance provisions in the bidding document generally set forth the insurance coverage that the contractor is obligated to provide. The company should read these clauses carefully and identify the following:

- The types of insurance policies that are to be provided.
- The duration for which the insurance coverage is to be obtained.
- The monetary limits for the policy.
- The form of the policy and the hazards to be covered under it (say, completed operations vs. ongoing operations).
- Evidence of compliance needed to be supplied to obtain these policies.

The most common forms of insurance under construction contracts include:

Contractor's All Risks Insurance: Insurance against physical damage to the works, equipment and materials on site. It typically covers the full reinstatement value of the works (100% contract value) and a markup for ancillary costs that are incurred, valid until the issue of the 'Taking over Certificate'.

Worker's Compensation Insurance: Insurance against compensation to workers and their dependents in case of death, occupational disease or injury (usually on site) caused at work or due to the nature of work.

Professional Indemnity Insurance: While insurance hedges against direct claims, indemnification involves three parties: the indemnitor, the indemnitee and the third party. Here, the indemnitor promises financial protection to the indemnitee for any potential legal liabilities and claims issued by a third party. Professional indemnity is an important policy taken in design-and-build contracts. It insures contractors and designers against claims made by clients based on alleged losses due to inadequate advice, services, or designs. The insurance amount (approx. 5%–10% of contract value) varies on a case-to-case basis depending on the extent and complexity of design works in the project.

Contractor's Plant and Equipment Insurance: This covers damage or loss associated with equipment and machinery (e.g., excavators, cranes, compressors, drilling machines and other heavy machinery) at the construction site.

Public Liability Insurance: This insurance provides the legal ability to pay compensation to third parties arising out of the business and construction activities of the insured. It is the insurance that protects the contracting party if a member of the public sues them.

Commercial General Liability (CGL) Insurance: It is a class of insurance that provides liability protection to companies in the case of bodily harm or property damage during the course of business. Most construction contracts require the main contractor to purchase CGL coverage to assure that the contractor has the means to pay for liability arising out of its construction operations. It usually covers a broad range of damages, such as faulty work, job-related injury and defamation.

Automotive Liability Insurance: It is the insurance for protection against claims for damages (physical injury or property damage) resulting from the ownership, operation, maintenance or use of all owned, hired and non-owned autos.

Pollution and Environmental Liability Coverage: These policies cover injuries, damages, defence costs and clean-up costs resulting from a pollution condition, sudden or gradual, arising from the business or construction activity. It also includes the emergency response cost coverage, usually extending to the subcontractors. Most pollution claims are not covered by CGL insurance, making it a requirement to obtain a separate pollution coverage.

Decennial Liability Insurance: In many contracts, the contractor is liable to the client for a particular period (usually 10 years from the taking over) for any defect that threatens the safety or stability of the building. Contractors take out insurance against this liability in some cases.

Delay in Start-up Insurance: The contractors are normally required to pay liquidated damages for the delay in work completion, usually capped at 10% of the contract price. This cap on the liquidated damages could mean that the client may not be fully compensated for delay damages and may suffer losses. Considering this, sometimes clients take out delay in start-up insurance for compensation for losses, including loss of anticipated revenue and other consequential damages arising out of the delay.

4.6 Mandatory Qualification Criteria

In many instances (often for tenders of higher value), the client specifies certain mandatory requirements to target a particular group of suppliers/contractors. The company must strictly evaluate these requirements, as a failure to comply with any of those requirements will immediately invalidate a submission. It should be ensured that the company qualifies them without any ambiguity in wording or criteria.

The mandatory qualification criteria often include International Organisation for Standardisation certification of quality, safety certifications, environmental management systems, minimum insurance requirements and proven experience in the delivery of similar projects. If the company does not qualify for any of these criteria or cannot provide proof for the same, then it should not invest time, money and other resources in tender preparation and submission.

Sufficient time and resources should be allocated for bid preparation, proofreading of documents, internal approvals and other bidding processes. Therefore, the availability of time and resources to make a high-quality tender submission must be kept in mind. The quality of the bid submitted also represents the reputation of the company, and poor-quality bids will leave a bad impression on the client's mind that may affect the company's chances of receiving more business opportunities from the same client in the future.

4.7 Competitive Bidding Strategies for an Optimum Markup

Competitive bidding has been widely used as a method for allocating and procuring construction contracts. Especially in public sector projects, it is often a legal requirement in most cases to use competitive bidding to reduce the chances of corruption and save public money. The process starts with contractors submitting their bids for the project with their respective quotes. The submitted bids are evaluated technically, and the bids that qualify for the requirements are further evaluated based on the financial bid submitted. There are many methods to make the decision to select financial bids, such as the lowest bid, the average bid, the second lowest bid and the below average bid. Usually, the lowest financial bid that is also technically qualified is selected and awarded the contract. The winner is expected to execute the project on the agreed price quoted and within schedule without compromising the quality.

The two main factors that affect the contractor's decision corresponding to a bid submission are (a) the bid/no-bid decision, where the contractor considers several aspects to make the decision, as discussed earlier in the chapter, and (b) if the contractor decides to bid, then the construction company's bidding strategy becomes the key factor in deciding the markup price. An optimum markup is crucial to ensure the company's survival, growth and success in the highly competitive and slim-profit construction industry.

An accurate estimation of the price of work, market competition and the type of work are some significant factors affecting the bid markup decision. There is an inverse relationship between the bid markup (level of profit margin) and the probability of getting the contract. A low bid would increase the chance of being selected for the work. However, if a bidder bids at a less-than-reasonable price to get the contract, it could have serious financial consequences and may lead to the project's failure. Clients may also reject bids that are not reasonably priced. Therefore, the bidding strategy adopted by the company is critically important to balance profitability and the chances of winning the bid.

Construction companies use various bidding strategy models, such as statistical models, cash flow–based models and game theory–based models, to derive the optimum markup for the bid. Statistical models such as Friedman's model (Friedman, 1956) and Gates' model (Gates, 1967) are most commonly used for decision-making. The objective of these models is to determine the optimal bid price that a risk-neutral contractor should submit to maximise the expected profit from the project.

The decision-making based on these models also depends on the construction company's objective, which could be the following:

1. *Maximise the expected profit or expected monetary value:* Here, companies focus on projects with a high return on investment, selecting projects that align with their company goals and offer favourable financial returns.
2. *Win the contract, even at a loss:* This approach is employed to attain strategic benefits, such as expanding into new markets or establishing a customer base, with an emphasis on long-term goals as opposed to short-term financial gains.
3. *Minimise the competitor's profit:* This approach is employed by companies that adopt aggressive pricing strategies to reduce competitor's profitability, aiming to capture a larger market share or weaken competitors' positions.

Additionally, as highlighted by Dulaimi and Shan (2002), the bidding strategy for determining the optimal markup is also influenced by other important factors, such as the company's perspective or outlook on the bid result (optimistic/pessimistic), cash flow, risk appetite, competition, inflation and overall economic environment and project characteristics (e.g., duration, location, complexity, etc.).

4.8 Bid Submissions

Online bidding systems have now replaced traditional paper-based systems of bidding in most countries. Online bidding systems are generally more secure and efficient for submitting, evaluating and managing bids. Bids are now placed in virtual lockboxes on digital portals, ensuring security and confidentiality. The bid preparation process is streamlined through standard formats and reusable templates, enhancing ease and efficiency. To participate, bidders are required to register using a digital identity and password, providing an added layer of security. It is mandatory for each bid to be securely stored to prevent tampering and to be duly signed by an authorised representative of the bidding company, accompanied by a written confirmation of the signatory's authority. Each bidder can submit only one bid per project, work or package to maintain fairness and transparency, and submitting more than one bid leads to disqualification.

The bidding documents required to be submitted for the bidding process are mentioned in detail in the tender document by the client. These documents can now, in most cases, be downloaded online from the respective portal or website after paying the cost for the bidding documents. For bids submitted by a joint venture or consortium, a copy of the agreement entered into by all partners shall be submitted. As an alternative, a 'Letter of Intent' can be signed by all partners to execute the joint venture agreement in the event of a successful bid, which can be submitted along with a copy of the proposed agreement.

Table 4.1 lists the typical documents required in a typical construction bid submission.

Table 4.1 Documents required for bid submission

Category	Typical documents
Documents establishing the qualifications and eligibility of the bidder	• A statement from the bidder maintaining that 'the bidder is neither associated, nor has been associated, directly or indirectly, with the consultant or any other entity that has prepared the design, specifications and other documents for the project/work or being proposed as project/work manager for the contract'. • A copy of the valid enlistment of the bidder with the government under the appropriate class and category of work. • Documents defining the legal status, principal place of business and place of registration of the bidder. • Written power of attorney to the signatory of the bid. • Details of similar works completed during previous years duly supported with performance certificates from the clients. • Details of equipment and machinery available, along with documentary proof of ownership or lease deed. • Document mentioning the experience and qualifications of the key personnel. • Reports on the financial standing of the bidders, stating their profit and loss statements and auditor's reports for the past few years. • Credit lines/letter of credit/certificates from banks for meeting the fund requirements. • Letter of authority given to the client or its representatives to seek references from the bidder's bank. • Affidavit/undertaking of not having been black-listed or debarred by any organisation at any stage. • Affidavit/undertaking that information being submitted is correct and true and that any false information shall lead to disqualification. • Details of the available bid capacity with an undertaking that the available bid capacity is more than the estimated value of the project/work. (In order to estimate the bidding capacity of the bidder, a statement is usually submitted showing the value of ongoing works and existing commitments, as well as the stipulated period of completion remaining for each of the works listed, duly signed by the respective authority.)
Technical bid	• Letter of technical bid. • A detailed construction planning and methodology. • Plans, specifications and drawings. • Proposed work method and schedule. • Equipment planning and deployment. • Quality assurance procedures. • The types, numbers and capacities of each machinery/equipment, along with the cycle time and production capacity, in order to match the project requirements. • Alternative technical bids, if permissible in the tender.
Financial bid	• Letter of financial bid. • Bill of quantities of the various classes of work to be done is to be submitted in a document called the bid template or the bid form. It provides a breakdown of costs, materials and labour, along with their respective rates. • The currency of the bid and that of the payments shall be as specified in the tender. • Terms and conditions of the contract that are to be complied with by the contractor.
Bid security/earnest money deposit	• The bidder shall furnish a bid security amount as specified in the tender document in the form and currency specified by the client. The amount could be submitted as an unconditional bank guarantee issued by a bank or surety, a certified check or cashier's check, an irrevocable letter of credit or another security indicated in the tender document.

4.9 Construction Contracts

A contract is an agreement (oral or written) that enables different parties to come together and collaborate towards their specific desires and needs. It creates, defines and governs mutual rights and obligations between the parties and how each party's duties and responsibilities will be enforced. It generally involves the transfer of goods, services, money or a promise at a future date. Contracts can be in written or oral form as long as there is mutual consent and consideration. However, to limit misunderstandings and make them more enforceable, contracts are usually written documents containing the obligations of both parties.

To be precise, a contract is an 'agreement' – 'enforceable by law'. Essentially, all agreements are contracts if (a) they are formulated by the free consent of the competent parties, (b) they are meant for lawful consideration and with a lawful object and (c) they are not expressly declared to be void. The three cardinal elements of a legal agreement are proposal, acceptance and consideration.

- *Proposal/Offer:* When a person signifies to another their willingness to do or to abstain from doing anything, obtaining the assent of that other to such act or abstinence, they are said to make a proposal.
- *Acceptance:* When the person to whom the proposal is made signifies their assent thereto, the proposal is said to be accepted. A proposal, when accepted, becomes a promise. The person making the proposal is called the 'promisor', and the person accepting the proposal is called the 'promisee'. Every promise and every act of promise, forming consideration for each other, is an agreement.
- *Consideration:* When, at the desire of the promisor, the promisee or any other person has done or abstained from doing, or does or abstains from doing, or promises to do or to abstain from doing something, such an act, abstinence or promise is called a consideration for the promise. Consideration must be mutual. For a promise to be legally binding as a contract, it must be exchanged for adequate consideration.

4.10 Classification of Contracts

There are a variety of construction contracts. The choice of a contract for a specific project depends on the value of the work and the contract strategy considered to achieve the project objectives. The various types of contracts offer different ways of managing aspects of pricing, risk transfer, responsibility for performance, cost certainty and complexity. Based on the nature of the project, complexities, timeline constraints and owner/contractor expertise, the delivery method is selected accordingly. Broadly, contracts can be classified based on the following five criteria.

4.10.1 Scope and Nature of Work

A person or a company enters into a contract to do some work under specified terms and conditions. The work may be for the construction and maintenance, supply of materials, labour, transport of materials, etc. The classification of contracts based on the scope and nature of work is shown in Table 4.2.

4.10.2 Mode of Project Delivery

The client uses a project delivery method for managing the design, construction, operation and maintenance of a structure or facility. The contract agreement process sets distinct expectations

Table 4.2 Examples of various construction contract types based on the scope and nature of work

Type of contract	Description
Supply-only contract	An agreement between parties where one party supplies goods or services to the other party for an agreed-upon length of time.
Supply and erection contract	The contractor is responsible for the supply, erection and installation of the end product or facility.
Erection and commissioning contract	Contract for providing service in relation to erection, commissioning or installation of plant, machinery, equipment or structures, whether prefabricated or otherwise. The contractor here must ensure that the owner receives a functioning facility or system that meets the end needs.
Civil works contract	The contract is between the property owner and the general contractor to complete the construction of the required facility or building in the stipulated timeline based on the terms and conditions mentioned in the contract.
Consultancy contract	It is a contract between the consultant and the client that outlines the terms and conditions regarding the services required, their description and payment terms.
Service contract	It is an agreement to perform the repair, replacement or maintenance of the asset due to a defect or wear and tear, with or without additional provision for indemnity payments.

and responsibilities for the involved parties, including payment processes and schedule details (Ahmed & El-Sayegh, 2024; Hosseini et al., 2016). A clearly defined process before construction begins can curtail potential misunderstandings and help with smooth project progress. The extent of involvement of the various service providers and the financial structuring of the project are the two key variables based on which the bulk of delivery methods are categorised, as discussed below.

General Contracting or Design–Bid–Build

Design–bid–build is a traditional form of contracting where the client is dealing with at least two parties (or more) for project delivery. This method consists of three distinct project phases: the design phase, the bid phase and the build phase. The client is responsible for the design finalisation before the project is put out for bids. A design professional or an architect finalises the design, which includes a set of plans, drawings and specifications. Based on these plans and documents, bids are solicited from multiple qualified contractors for the execution of work. The contract is then awarded to a contractor. This model has been traditionally used in public sector infrastructure projects. It offers the client control over design decisions as they work directly with the designer. However, in this method, there is generally a lack of collaboration between the designer and the contractor, and the onus of any execution and design misalignments falls on the client, leading to variations and delays. Also, the project duration tends to increase due to the 'in-series' or sequential nature of project delivery; that is, the design is first finalised before the construction process starts.

Design–Build

Under this type of contract, the client engages a single entity or joint venture of architects, designers, engineers and builders to perform design and construction under a single contract instead of soliciting separate bids from each one. Here, generally, the contractor owns the design responsibility in consideration of the client's requirements that specify the purpose, scope, standards and performance criteria for the work. The contractor may perform all the design and construction work or subcontract parts of the work to other companies. The construction process could begin before the final design is completed, saving the client time and money. However, the method requires high levels of collaboration between various parties. While the client gains greater predictability of price and time for completion, the contractor assumes a higher risk, which is accounted for in the bid price as a risk surcharge. In recent years, the DB method has been increasingly used. While this method reduces the client's risk in the project, it also minimises their role in decision-making and project management. However, since design and construction processes overlap, the project could be completed at a faster pace. Design–build contracts may suit projects that involve high risk, many unknowns and limited time to complete.

Engineering, Procurement and Construction Contracts

In the case of engineering, procurement and construction (EPC) contracts, the contractor is responsible for engineering, including design, procurement of works, plants, materials and services, and execution of the construction works. An EPC project typically results in a turnkey facility, where the contractor hands over a working facility that is ready to go. This delivery method is generally used for large-scale developments such as power stations, process plants, other major plants, oil and gas projects and the delivery of mining infrastructure. It allows for better overall prediction of price and completion time (Klee, 2015). Compared to the DB method, in the EPC method, far more liability and risk lie with the contractor. The expectation is that the contractor will be able to control and assess risks in their bid price. EPC contracts provide the most suitable arrangement for projects in which (a) substantial engineering expertise is required, (b) design is determined by functionality, (c) greater focus is on performance requirements and (d) the client does not need to control the design.

Engineering, Procurement and Construction Management

The engineering, procurement and construction management (EPCM) mode of project delivery involves a professional engineering services contract. This contract is popular for the construction of heavy engineering facilities and manufacturing plants across many industries, such as energy, chemical, agriculture and pharmaceutical. Here, the client concludes direct contracts with particular contractors on a lump-sum basis. For their coordination, the client hires construction or project manager(s) on a professional service agreement basis and pays them on a cost-plus basis, so the general contractor's surcharges are restricted. The project manager or consultant is involved in managing all the stages of engineering, from planning and engineering to coordinating all contractors and vendors on the client's behalf, and provides a single point of contact. They act as an intermediary between the client and their contractors. The client stays in complete control of their project. So, EPCM is different from the EPC mode of project delivery, where the client is completely 'contracting out' the project rather than having consultants manage it. In the case of EPCM mode, the client holds control of all the contracts and purchase orders with third parties (suppliers and vendors and has overall responsibility and control here). In contrast, the contracts are made directly between the third parties (subcontractors and suppliers) and the main contractor, with minimum involvement of the client in EPC delivery mode.

Integrated Project Delivery Contract

The IPD contract is a multi-party agreement with a shared risk/reward model between the design firm, the contractor and the client. It may also include trade partners. Subcontractors typically fall under the contractor's part of the agreement. Like the DB contract, it brings all deliverables into one contract. A single contract is established between the client, designer and general contractor. It is a popular and novel project delivery method built on ideas of teamwork and shared responsibility. It pairs well with the lean construction management approach to projects. Here, all contractors, subcontractors, suppliers and service providers are in communication with the client from the outset. The client often sets incentives for design and construction teams, intending to foster collaboration and goal alignment. The IPD method is the latest one being explored in the industry, especially for large, complex and repetitive project types. As liability, responsibility and risk/reward are shared among the stakeholders in the project, the method requires extensive collaboration among parties, heavy technology integration and an up-front cost investment. Even though advanced and costly technologies are an added cost compared to traditional projects, when delivered successfully, IPD could provide some of the most cost-effective and fastest projects.

4.10.3 Pricing

There are various contracts that are based on pricing structures. The common ones are explained below.

Lump-sum Contract

A lump-sum contract is a fixed-price contract, quite common in the construction industry. As the name suggests, the contract document outlines the work to be done for a fixed amount. This type of contract is usually considered for smaller projects that are not likely to have much variance in the project scope or cost.

Item Rate Contract

Major public work contracts follow this form of contract. Here, the rates are quoted by the contractor item-wise in the bill of quantities. The contractor carries out the work as per the drawings and specifications approved by the client, and the payment is made entirely on measurements of the work executed. It is also called a unit price contract, as the contractor quotes the rate or price per unit for each item of the construction work, including materials and labour.

Cost-Plus-Profit Contract

The cost-plus-profit contract includes two elements of payment for the contractor: the cost of the materials and labour plus the contractor's profit. The profit can be predetermined either as a set figure or as a percentage of the final project costs. These contracts are best suited for projects with an unclear scope or with many variables. However, they include a 'not to exceed' limit, which sets some budgetary boundaries. In such contracts, the focus is on quality over costs, and the contractor's risks are greatly reduced. The client is responsible for the costs and all connected expenditures. Depending on different situations and project preferences, there are different variations of cost-plus contracts: cost-plus percentage of cost, cost-plus fixed fee, cost-plus incentive fee, and cost-plus award fee.

Guaranteed Maximum Price Contract

A guaranteed maximum price (GMP) contract sets a maximum price for a construction project that the client will pay to the contractor. Any amount beyond this is covered by the contractor, whether in the form of labour or materials. A GMP contract is often used for large and complex projects where it minimises financial risk for the client. However, it is important for projects using GMP contracts to have a well-defined scope of work, timelines, drawings and specifications and site-specific information. The guaranteed maximum price consists of the direct project costs, the indirect costs or overheads and the contractor's profit. If the project costs go above this amount, then the contractor is responsible for the increased amount unless there is a change order request or change in scope approved by both parties. Since contractors assume a lot of the financial responsibility and risk in this case, they are likely to inflate rates and markup so that they can build in a financial cushion for the unknown risks in the project, increasing the overall project cost. A GMP contract can be a standalone contract or a different type of contract may incorporate a GMP and other specific terms.

Incentive Contract

The incentive contract is a type of construction contract that offers rewards or incentives to the contractor based on their performance and accomplishments. Incentive contracts are designed to motivate the contractor to achieve specific goals, such as the timely completion of a project within budget or with high quality or safety (Kwawu & Laryea, 2013; Suprapto et al., 2016). These incentives are generally supplementary remuneration according to the contract terms and conditions.

4.10.4 Ownership and Financing

In the case of complex public sector infrastructure projects, some of the contract forms are increasingly used worldwide based on the ownership of assets, project financing, risk-sharing and incentive mechanisms. These are discussed below.

Public–Private Partnership

The PPP method involves a long-term arrangement between the parties in the public and private sectors. It was developed to enable and make optimal use of private funding and expertise for public projects. The public and private stakeholders agree to develop, finance, execute and operate a project (mostly an infrastructure project) in a joint manner (Almaaz et al., 2020). An entity or a special purpose vehicle is created to manage and execute the project. After an agreed-upon period of operation and maintenance, the facility is passed over to public ownership.

The contract delineates the responsibilities of the partners involved. In most cases, the public partner assumes work like statutory approvals, political resolution of issues and land acquisition, in addition to overall tracking of the work. The private party is responsible for obtaining the project finance and its execution and, thereafter, for running the assets thus created for a pre-defined period to realise a return on its financial investments. Public clients have chosen this approach in the past few years as it does not directly burden the budget or increase public debt. The positive aspect of the arrangement is that it provides the opportunity to start new projects that could not have been financed otherwise without private resources. The negative aspect is higher overall transaction costs and reduced client control. However, the long-term effects are hard to foresee.

PPP is, in practice, more of an umbrella term for different kinds of project delivery methods, including BOT (build–operate–transfer), DBO (design–build–operate), BFOT (build–finance–operate–transfer), BOTM (build–operate–transfer–maintain), BOOT (build–own–operate–transfer), DBFO (design–build–finance–operate) and DBFOM (design–build–finance–operate–maintain), where the abbreviations signify the different phases and responsibilities involved in these projects. While the PPP model is now being widely used in infrastructure projects in developing countries, it faces many challenges. Misaligned organisational goals, poor collaboration, improper risk management and allocation, unreliable risk- and responsibility-sharing mechanisms and a lack of transparency are some of the key barriers to its successful implementation (Batjargal & Zhang, 2021; Jayasuriya et al., 2019). Alternate methods of collaborative and relational contracting, such as alliance contracts, are gaining more prominence in many countries, such as the United Kingdom and Australia.

Alliance Contracting

Alliance contracting could be described as collaborative relational contracting typically used in long-term multi-party infrastructure projects. Here, the client (usually a public sector entity) enters into a contract with an alliance of delivery partners, where project risks are shared collectively by the parties (Gransberg et al., 2015; Suprapto et al., 2016). The traditional contract models have historically lacked incentives to promote client–contractor collaboration, and risks are often transferred to private contractors. In the case of an alliance, risks are jointly shared and managed via a risk–reward framework included in the terms of the agreement. As a model, alliance contracting operates with key features, including a cost-plus-fee structure, risk- and opportunity sharing, commitment to collaboration and no disputes, transparency through open-book project reporting and a joint management structure. Some commonly used alliance contract forms are Framework Alliance Contract (FAC-1), Term Alliance Contract (TAC-1) and NEC4 Alliance Contract.

4.10.5 Other Perspectives

From a contract management perspective, contracts are classified based on their legal validity, mode of formation and performance. The different categories are summarised in Table 4.3.

Table 4.3 Examples of various construction contract types based on different perspectives

Type of contract	Description
Bilateral and unilateral	In a unilateral contract, the contract is drafted by one party, and an express offer is made to the other party, usually the general public or a specific group of people, as prescribed in the offer. Here, the other party is in no way obligated to perform upon the offer (e.g., reward contracts or insurance contracts). In bilateral contracts, both parties promise to perform their contractual obligations as decided and negotiated in the contract document.
Void, voidable and unenforceable	A void contract is not a contract but instead an agreement that cannot be enforced by either party because it fails to meet the standards of a legitimate contract. Voidable contracts are those where minor breaches exist, and the aggrieved party has the option to decide whether the contract is to be treated as valid or void. A voidable contract remains valid until it is declared void by the party who has suffered due to the breach. Unenforceable contracts are those contracts that are very much valid as contracts but cannot be enforced in a court of law due to the absence of some evidential features or essential legal requirements.

(Continued)

Table 4.3 (Continued)

Type of contract	Description
Formal and informal	Formal contracts are written and signed by the contracting parties and contain all the contract conditions, timelines and work details. They are generally used for big projects involving a large amount of money. Informal contracts can be oral or written and are much simpler. Most informal contracts are also legally binding as long as they work in accordance with the law of the land.
Executory and executed	An executed contract is one that is completely performed by all the parties to the contract. By contrast, an executory contract is one that is yet to be fulfilled by one or more parties.
Contingent contract	The word contingent here stands for an event or situation that is contingent, meaning it depends on some other event or fact. Therefore, contingent contracts are those where the promisor performs their obligations only when certain conditions are met. Contracts of indemnity, insurance and guarantee are some examples of contingent contracts.
Express and implied	An express contract is one where the terms and conditions of the contract are entirely and explicitly stated orally or in writing in the contract. An implied contract is a contract formed in whole or in part by the conduct or actions of the parties involved.
Quasi-contract	These contracts outline the obligation of one party to another when the latter is in possession of the original party's property. These parties may not necessarily have had a prior agreement with one another. The agreement is imposed by law through a judge as a remedy to achieve a fair outcome in a situation where one party has an advantage over the other.
International contracts	These contracts are legally binding agreements between parties based in different countries or projects being executed in different countries, stating the obligations of each of the parties. These contracts specify which governing law and jurisdiction will be applicable in case of disputes or non-performance.

4.11 Contract Structure

A construction contract is a legal agreement that defines the specifications and terms of a project. Unlike other industries, a construction contract is not one single document. Instead, it is a collection of documents that present the terms and specifications of the project.

General conditions of the contract are standard terms that suit the majority of the projects and are an inherent part of the construction contract. They are strategic in nature and not advised to be amended or modified. They include project definition, scope, roles and responsibilities, payment schedule, delay penalty and damages, jurisdiction and modes of dispute resolution, process of variations and change orders, insurance of works, quality assurance programme, applicable laws and regulations, contract coordination procedures and progress monitoring, completion certificate and completion plans, etc.

Special or particular conditions of the contract are specific to a particular project or bid and are used as a supplement or amendment to the general conditions of the contract. These are the conditions that make the contract flexible enough to achieve the project objectives. They specify the modifications required to suit the uniqueness of the project. Whenever there is a conflict, the provisions in the special conditions of the contract prevail over those in the general conditions of the contract.

A *letter of tender* or *letter of bid* is a document stating the signed offer made by the contractor to the client for the execution of the works, whereas a *letter of acceptance* is the formal acceptance of the contractor's bid and usually presents the point in time when the contracting parties enter into the contract.

Plans and drawings are a set of drawings (architectural, electrical, mechanical, landscape or construction detail drawings) that provide an overview of the project and define what is to be built. The *bill of quantity* is an itemised document that lists all the components of the work, along with the quantity and price for each of them. It is provided by contractors at the time of bidding, and once the bid is accepted, it becomes a part of the contract document.

Technical specifications contain all the technical data and performance requirements pertaining to the construction materials, workmanship and equipment required for the project. Typically, technical specifications outline the quality standards, acceptable materials, deviation limits, quality testing and compliance requirements.

4.12 Risk Allocation in Construction Contracts

Due to their complex and unique nature, construction projects are inherently prone to risks such as environmental risks (natural disasters, adverse weather conditions, etc.), technical risks (inadequate site investigation, incomplete designs, etc.), financial risks (inflation, fluctuation in estimates, etc.) and construction risk (disputes, damages to third-party properties and persons, non-availability of resources, health and safety issues, etc.). Projects may also have additional risks specific to the nature of work and other circumstances.

An integral part of risk management is risk allocation in the contract document. A construction contract allocates risks between the contracting parties in such a way as to enable risks to be managed efficiently and effectively throughout the project. The intent should be to impart a 'fair and equitable' allocation of risk to create the best chance of successful project delivery. Risk allocation involves more than merely assigning risks to a particular party; it requires a balanced distribution of these risks, taking into account the potential consequences, such as additional time or costs. The client is responsible for initially identifying the risk, followed by its assessment and analysis, before determining which party should bear the risk. This decision is made based on available information and thorough analysis.

Typically, the client is responsible for preparing the contract bid package, except in cases where the contractor assumes this role. Consequently, the responsibility for risk allocation often falls on the client, who tends to shift more risks onto the contractor while accepting as little risk as possible. In the case where a risk is transferred to the contractor, who has no means by which to control the occurrence or outcome of the risk, the contractor either insures against it or adds a contingency to the bid price. This approach induces several hidden costs into the project, including restricted bid competition, increased potential for claims and disputes and, above all, more adversarial client–contractor relationships, ultimately affecting the financial viability and success of the project itself.

Several important studies and surveys have been conducted to establish the criteria of risk allocation (Abrahamson, 1984; Barnes, 1983; Thompson & Perry, 1992). One of the most important works on the topic is by Max Abrahamson (1973), who defined these risk allocation principles in construction projects, known today as Abrahamson's principles. Based on these principles, risk should be allocated to the party when:

1. It is in their control.
2. They can transfer this risk in an economically beneficial way.
3. The economic benefit of the risk rests with that party.
4. It is more efficient to put the risk on that party.

5. If the risk eventuates, the loss falls on that party, and there is no valid reason to try to transfer it.

The criteria mentioned above present a road map to determine the party expected to manage the risk effectively. Nevertheless, their application in the real world relies heavily on the subjective judgement of the decision makers. For example, one of the principles states that 'a party should bear a risk where it is in their control'. The term *control* is difficult to interpret precisely, as the control could be partial or limited for the contracting party. The problem with this kind of decision-making process is its implicitness. Factors such as attitude and bias in personal judgements may significantly influence the outcome of the decision (Lam et al., 2007). While the principle of control of risk is a decisive factor in determining risk allocation, it is not always comprehensive. It may not necessarily be suited for each project or project participant. A party may be best placed to control and manage risk, but this does not mean that allocating the risk to that party is always equitable. Hence, it can be inferred that effective risk allocation requires thoroughly considering each party's capability, resources, awareness and authority to effectively bear and manage the risk, along with assessing any potential repercussions arising from these allocations.

While risks and their allocation depend on the contract structure, project delivery methods and market conditions, the risks in general could be allocated into three broad categories: (a) contractor-borne risks, (b) client-borne risks and (c) shared risks. *Contractor-borne risks* refers to all risks associated with the shortage of resources, labour, equipment and machinery required to complete the project and are usually attributable to the contractor. Also, the contractor must take care of the change in quantity, availability and transportation of construction material, except in cases where the material is to be supplied by the client.

Client-borne risks are associated with the client's responsibility for encumbrance-free access to the site, environmental clearances and timely contractor payments. Apart from these, the project-related information provided to the contractor may be incorrect or inadequate. In such cases, the client may assume some or all of such risk by allowing the contractor time and cost relief. Alternatively, the risk may wholly sit with the contractor in case the information is provided on a 'non-reliance' basis.

Finally, *shared risks* may include *force majeure* risks, such as natural disasters, war-like situations and pandemics, which usually provide for a temporary suspension or premature termination of the contract, depending on the severity of the event. In such cases, the damages and losses are shared by the parties; however, the specific share of each party is different based on the circumstances and the contract. Apart from these, there are certain third-party or residual risks that are not clearly attributable to any party. Their allocation is mostly debated and disputed and is attributed based on the contract clauses on risk and damages, such as the liquidated damages clause, limitation of liability clause, residual risk clause, apportionment of liability clause and insurance clauses.

It needs to be noted that equitable risk allocation in the contract document is essential; however, it does not resolve the problem on its own. In practice, the allocation is not always equitable, and risks are transferred to the parties least able to refuse them. The unequal distribution of power and decision-making makes vulnerable parties more prone to accepting certain risks, even when they are not in the best position to bear them. In many cases, when risks are transferred to the contractors, they may further pass them down the chain to subcontractors and vendors, who may not be capable of bearing such risks.

Therefore, it is evident that achieving equitable risk assignment and allocation necessitates cooperation and collaboration among the contracting parties. Risk allocation should be strategically addressed during the initial phase of contract drafting, allowing contract parties to negotiate and deliberate on the contract terms. This proactive approach reduces potential disputes and conflicts at a later stage. Beyond the mere structure of the contract, a fair distribution of risk and a risk–reward system tailored to the unique nature and specific considerations of the project can be accomplished through the implementation of practices such as a clear understanding of the risks being borne by each party and who owns or can manage the risk, maintaining active communication, instituting a system for early risk identification, providing substantial project information to the contractor to minimise the 'unknowns', building trust and enabling proactive dispute prevention and resolution mechanisms.

4.13 Conclusion

Understanding the bidding process in construction is crucial, as it determines a company's project pipeline and profitability. Companies use specific criteria for bid/no-bid decisions, such as their capability to deliver the projects, financial viability and risk considerations and the alignment of the projects with their expertise and resources. Similarly, construction companies need to have a comprehensive understanding of contract structures and the risks and rewards associated with them. This chapter provided a detailed overview of the bidding processes and decision criteria from the viewpoint of construction companies. The key elements of a construction contract and their classification based on key criteria were also presented. This knowledge is important for construction professionals and stakeholders to make informed decisions, ensuring that they undertake projects that are not only profitable but also well within their operational capabilities and strategic objectives. This knowledge is instrumental in effectively managing contractual issues and ensuring a fair and equitable distribution of risks. Achieving a balanced allocation of risk is crucial for the timely and successful completion of a project and for managing a successful construction company. The chapter also outlined specific criteria to identify which party should manage and bear specific risks to facilitate smoother project execution and reduce disputes.

References

Abrahamson, M. (1973, November). Contractual risks in tunnelling: How they should be shared. *Tunnels and Tunnelling*, 587–598.

Abrahamson, M. (1984). Risk management. *International Construction Law Review, 1*(3), 241–264.

Ahmed, S., & El-Sayegh, S. (2024). Relevant criteria for selecting project delivery methods in sustainable construction. *International Journal of Construction Management, 24*(5), 512–520.

Almaaz, R., Zakaria, I., & Odimegwumas, T. (2020). Comparative analysis of alliancing and public private partnership cost performance in Australasia infrastructure projects. *IOP Conference Series Earth and Environmental Science, 498*(1), 012102.

Barnes, M. (1983). How to allocate risks in construction contracts. *International Journal of Project Management, 1*(1), 24–28.

Batjargal, T., & Zhang, M. (2021). Review of key challenges in public–private partnership implementation. *Journal of Infrastructure Policy and Development, 5*(2), 1378.

Benjamin, N. B. (1969). *Competitive bidding for building construction contracts* (Technical Report No. 106). Department of Civil Engineering, Stanford University.

Dulaimi, M., & Shan, H. (2002). The factors influencing bid mark-up decisions of large- and medium-size contractors in Singapore. *Construction Management and Economics, 20*(7), 601–610.

Friedman, L. (1956). A competitive-bidding strategy. *Operations Research, 4*(1), 104–112.

Gates, M. (1967). Bidding strategies and probability. *Journal of the Construction Division, 97*(1), 75–107.

Gransberg, D. D., Scheepbouwer, E., & Loulakis, M. C. (2015). *Alliance contracting – Evolving alternative project delivery* (NCHRP Synthesis 466). The National Academies Press.

Hosseini, A., Lædre, O., Andersen, B., Torp, O., Olsson, N., & Lohne, J. (2016). Selection criteria for delivery methods for infrastructure projects. *Procedia - Social and Behavioral Sciences, 226*, 260–268.

Jayasuriya, S., Zhang, G., & Jing Yang, R. (2019). Challenges in public private partnerships in construction industry: A review and further research directions. *Built Environment Project and Asset Management, 9*(2), 172–185.

Klee, L. (2015). *International construction contract law*. Wiley.

Kwawu, W., & Laryea, S. (2013, September 2–4). *Incentive contracting in construction* [Paper presentation]. 29th Annual ARCOM Conference, Reading, UK.

Lam, K. C., Wang, D., Lee, P. T. K., & Tsang., Y. T. (2007). Modelling risk allocation decision in construction contracts. *International Journal of Project Management, 25*(5), 485–493.

Li, G., Chen, C., Martek, I., & Ao, Y. (2021). Bid or no-bid decision model for international construction projects: Evidential reasoning approach. *Journal of Construction Engineering and Management, 147*(2), 04020161.

Murdoch, J., & Hughes, W. (2002). *Construction contracts: Law and management*. Routledge.

Skitmore, M. (1989). *Contract bidding in construction: Strategic management and modelling*. Longman Scientific & Technical.

Suprapto, M., Bakker, H. L. M., Mooi, H. G., & Hertogh, M. J. C. M. (2016). How do contract types and incentives matter to project performance? *International Journal of Project Management, 34*(6), 1071–1087.

Thompson, P., & Perry, J. (1992). *Engineering construction risks: A guide to project risk analysis and risk management*. Thomas Telford.

5 Strategic Management

Kumar Neeraj Jha and Asheem Shrestha

5.1 Introduction

The construction industry is inherently risky and is strongly influenced by economic conditions and changing demand. Construction companies operating in this environment require an adaptable and resilient approach that helps them sustain their business in the long term. So, they must plan well to adjust to shifting market conditions, technological progress, growing customer requirements and changing regulatory environments (Langford & Male, 2008).

Construction companies have undergone significant changes in response to the industry's shifts over the past few decades. These changes have included developing strategies and restructuring their strategic management procedures to adapt to the dynamic environmental conditions. The concept of strategic management is particularly relevant for construction companies because it involves navigating the complexities and uncertainties of the industry while being able to compete effectively, manage risks and sustain growth. But despite this, the focus of construction companies generally lies on the execution of projects, while strategic management at a business level does not get enough attention (Kazaz et al., 2014). Therefore, it is critical for construction companies to pay particular attention to their strategic thinking and direction to achieve long-term success.

In this chapter, the fundamental concepts of strategic management are presented in the context of construction companies. It begins by explaining the process of strategic management in line with the objectives of construction companies. It then provides insights into the construction industry, with a particular focus on construction markets, the barriers to entry and exit and how they play a role in the firm's strategy development. The chapter then focuses on some of the common tools that construction companies can use to develop and manage their strategies. Finally, portfolio management and its importance for construction companies are discussed.

5.2 Concept of Strategic Management

Goldsmith (1995, p. 1) defined *strategic management* as

> the process by which managers set an organisation's long-term course, develop plans in light of internal and external circumstances and undertake appropriate action to reach those goals. . . . [It] is a broad activity that encompasses mapping out strategy, putting strategy into action, and modifying strategy or its implementation to ensure that the desired outcomes are reached.

DOI: 10.1201/9781003223092-5

The concept of strategic management is believed to originate from military strategy. However, it evolved during the mid-20th century and gradually transitioned into the business domain over time. According to Goldsmith (1995), the concept has its roots in teaching and research in business administration, which dates to the 1960s. Business policy and management courses taught in the 1960s focused on identifying the decision-making processes and strategies adopted by top executives of Fortune 500 companies. Because these top executives spent much of their time strategising about the long-term direction, goals and policies of their businesses, business policy academics and researchers were forced to undertake systematic analyses of the companies' strategies, which ultimately resulted in the self-styled study of strategic management. Research findings during this time also revealed that successful companies in the same industry were those that adopted different approaches, while the less successful companies adopted similar approaches. Traditional economic theory was inadequate in explaining these anomalies. Therefore, the concept of strategic management was seen as a better way to understand how organisations formulate strategic plans to achieve successful businesses.

In the annals of strategic management, 1980 stood as a pivotal year due to the release of Porter's seminal work, *Competitive Strategy*. Porter (1980, p. 6) established a fundamental inquiry into strategic management theory when he wrote, 'The basis of strategy formulation is establishing a relationship between a company and its environment'. The linkages between economics and strategic management become clear through three key interconnected concepts emerging from this statement: strategy, the organisation and the environment (Pamulu, 2010). So, it is inferred that a strategy is created to align with and respond to different organisational and environmental factors.

In essence, strategy refers to setting long-term objectives and determining the best route to achieve them. It is also about effectively competing and positioning oneself to achieve a unique or leading position in the market. One of the fundamental concepts of strategic management is the notion that a company must be distinctive to maintain long-term competitiveness. *Competitive advantage* is a concept popularised by Porter (1980) and pertains to the qualities that enable an organisation to surpass its rivals in terms of performance.

Competitive advantage can come from resource accessibility, geographical location, entry barriers to the market, distinctive technologies, reputations and relationships with other stakeholders. A competitive advantage enables businesses to establish a sustained market position by facilitating the creation of value in a manner that is difficult for rivals to imitate or exceed. To utilise these advantages effectively, companies must integrate them into their overall business strategy, ensuring that they align with their long-term organisational goals and are consistently applied across all projects and operations.

5.3 Porter's Generic Competitive Strategies

Porter's generic strategies framework outlines three primary ways companies can achieve sustainable competitive advantage: cost leadership, differentiation and focus (see Figure 5.1). These generic strategies are relevant to construction companies as they navigate a complex and competitive market environment.

5.3.1 Cost Leadership

With this strategy, a construction company seeks to become the lowest-cost producer or service provider in the industry. It can be achieved through economies of scale, efficient utilisation of resources, effective procurement and streamlined management practices. By minimising costs,

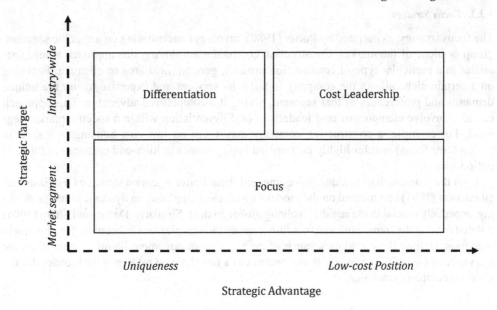

Figure 5.1 Porter's generic strategies.

Source: Porter (1980)

the company can offer competitive pricing to its clients. This approach is particularly effective in markets where price is a major factor for clients in selecting the service provider or for price-conscious customers. However, it requires maintaining rigorous cost control and often involves continuous investment in process optimisation and cost-saving measures. An example of cost leadership in the context of construction may include larger companies leveraging economies of scale, reducing the overall project cost and offering competitive bids. Similarly, volume home builders do not provide customisation opportunities like custom-home builders. Still, due to their standardised approach to construction based on a few fixed house plans, they offer more affordable housing options and, thus, take a major share of the housing market.

5.3.2 Differentiation

Differentiation is where the company seeks to establish itself as unique among its competitors. Adopting this approach, the company will focus on offering unique services or products that are perceived as superior and/or distinct in the market. It could be through innovative construction techniques, exceptional design quality, sustainable materials or practices or excellent customer service. By differentiating its products or services, the company can also create a niche market, allowing it to charge a premium price for its unique offerings. This strategy requires a strong focus on innovation and quality standards and may require a significant effort in the acquisition of relevant resources and specialised knowledge. An example of a differentiation strategy is a construction company specialising in high-tech, sustainable building practices, which might include innovative designs, energy-efficient materials and smart technology integrations, catering to clients who prioritise environmental sustainability.

5.3.3. *Focus Strategy*

The focus strategy, as outlined by Porter (1980), involves concentrating on a specific segment, group or niche of the market. Construction companies employing this approach would specialise in a particular type of construction product, geographical area or clientele. Focusing on a certain niche allows the company to tailor its services and expertise to suit the unique demands and preferences of that segment, giving it a competitive advantage. This approach can also involve elements of cost leadership or differentiation within a specific market segment. For example, a construction company may focus on low-cost housing in a specific region (cost focus) or offer highly customised luxury homes to high-end customers (differentiation focus).

Over the years, various critiques have emerged about Porter's generic strategies. For instance, Ghemawat (1991) commented on the model's insufficient emphasis on dynamic strategic thinking, especially crucial in the rapidly evolving global context. Similarly, Dawes and Sharp (1996) pointed out that the framework may not fully capture the complexities encountered in real-world scenarios. Despite these critiques, much of the strategic management literature recognises the relevance of Porter's framework. It often serves as a foundational reference for broader discussions on company strategies.

5.4 Strategic Management Process

Strategic management is a systematic process by which a construction company competes and positions itself to continue the success of the projects that it takes on. It involves two stages: (a) strategy formulation, where strategies are developed and finalised, and (b) strategy implementation, where the strategies are put into action.

Developing strategies begins with defining the mission and objectives of the company. During this stage, the organisational objectives are analysed in line with the company's core missions and its internal and external environments. Based on this comprehensive analysis, the company may need to revise its objectives and make informed decisions about which strategies to implement for optimal outcomes. There are several factors that need to be considered during this stage, including the environmental pressures of operating in the market, the resources that are available to the company and the value system of the people making the decisions, as well as the characteristics of the company (Langford & Male, 2008).

In the strategy implementation phase, strategies are brought to life through actionable steps across different levels of the organisation. The implementation is multifaceted, involving areas such as corporate governance, management systems and practices and strategic leadership. Each of these areas plays a vital role in ensuring that the strategies are executed effectively and align with the company's missions and objectives. As the strategies are implemented, they are also carefully evaluated to see whether the desired objectives have been achieved. It also involves renewing and refining the strategy to enable the company to stay agile and responsive to changing market conditions, internal dynamics and emerging opportunities.

For construction companies, the process acts as a guide for pursuing overall development and growth in the market. Strategic management is crucial for construction companies to implement due to the dynamic and evolving nature of the construction industry. Companies need to carefully plan for potential changes that could impact their operations and competitiveness in the future. Additionally, both the internal and external environments require a thorough evaluation, as understanding the risks and opportunities is critical for the survival and success of construction companies in highly competitive construction markets. For larger construction companies, adapting to changes is a complex process that must be systematically planned and executed to ensure organisational coherence and efficiency. Additionally, realistic forecasting of budgets and revenues is pivotal for maintaining financial health and consistency.

Strategic management is generally conducted at three levels in the construction company, namely:

1. *Corporate or strategic level:* Here, the company focuses on identifying and evaluating potential businesses or markets. This level involves crafting a comprehensive plan that outlines the overarching direction for the entire company.
2. *Organisation or business level*: At this level, the company formulates strategies for competing in specific businesses or markets. This involves strategic choices tailored to the unique challenges and opportunities of these selected areas.
3. *Operational or functional level*: The company focuses on detailed, day-to-day operational decisions at this level. These include choices between hiring or purchasing equipment, resource allocation, budgeting and other critical actions that directly influence project execution, turnover growth and profit maximisation.

To achieve their long-term goals, companies need to make important strategic decisions. This involves choosing from options that shape the company's future position in the market. Ansoff (1987) identified three types of decisions that companies make. Firstly, there are operating decisions, which are primarily concerned with the transformation of inputs into outputs. These decisions occupy the majority of a company's time and focus on the resource conversion process. In the context of construction companies, such decisions include setting pricing for bids, developing market strategies for selecting project types or services, scheduling production through site planning and allocating budgets among various functions like departmental and project budgets. Secondly, administrative decisions revolve around the structuring of the company and the allocation of resources within its structure. These decisions are crucial for the efficient and effective functioning of various organisational departments and teams.

Finally, there are strategic decisions that connect the company with its broader business environment. These decisions have a widespread impact on the organisation, influencing both operating and administrative decisions. Strategic decisions are a critical component of business planning and management, marked by their focus on the company's long-term success. They are typically made under conditions of uncertainty, which requires careful consideration and weighing potential outcomes. These decisions involve evaluating various alternatives, understanding the potential consequences of each and making a deliberate choice. Given the significant commitment of resources that strategic decisions entail, they demand thorough analysis and thoughtful deliberation. Furthermore, once made, these decisions are not easily reversible, often setting a course that can define the future trajectory of a company for years to come. Therefore, strategic decisions carry a weight that underscores their significance in determining the overall trajectory and long-term viability of a company.

Besanko et al. (2009) highlighted that for a company to successfully formulate and implement long-term strategies, it must address four key areas. First, the firm must define its boundaries, deciding on the scope, size and specific business areas it will operate in. Second, it should conduct a thorough market and competitive analysis to understand the unique characteristics of its operational markets and the competitive dynamics that prevail within them. Third, the company needs to determine its positioning and dynamics by deciding how it should compete, what the basis of its competitive advantage will be and how it plans to adapt these strategies over time. Lastly, the company's internal organisation must be considered, with the establishment of the most effective internal structure and systems to support its strategic objectives. These four dimensions are interrelated and critical to the holistic strategic planning process.

5.5 Objectives of Construction Companies

Every company has objectives they strive for. Objectives provide a clear focus or targets for the companies. Some objectives focus on the short term, while others may focus on the long term. These objectives may vary between companies and are based on several factors, including the nature of the business the company is in, the environment in which the business operates and the characteristics of the company itself. Generally, in the short term, companies may focus on revenue generation, profitability and building relationships and reputations. For construction companies, the short-term focus would be on winning and delivering projects efficiently by managing their cash flow well. They would also want to adhere to all the legal requirements and deliver projects to the satisfaction of their clients. For newly established companies, survival might be their main priority in the short term, while more established companies may look for ways to optimise their operations, increase efficiency and strengthen their relationships.

A company's long-term objectives are generally strategic in nature and involve factors that determine the company's long-term success. Often, companies look for sustainable growth. They want a consistent return on their assets and to satisfy their shareholders. Construction companies typically pursue growth through one of two main strategies: (a) organic growth or (b) mergers and acquisitions (Lavender, 2014). Organic growth, a steady and controlled approach, involves reinvesting profits and maintaining a balanced level of borrowing. This method focuses on internal expansion and gradual market development. On the other hand, growth through mergers and acquisitions can lead to more rapid expansion, offering immediate access to new markets and resources. However, this approach often involves higher levels of borrowing and carries greater risks, including the challenges of integrating different corporate cultures and systems.

In any case, one of the primary motivations for growth for construction companies is to generate higher profits, which occurs through increased revenue and cost reduction. Revenue may increase through business expansion and market share growth, scaling operations and entering new markets or sectors while ensuring survival. Costs may be reduced through greater efficiencies from economies of scale, efficient utilisation of resources and better supply chain management.

However, to achieve their objectives, companies must surpass others with similar focus in a highly competitive landscape. Gaining a distinct advantage over competitors offering similar products or services is essential for success in the construction market. However, compared to other sectors, such as manufacturing, it is difficult for construction companies to maintain a sustained competitive advantage, primarily because the 'product' is generally homogenous in nature. The buildings and structures may vary in terms of design and the way they are produced, but the final products are not that different and can be easily imitated by competitors. So, the differentiation strategies in the construction sector are more about process and efficiency than the delivery of unique products.

5.6 Understanding Construction Markets

Langford and Male (2008) highlighted that construction companies operate in a very distinct market where there are economic forces at play that are unique to the industry. Firstly, unlike other industries, contractors set prices before production, and the built structures are pre-demanded by the client. Secondly, it also involves speculative construction, a process that either anticipates, responds to or generates demand, much like traditional manufacturing methods. An example of this is speculative house building.

In light of these industry-specific characteristics, it becomes imperative for construction companies to gain a thorough understanding of the construction market. This involves not only keeping informed on current trends and customer needs but also anticipating future demand and shifts in the market. This allows construction companies to strategically position themselves to remain competitive and meet their strategic objectives.

5.6.1 Market Structure and Competition

Market structure plays an important role in determining the level of competition in the construction sector. Strategies to be executed depend on the competitive environment in which the construction companies operate (De Valence, 2010). The competitive environment, in turn, depends on the demand and supply dynamics of the construction sector.

Demand in the construction sector is driven by prices and non-price determinant factors such as government policies, the income of individuals, the quality and age of the existing infrastructure, etc. Supply in the construction sector is driven by prices and non-price-determinant factors such as supply chain management, adoption of technology, etc. These factors typify the market structures prevalent in the construction sector (Myers, 2022). There are four market structures that exist in the construction sector: perfect competition, monopolistic competition, oligopoly and monopoly.

Perfect Competition

Perfect competition is when there are a very large number of companies producing the same product. In this market structure, there are no price makers, and every company is a price taker. The products in this market structure are standardised, and every product has a substitute for its replacement. There are low barriers to entry for construction companies in this market. The products in this market are homogenous in nature. A close example of perfect competition in the construction market can be seen among small-scale home builders in local housing markets. These builders compete against numerous similar builders, each with a small market share. In this scenario, no single builder has the capacity to notably impact the pricing of new homes, as the market is characterised by a high level of competition and a lack of dominant players. However, it must be noted that it is rare to find perfect competition in a real-life market, especially in the construction market. The construction industry is predominantly made up of small companies; however, it is also true that the largest companies in the industry in many countries account for a significant share of industry turnover (De Valence, 2010).

Monopolistic Competition

Companies operating in this market structure tend to differentiate themselves from other companies to create monopolies. They differentiate similar products with different attributes such as brand name, quality, durability of the product, etc. In monopolistic competition, companies have the freedom to set their prices and may have some market power. However, they must be careful not to overprice their products or services, as consumers have alternative options and may switch to a competitor. Therefore, companies do not have excessive control over the price. If they tend to increase the price, the product becomes uneconomical. In the construction market, examples of monopolistic competition could include custom home builders operating in a local housing market with many custom home builders, where each builder offers slightly different designs and features and buyers have a range of options to choose from. However, each builder has a limited market share, and no individual builder can significantly influence the prices of new homes.

Oligopoly

In this market structure, there are very few companies operating in the industry with either identical or differentiated products. It is not easy to enter the oligopoly market due to its high fixed costs. Control of prices could vary depending on the relationship between the companies operating in the oligopoly market. If the companies compete to operate, then the prices will be competitive. If the companies cooperate or collude, then the prices can be interdependent and could lead to rigged pricing in the market.

In the construction market, examples of oligopoly could include large-scale home builders operating in a local housing market where a few large-scale builders control a significant portion of the market share and have the ability to influence the prices of new homes. Additionally, infrastructure markets are akin to oligopolies, particularly for large projects where few large companies have the capacity, expertise and resources to undertake and deliver these projects.

It is important to note that in an oligopoly market, companies are interdependent on each other, and their decisions on prices, output, and marketing strategies will affect the market. Oligopolies may engage in price collusion or tacit collusion to keep prices high and restrict output. Also, barriers to entry are very high, making it difficult for new companies to enter the market. Collusion in the oligopoly market refers to a situation where companies in the market work together to restrict competition and maintain higher prices. This can take different forms, such as price fixing, market sharing or bid rigging.

Monopoly

The term *monopoly* signifies that there is only one company producing the product. In this market structure, companies are the price makers, unlike in a perfectly competitive market. An example of a monopoly could include public utility providers, which may be a single company holding the monopoly on certain utilities, such as electricity or water, for a specific region and, therefore, having complete control over the prices for those services. In the defence sector, a government-approved company could be the only company eligible to execute the projects due to security considerations. Monopolies may restrict output to increase prices, and companies may not have any incentive to innovate or improve their product or service. Also, barriers to entry are high, making it difficult for new companies to enter the market.

5.6.2 Entry and Exit Barriers

Entry and exit barriers refers to structural principles within an industry that governs the degree of ease with which a company can enter or exit the industry. In general, if an industry has higher barriers to entry, it typically means that new competitors cannot easily enter and disrupt the market, which could lead to longer periods of prosperity for existing companies. On the contrary, if an industry is difficult to exit for existing companies, there will be a greater amount of rivalry in the industry. This can result in overcapacity, where too many companies compete for the same customers, leading to increased competition and potentially lower profitability. Entering the construction industry is easy for companies involved in small-scale work. However, as projects become larger, more complex and technologically demanding, the number of companies capable of handling such work diminishes.

Gruneberg and Ive (2000) identified six barriers for companies to enter the construction market, namely, (a) economies of scale, (b) supply chains, (c) incumbent cost advantages, (d) private information and client relationships, (e) contestable markets and (f) client-imposed barriers. These barriers are explained below in detail.

Economies of Scale

Higher levels of construction output would result in a reduction in average unit costs, reflecting the condition termed *economies of scale*. These include technical economies of scale (exploiting the equipment), managerial economies of scale (exploiting the potential of the management and supervisory team), commercial economies of scale (buying in bulk to reduce the unit cost of materials), financial economies of scale (greater availability of funds for larger companies) and risk-bearing economies of scale (market diversification). Large companies with high output can lower their costs, making it difficult for smaller new entrants to compete. New entrants may not be able to exploit economies of scale until and unless their production reaches higher levels. The condition of economies of scale is not a permanent phenomenon, as construction companies tend to observe diseconomies of scale due to ineffective management and the fragmented nature of the industry (Myers, 2022).

Supply Chains

Existing construction companies often have strong relationships and long-term agreements with suppliers and subcontractors. New entrants, on the other hand, may struggle to establish a reliable supply chain and may have to face higher costs or less favourable terms, which can be a significant barrier to entry. Establishing credibility and securing business with partners would require a substantial investment of time and effort for new entrants.

Incumbent Cost Advantages

Existing companies may have cost advantages due to their established position in the market that are not available to new entrants. These advantages can be in the form of lower production costs due to more efficient processes, better economies of scale or favourable contracts with suppliers. Incumbents can also access cheaper financing due to their proven track record. Matching these cost efficiencies can be challenging for new entrants, making it harder for them to compete on price. Larger projects are also capital intensive, presenting high initial investment requirements where companies must not only accumulate substantial capital but also acquire the necessary equipment and skilled human resources. However, the considerable sunk costs associated with these preparations in large projects often pose a significant challenge for new entrants.

Private Information and Client Relationships

Established construction companies often have access to private information and have built strong relationships with clients over time. Gruneberg and Ive (2000) highlighted this as one of the biggest barriers to entry into the construction industry. Private information may consist of proprietary knowledge pertaining to certain technologies, procedures or insights into customer preferences and demands, among other things. Such information is often safeguarded. This knowledge can be a significant barrier to new entrants who do not have access to the same level of information or have not yet built trust with potential clients.

Contestable Markets and Contacts

A contestable market is one in which entry and exit are unrestricted, and new businesses can enter and compete with established enterprises with relative ease. In a contestable market, new entrants can easily replicate existing companies, so, in response, the existing companies modify their behaviours to deter new companies from entering the market. For example, an established company might respond to the threat of new entrants by leveraging its existing relationships with suppliers to negotiate lower material costs, making it harder for new companies to compete on price. Within the construction industry, markets are frequently not entirely contestable with various entry and exit barriers making it difficult for new companies to enter the market.

Client-Imposed Barriers

The clients themselves can impose barriers to entry. According to De Valence (2010), this barrier is unique to construction, where the clients put several restrictions in place that make it difficult for new entrants to bid on projects. For instance, they might prefer working with established companies with whom they have a long-standing relationship or might require contractors to have specific qualifications, experience or scale that new entrants do not possess. When clients prefer contractors who have prior experience with similar projects, it significantly limits the growth opportunities for newer construction companies.

Exit barriers in the construction market are significant and multifaceted, involving aspects like investment in specialised equipment, the employment of skilled labour and substantial fixed costs. The first major exit barrier concerns the specialised equipment used on construction sites. If a firm decides to leave the construction market, repurposing or selling specialised equipment can be challenging due to its limited applicability in other industries. Various assets represent a significant capital investment that may not be easily recoverable or transferable to other sectors.

The second key exit barrier is the specialised workforce that construction companies employ. These professionals, trained specifically for construction tasks, may find their skills less transferable to other industries. In the event of a market exit, this workforce becomes redundant, leading to issues like unemployment or the need for retraining, which can be both costly and time-consuming.

Thirdly, the fixed costs associated with setting up and managing construction projects present another significant barrier. These costs include the mobilisation of resources, advance payments to subcontractors and suppliers and investments in new technologies. These technologies often require substantial outlays for licenses, training and implementation. When a company exits the construction sector, these fixed costs transform into sunk costs, representing financial resources that the company cannot recover. This adds a financial dimension to the exit barriers, making the decision to leave the market complex and potentially very costly.

5.7 Strategic Analysis Tools

There are various tools that can help companies develop their strategies. In recent years, complex models have been developed. However, many of the new models are grounded in some of the classic, fundamental strategic analysis methods, which provide the underlying principles for understanding strategic models. Some of the important strategic analysis tools are discussed next.

5.7.1 Porter's 5 Forces Analysis

Porter's 5 forces analysis is a framework that allows companies to examine the level of competition within a market, based on which they can formulate their strategies. Porter (1980) argued that the structure of an industry defines the competitive rules of the game as well as the strategies available to a company. The structure is influenced by the interplay of five competitive forces:

1. *Threat of New Entrants:* This force is related to entry barriers. The intensity of competition will increase according to the ease with which new companies can enter the market. As new entrants introduce their products or services, they become rivals to established companies. A proliferation of competitors will dilute the market share and weaken the standing of existing companies.
2. *Threat of Substitute Products or Services:* A significant competitive threat is posed to the company when alternative products or services enter the market, serving as substitutes for the company's offerings. When there are options in the market for offerings that perform the same functions, customers may switch to alternatives. The more viable and attractive these substitutes are, the more they limit the potential returns for the existing companies.
3. *Bargaining Power of Suppliers:* This force analyses the extent of influence suppliers have on companies. When suppliers have substantial bargaining leverage, they can impact the attractiveness of a market through their ability to increase the prices of goods and services. This influence is determined by the number of suppliers providing essential components of a product or service, the distinctiveness of these components and the cost incurred by a company to shift between suppliers. A limited pool of suppliers, coupled with the high dependency of a company on a particular supplier, amplifies the supplier's power.
4. *Bargaining Power of Buyers:* The power of buyers determines how much pressure customers can place on the margins and volumes. Buyers who make large purchases or those who have the option of switching to other sellers hold considerable influence, as they can negotiate for reduced prices or improved product quality, thereby affecting a company's profit margins. This force is dependent on the number of customers a company has, the importance of each individual customer and the expense involved for a customer to transition from one company's product or service to another's. The smaller and more powerful a client base, the more power it holds.
5. *Intensity of Competitive Rivalry:* The core of Porter's model is competitive rivalry, which is influenced by all the four forces discussed above. This force examines the degree of competition among existing companies. A greater number of rivals and the array of comparable products and services they provide limit the company's market influence. If suppliers and buyers find the deals offered by a company unsatisfactory, they may turn to its competitors, underscoring the competitive pressure within the industry.

According to Porter, the ultimate success of an industry in terms of long-term return on invested capital is determined by the combined strength of these forces. An industry where the five forces are strong will be less attractive because the intense competition will drive down profit potential. In contrast, industries with weaker forces can be more attractive. Understanding these dynamics helps businesses identify the most advantageous positions within their industry. For example, if a company is in a highly competitive industry, it might seek to improve its competitive position through strategies like innovation and cost leadership or by focusing on niche markets where it can have a competitive advantage.

5.7.2 SWOT Analysis

SWOT analysis is a common strategic analysis and planning tool that is used to understand a company's capabilities in line with opportunities to develop strategies. It enables companies to develop informed strategies by analysing their operational environment and aligning their resources and actions accordingly. SWOT stands for strengths, weaknesses, opportunities and threats. It was developed in the 1960s to evaluate the strategic fit between a company's organisational strategy and the competitive environment in which it operates. Strengths and weaknesses are generally internal to the company, while opportunities and threats are external.

According to Langford and Male (2008), a company faces two types of strengths and weaknesses: structural and implementational. Structural strengths and weaknesses are related to external factors that might affect the company's strategic position. They can range from the economy and market conditions, the technological and societal developments or the industry structure and competition. This type of strength or weakness is difficult to change or influence. Using SWOT analysis, these factors are considered during the company's assessment to identify opportunities and threats. On the other hand, implementational strengths and weaknesses are related to the company's capacity to execute its strategies, which comprises skills, resources and management capabilities. These are usually assessed as part of the company's internal appraisal of its strengths and weaknesses.

Strengths could include experience, expertise, reputation, established networks, financial stability, technological advancement, etc. Weaknesses could include dependency on certain markets, a low skill base, challenges in supply chain management, safety and quality concerns, etc. Opportunities could include emerging technologies, the potential to incorporate sustainable practices, expansion into new markets, etc. Threats could include high competition, regulatory changes, recession, inflation, etc.

A SWOT analysis effectively maps out the environmental factors and capabilities but does not directly prescribe strategies. It requires a careful synthesis of the insights to formulate coherent and applicable strategies. A common method that is used is the TOWS matrix (see Figure 5.2), where the strengths are mapped along with opportunities to identify how the company's strengths can be leveraged to capitalise on these opportunities, creating strategies that align internal capabilities with external possibilities for growth and success. Similarly, weaknesses can be mapped to threats to devise mitigation strategies that address vulnerabilities and safeguard the company against potential challenges and risks in the external environment.

Often, SWOT is also combined with the PESTEL framework to get a comprehensive view of the external environment. PESTEL stands for political, economic, social, technological, environmental and legal factors (Buys & van Rooyen, 2014). The combined analysis facilitates understanding broader macro-environmental factors that could impact strategic planning.

External Factors

		Opportunities	Threats
Internal Factors	Strengths	**Strengths – Opportunities** *How can we use our strengths to take advantage of the opportunities?*	**Strengths – Threats** *How can we use our strengths to manage the threats?*
	Weaknesses	**Weaknesses – Opportunities** *How can we overcome weaknesses that prevent us from taking advantage of the opportunities?*	**Weaknesses – Threats** *How can we overcome weaknesses that make threats a reality?*

Figure 5.2 TOWS matrix for SWOT analysis.

5.7.3 Value Chain Analysis

Another tool particularly relevant in the construction industry context is value chain analysis, which Porter developed in 1985 (Porter, 1985). It helps businesses dissect their internal activities and understand how they contribute towards creating value, which becomes the source of competitive advantage. The value chain involves primary activities and support activities. Primary activities are the core activities that are directly involved in producing and delivering products or services. Support activities are those that are not directly involved in production, but they support the primary activities to enhance the production process.

Value emerges from the way the company manages its resources and processes, including how inputs from suppliers are transformed into outputs valued by customers. The company's strategic management process impacts the formation of its value chain. Retaining competitive advantage depends on the company's capacity to provide distinctive value via its value chain management. A value chain analysis involves a systematic process where a company identifies and reviews its primary and support activities. Each activity is scrutinised for efficiency, cost-effectiveness and role in adding customer value. The company also examines how these activities are interrelated, seeking optimisation opportunities within these linkages. The analysis helps in identifying potential areas for competitive advantage. The insights gained from the analysis help develop a strategic plan for areas with potential for innovation or improvement.

Porter (1985, p. 46) presented the generic value chain, which includes 'inbound logistics', 'operations', 'outbound logistics', 'marketing and sales' and 'service' as primary activities. The support activities include 'firm infrastructure', 'human resource management', 'technology development' and 'procurement'. In the construction industry, the value chain can vary depending on the perspective from which we are examining it. For instance, from a contractor's viewpoint, the value chain may include material procurement, workforce coordination, operations, project management and on-site logistics as key activities. In contrast, a real estate developer might focus more on financing, land acquisition, pre-construction planning, etc.

5.7.4 Strategic Group Analysis

Companies need to develop strategies to position themselves competitively relative to the other companies operating in the same market. While competition is perceived as a threat, it also serves as a catalyst for companies to innovate and evolve. Strategic group analysis is a useful instrument that aids companies in understanding their competitors. It focuses on distinct clusters of competing companies that share a common strategic perspective or competitive approach (Fleisher & Bensoussan, 2000). It is used as a tool to ascertain the distinct strategic positions held by competing companies, the level of competitive rivalry between companies, the profit potential of the different strategic groups operating within an industry and the implications for the competitive position of the analysed firm (Dikmen et al., 2009).

To perform strategic analysis, the competitors first need to be identified through competitor analysis. Key aspects could include the types of clients these competing companies deal with, the services they offer and their pricing patterns, as well as their strengths and weaknesses. The analysis would identify the primary, secondary and tertiary competitors upon which strategic group analysis is applied. A strategic group consists of companies within the same industry with similar business models and strategies. The construction companies in a strategic group have similar characteristics, offer similar services, have similar market shares, etc. Strategic group analysis explores the entire construction market's competitiveness and the amount of rivalry present. This can help a construction company in various ways, such as:

- Examining the areas and services where the other companies are operating helps it devise better strategies where there is more competition.
- Identifying the paths/solutions taken by the other construction companies in a strategic group that the company can explore.
- Identifying gaps in the services offered in the market that the company can capitalise on.

A strategic group map is created with the identification of similar and distinguishing attributes that the companies in a strategic group exhibit. For example, Dikmen et al. (2009) identified

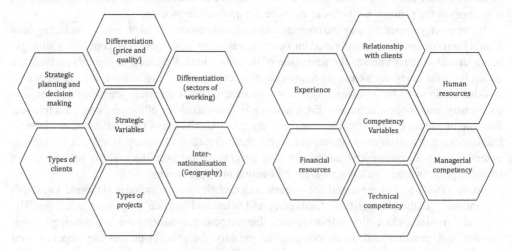

Figure 5.3 Strategic and competency-based variables identified for construction companies.
Source: Based on Dikmen et al. (2009)

12 variables for analysing Turkish construction companies, which include strategic and competency variables (see Figure 5.3), based on which direct rivals can be identified and strategies can be developed to gain competitive advantage.

5.8 Business Portfolio Management

Portfolio management refers to the systematic approach to managing a company's investments to achieve strategic objectives and maximise overall performance. It involves carefully selecting, prioritising and supporting projects and investments based on their alignment with the company's strategic goals, their expected returns and associated risks. Portfolio management also helps companies decide how to mix their investments optimally and allocate assets to balance corporate risks.

According to McNamee (1985), portfolio management requires identifying three key aspects of the product or the strategic position of the company: (a) its market growth rate, (b) its relative market share in comparison to the market leader and (c) the revenues generated from the product sales or the company's activities. The Boston Consulting Group matrix, developed in the 1970s for the Boston Consulting Group, has since become a fundamental tool in portfolio management. Its primary purpose is to help companies evaluate their portfolios in terms of their market growth rate and relative market share. This analysis enables businesses to categorise their business units or product lines into four distinct quadrants – stars, cash cows, question marks and dogs – each representing a different strategic scenario, as shown in Figure 5.4.

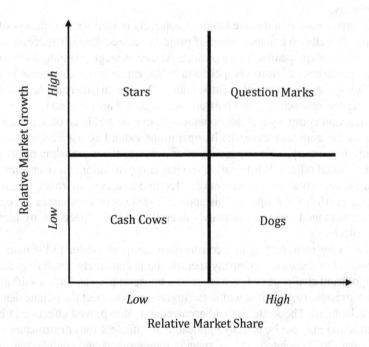

Figure 5.4 Boston Consulting Group growth-share matrix.
Source: Adapted from McNamee (1985)

Stars are those units with a high market share in a growing market. These units generally strategise to allocate resources to maintain their market position while looking to expand further. Cash cows have a high share operating in a market that is not growing. They generate steady, reliable profits, and their strategies are related to maintaining stability and modest growth while not putting further investments into the business. They may also support their sister business units that are in other quadrants or look to invest in question marks. Question marks are units operating in a high-growth market but with a low market share. These companies, particularly the ones that show promise, are seen as potential future stars but require significant investment to grow. Their priority strategy would be to grow to capture more market share. Those that do not show promise would think about divestiture. Finally, dogs are units that operate with a low market share in a low-growth market; for them, the best option is to divest.

For large, diversified companies involved in construction, especially those with subsidiaries in other business areas, the traditional sense of portfolio analysis applies well. These companies often manage a range of investments across different business sectors. In such cases, portfolio analysis is used to assess and balance these diverse investments, which involves evaluating each business unit or subsidiary's performance, market position and growth potential. For most construction companies, however, especially those not part of larger, diversified conglomerates, the portfolio typically refers to the collection of projects they are involved in. In this context, portfolio analysis is more project centric. It involves assessing the risk, profitability, resource allocation and strategic value of different construction projects. Unlike diversified companies, where portfolio analysis might span across different industries, in most construction companies, the analysis is focused on the various construction projects they undertake, whether these are in different geographical locations, involve different types of construction (e.g., residential, commercial, infrastructure) or vary in size and complexity.

The effective management of a diverse range of contracts is vital for the success of a construction company. The effective management of projects secured through successful tenders is crucial for construction companies. These projects require strategic planning to enhance the company's overall performance. However, portfolio management is not a common practice in the construction industry due to its project-centric nature. Despite this, there are several compelling reasons why regular restructuring and portfolio management are necessary.

Firstly, a construction company typically operates from a central head office that oversees all project sites. As the company expands, its operations extend across various geographical regions within the country. This expansion often leads to the establishment of different business units or regional offices. While part of the larger organisation, these entities operate semi-autonomously, necessitating a more coordinated and strategic approach to effectively manage the diverse portfolio of projects. This approach ensures that resources are optimally allocated, risks are managed and the company's overall strategic objectives are met across all regions and projects.

The second reason for restructuring in a construction company relates to the nature of the projects it undertakes. For instance, a company specialising in constructing buildings and factories might face significant challenges when it takes on a transport project. Such a shift in project type necessitates a strategic realignment within the organisation to meet the unique demands of infrastructure development. The strategies and arrangements that proved effective in building and factory construction may not be directly applicable or sufficient for infrastructure projects. As Smyth and Yanga (2021) pointed out, the specific requirements and complexities of infrastructure work require a tailored approach, differentiating it from other types of construction projects. This underscores the importance of adaptability and strategic reorientation in response to the changing nature of the company's project portfolio.

Moreover, portfolio management plays a crucial role in balancing profitability and strategic growth. Ball (2014) highlighted the necessity of maintaining a strategic equilibrium between projects, advocating for the diversification of projects as a key to achieving scale economies. This diversification not only allows construction companies to distribute risk across multiple projects, thereby reducing the impact of any single project on their overall financial stability, but also provides opportunities to enter new markets with lower resistance.

Furthermore, by expanding their project portfolios, either organically or through acquisition, companies can access new customer bases, tender opportunities and networks, enhancing their market presence. For larger construction companies, this diversified portfolio strategy also offers the flexibility to withdraw from specific markets with relative ease, as their commitment is often project specific. Additionally, a wider range of projects and a broader client base can significantly strengthen a company's bargaining power with clients, leading to potentially more favourable terms and increased profitability. This approach to portfolio management underscores its importance as a strategic tool for growth and stability in the construction industry (Langford & Male, 2008).

As a construction business experiences growth in turnover, it faces the challenge of integrating and strategically managing an increasing number of profit centres within the construction market. A widely used approach to effectively navigate this complexity is product market portfolio analysis. It allows businesses to evaluate and organise their diverse range of projects and market segments, ensuring that each profit centre is aligned with the overall strategic objectives and contributing optimally to the company's growth. Implementing this analysis aids in making informed decisions about resource allocation, market positioning and strategic planning, which are essential for maintaining growth momentum in a competitive industry.

5.9 Conclusion

The chapter has provided an overview of the concepts and processes of strategic management, particularly from the perspective of construction companies. It has highlighted the role of strategic management in guiding these companies towards achieving their objectives and sustaining competitiveness in the construction market. Strategic management in the construction industry is a dynamic process. It requires continuous adaptation, keen market insights and a strong focus on long-term goals.

The tools and concepts discussed in the chapter are not just theoretical constructs but practical instruments that can help construction companies navigate the evolving challenges and opportunities of the industry. As the industry continues to evolve with technological advancements and shifting market dynamics, the principles of strategic management will remain more relevant than ever. It is through strategic rigour and adaptability that construction companies will continue to lay the foundations not just for business continuity but for a resilient, prosperous future.

In the forthcoming chapter on business models, the ideas presented in this chapter are extended, bridging the gap between the theoretical underpinnings of strategic management and their practical application in the construction sector. This business model chapter will delve deeper into how the fundamental concepts of strategies discussed here can be operationalised into tangible, actionable tools. Various business models used by companies in the construction industry are explored, with a focus on examining how these models are designed and used to meet industry-specific challenges and opportunities.

References

Ansoff, I. H. (1987). *Corporate strategy*. Penguin.

Ball, M. (2014). *Rebuilding construction (Routledge revivals): Economic change in the British construction industry*. Routledge.

Besanko, D., Dranove, D., Shanley, M., & Schaefer, S. (2009). *Economics of strategy*. John Wiley & Sons.

Buys, F., & van Rooyen, R. (2014). The strategic management of construction companies during recessionary cycles. *Acta Structilia: Journal for the Physical and Development Sciences, 21*(2), 1–21.

Dawes, J., & Sharp, B. (1996). Independent empirical support for Porter's generic marketing strategies? A re-analysis using correspondence analysis. *Journal of Empirical Generalisations in Marketing Science, 1*(2), 36–53.

De Valence, G. (2010). *Modern construction economics: Theory and application*. Routledge.

Dikmen, I., Birgonul, M. T., & Budayan, C. (2009). Strategic group analysis in the construction industry. *Journal of Construction Engineering and Management, 135*(4), 288–297.

Fleisher, C. S., & Bensoussan, B. (2000). A FAROUT way to manage CI analysis. *Competitive Intelligence Magazine, 3*(2), 37–40.

Ghemawat, P. (1991). *Commitment: The dynamic of strategy*. Free Press.

Goldsmith, A. A. (1995). *Making managers more effective: Applications of strategic management* (Working Paper No. 9). USAID.

Gruneberg, S., & Ive, G. J. (2000). *The economics of the modern construction firm*. Springer.

Kazaz, A., Er, B., & Ozdemir, B. E. (2014). A fuzzy model to determine construction firm strategies. *KSCE Journal of Civil Engineering, 18*, 1934–1944.

Langford, D., & Male, S. (2008). *Strategic management in construction*. John Wiley & Sons.

Lavender, S. D. (2014). *Management for the construction industry*. Routledge.

McNamee, P. B. (1985). *Tools and techniques for strategic management*. Pergamon Press.

Myers, D. (2022). *Construction economics: A new approach*. Routledge.

Pamulu, M. S. (2010). *Strategic management practices in the construction industry: A study of Indonesian enterprises* [Unpublished doctoral dissertation]. Queensland University of Technology.

Porter, M. (1980). *Competitive strategy: Techniques for analyzing industries and competitors*. Free Press.

Porter, M. (1985). *Competitive advantage: Creating and sustaining superior performance*. Free Press.

Smyth, H., & Yanga, W. (2021). *Project portfolio management – Practice-based study of strategic project portfolio management in the UK construction sector*. University College London.

6 Business Models

Asheem Shrestha

6.1 Introduction

The construction industry is witnessing a transformation. Construction, which has generally been slow to change, is now transitioning at an accelerated pace, with the construction market seeing shifts in the use of new and improved materials, products and processes. These changes driven by industrialisation, globalisation and digitalisation (Ribeirinho et al., 2020) have created huge disruptions in construction markets worldwide, presenting construction companies with both formidable challenges as well as enormous new opportunities (Renz et al., 2016).

The construction industry has largely relied on site-based work but is now seeing more industrialisation with prefabrication and modular construction being employed; supply chains have expanded with increased global trade as a result of advances in transportation, logistics and information technology; and project operators are demanding more digital interactions, with companies beginning to employ new technologies to improve efficiency and environmental sustainability. The transformation is being further driven by changing customer expectations, including those of homebuyers and taxpayers, as well as users of non-residential buildings and infrastructure. So, the level of sophistication of construction projects is increasing, and we are seeing more and more complex projects being planned and delivered. Entrants with new specialisations are emerging in the market to produce innovative products or provide new services to meet the demands of the industry, while new international players are appearing in the global market.

Construction companies are vulnerable to disruptions and recognise that they must adapt to the new market conditions to remain competitive. New technologies, materials and processes are now seen as important means to improve productivity, enhance quality, improve safety conditions and reduce environmental impact. Most construction companies follow a business model (BM) to compete in the market and generate profits, usually knowingly but sometimes also unknowingly. BMs have a significant impact on a company's decision-making and are founded on a business logic derived from the core beliefs, value system, strategies and market knowledge. However, in times of major disruptions that change the market dynamics, companies need to reconsider their business logic so that they can keep delivering value to their customers and stakeholders (Jones et al., 2019). To this end, there has been a significant shift in the construction industry recently, where companies are placing their focus on new BMs as a means of addressing the inherent complexity and challenges of the sector (Abeynayake et al., 2022).

This chapter focuses on BMs for construction companies. However, before delving into the specifics of construction BMs, a thorough introduction to the concept of BMs and the function they serve is provided. Then, the relevance of BMs in construction is discussed, and some of

DOI: 10.1201/9781003223092-6

the common forms of BMs utilised by construction companies are presented. Next, the chapter discusses BM innovation and why it is crucial for construction companies, mainly focusing on two main areas: technological innovation and sustainability innovation. Finally, case studies illustrate the main ideas presented in BMs applied in the construction industry.

6.2 What Is a Business Model?

6.2.1 Key Concepts

The use of the term *business model* gained traction after the 1990s, when the internet became a focal point for discussions on BMs and the shifting boundaries of firms. According to Karam (2022), before this time, effective BMs emerged by chance, without deliberate planning. In recent years, however, the notion of BMs has been highly popular in a wide range of fields, being utilised as a broad framework for describing the interplay between a company and its suppliers, consumers and partners (Zott & Amit, 2007).

At the most fundamental level, a BM is seen as the description of an organisation and how it accomplishes its goals (Massa et al., 2017). A BM explains the design or architecture of the systems utilised by companies for value creation, delivery and capture by documenting their organisational activities, resources and processes (Pan & Goodier, 2012). In doing so, it seeks to answer questions such as who the customer is, what they value and how the company can make money in the business (Magretta, 2002). BMs are a fundamental component of any successful enterprise, as they serve as a blueprint for creating the company's unique value proposition for the target customers. It also helps determine the necessary resources and capabilities required to deliver the value proposition and direct them towards achieving the company's objectives through an effective profit formula (Osterwalder & Pigneur, 2010).

Johnson et al. (2008) explained that the four building blocks that form a BM are customer value proposition, profit formula, key resources and key processes. The company's value proposition is what it offers to its customers. It is most effective when designed to answer the customer's problems in a specific manner that competing alternatives do not. The profit formula encompasses various key components that collectively determine how a company generates value for itself. It comprises the revenue model, which centres around the interplay between pricing and production volume; the cost structure, encompassing both direct and indirect expenses and the influence of economies of scale; the margin model, designed to achieve desired profit levels; and the resource velocity, dictating the speed at which assets can be turned over to meet profit expectations. The third building block, key resources, is related to assets like people, technology, facilities, equipment, skills and relationships that are essential for delivering the value proposition to the customer. It emphasises how these elements interact to create value for both the customer and the company. The fourth building block, key processes, relates to the company's operational and management procedures and guidelines that are designed to be efficient, streamlined and scalable.

6.2.2 Business Model versus Strategy

We need to clarify how a BM differs from strategy. Many studies on BMs do not make this distinction, and, worse even, they often use the terms interchangeably. Though the two concepts are interrelated, they are not the same (Magretta, 2002). According to DaSilva and Trkman (2014), the term BM is often misused both in academia and practice, where it often appears to encompass everything from strategy, an economic model, a revenue model or sometimes even a business plan. Casadesus and Ricart (2011) highlighted that this misconception often leads to poor decision-making.

A company's strategy is its goal to achieve a distinguishable and advantageous position in the market. The strategy focuses on competition and includes plans for a wide range of potential outcomes, such as the actions of rivals or unexpected changes in the external environment, whereas a BM describes the business logic behind how the company functions and focus on how it can create more value for customers and stakeholders in a competitive marketplace while still being aligned with the specific strategy and culture of the company (Pekuri et al., 2013). The BM structures a profit formula for delivering value by organising its resources and developing processes that allow the company to generate a profit. An excellent case in point is Amazon, whose BM included bypassing traditional retail distribution channels in favour of reaching consumers directly via online activity.

As an illustration of how different BMs work, we can look at the study by Ferrero et al. (2015), which examines a BM in the context of car-sharing services. They explained that going from point A to point B could be achieved with a car that is either purchased, leased, rented or shared. While there is not much variation in the core offering, the underlying activity structure provides cover for a wide variety of possible BM implementations. The same idea can also be used to discuss BM and strategy. While the main strategic goal of the company will remain the same, the means to achieve that goal can vary. Thus, a BM may be thought of as a layer between business strategy and business processes and the means through which a company translates its strategic goals into actionable plans. Figure 6.1 shows how the BM fits within the company strategy.

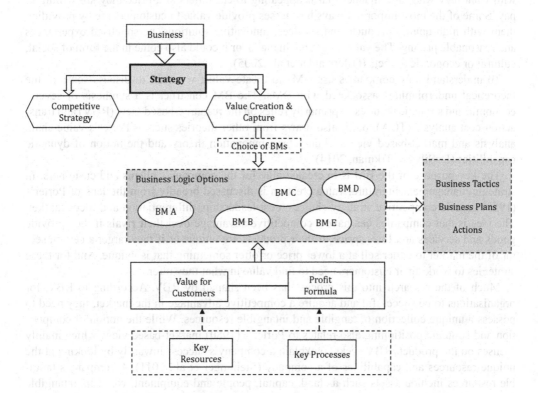

Figure 6.1 The role of a business model.

In past decades, companies have heavily focused on their 'company strategy' to survive and thrive in competitive markets. However, according to Casadesus and Ricart (2011), companies are now beginning to give more weight to BMs in the pursuit of company success. While strategy is essential for the company's success, particularly for creating a competitive advantage over the long term, BMs serve as practical instruments that enable the company to achieve its strategic goals, particularly by taking advantage of emerging market possibilities and counteracting the impact of disruptive technologies (Zhao, Chang et al., 2017). BMs become particularly relevant when companies are in the process of starting up, entering a new market or need to pivot their operations in response to changing market conditions.

BMs have recently received a good deal of interest in the construction industry, mainly as a result of the disruptions and changes the industry is presently undergoing. As Zott et al. (2011) highlighted, the BM has gone from being a supplementary part of strategy research to becoming the main emphasis of business research. For companies, BMs are easier to adjust or entirely replace compared to strategies because BMs can be tailored more easily to suit not only a specific industry, geographic location and customer segment but also to introduce new products, technologies or processes. A viable BM in these circumstances facilitates identifying a customer base and understanding what they value and then utilising resources and developing means for creating that value with a profit formula that benefits the company.

6.3 Value Creation via Business Models – Theoretical Foundations

The underlying premise of any company is that it has a value proposition, providing consumers with what they want in a manner that is appealing to customers at a price they are willing to pay. Some of the most important ways businesses provide value to customers are by providing them with high-quality products and services, innovative solutions, personalised experiences and reasonable pricing. The value might be financial, or it could also come in the form of social, cultural or economic aspects (Osterwalder et al., 2005).

To understand how companies use BMs to create value, we need to discuss some of the theoretical underpinnings associated with BMs. The BM construct is a synthesis of several economic and strategic theories. It primarily relies on the resource-based view (RBV) and transaction cost analysis (TCA), but it also draws from other theories such as Porter's value chain analysis and market-based view, Schumpeter's innovation theory and the notion of dynamic capabilities (DaSilva & Trkman, 2014).

The key purpose of the BM is to create value for the company's clients and customers. In strategic management literature, value creation is discussed broadly from the lens of Porter's (1980) generic competitive strategy choices: cost leadership, differentiation and focus market. The idea is that companies can have a competitive advantage over their rivals if they provide goods and services at a lower price, if their products are unique or if they target a certain sector of the market to either sell at a lower price or offer something that is unique. And for these strategies to work, their customers need to find value in what they offer.

Much of the research into this process has been shaped by RBV. According to RBV, for organisations to be successful and acquire a competitive advantage in the market, they need to possess a unique collection of tangible and intangible resources. While the notion of competition and strategic positioning also relates to Porter's (1980) market-based view, which mainly focuses on the product, RBV seeks to explain a company's success inwardly by looking at the unique resources and capabilities of a company (Steininger et al., 2011). A company's tangible resources include assets such as land, capital, people and equipment, while its intangible resources include assets such as reputation, relationships, knowledge and intellectual property.

Having a unique set of resources and skills and utilising them effectively also allows companies to develop BMs that deliver value to their clients. Hedman and Kalling (2003) used the example of IKEA's BM to illustrate how their resources, such as design capabilities, supplier relations and networks, as well as commitment and leadership, all lead to value creation for their customers.

However, while many scholars consider RBV as the primary theoretical foundation for determining how a BM creates value, others have suggested that RBV alone cannot adequately describe the complexity of BMs and, thus, a combination of theoretical applications is necessary (DaSilva & Trkman, 2014). For example, Morris et al. (2005) highlighted that while value is created from unique combinations of resources, it is TCA that helps identify transaction efficiency as a source of value. The core premise of TCA is that economic transactions are not costless and that these costs can significantly impact the efficiency and outcomes of transactions. So, BMs need to be devised so that these costs are minimised. According to Amit and Zott (2001), practitioners often use the theoretical foundations of both RBV and TCA when developing a BM.

A company's attempts to capture value or, in other words, create value for itself are generally based on reduced production costs, higher revenues and increased profits (Mangematin et al., 2017). Costs are reduced where inputs for production are sourced more cheaply or when better processes are utilised to make sourcing more efficient, revenues are increased by selling the products in larger volumes and/or at higher prices and profits are improved when the company can transform its inputs into more desirable outputs more efficiently (Liu et al., 2018). Porter's value chain concept is generally applied here to determine the internal processes where the value can be captured. Supply chain and distribution channel management are integral parts of this process for many businesses. Value capture is made possible by reducing transaction costs, which is related to the company's capacity to mechanise and standardise processes across the activity systems (Karam, 2022).

For example, when a company is considering whether to participate in market transactions or to internalise the production process, value chain analysis and TCA provide a useful framework for understanding how and where value can be captured. Similarly, standardising processes and procedures may reduce the need for negotiation and monitoring or help to develop standard contracts for suppliers or customers that reduce the need for negotiation. DaSilva and Trkman (2014) further extended RBV to incorporate dynamic capabilities by placing the focus on the company's ability to adapt and renew its resource base over time. Dynamic capabilities hold that businesses should be proactive and responsive to act upon emerging opportunities and risks, which requires a persistent awareness of the external environment and the development and use of current resources and skills.

6.4 Design and Application of Business Model

Even though the term *business model* is common and familiar in business and academia, its exact definition remains vague. There is a lack of consensus on what exactly constitutes a BM (Morris et al., 2005; Osterwalder et al., 2005). According to Pekuri et al. (2013), this lack of consensus is attributed to the BM concept being drawn from a wide array of academic and practical domains. However, in more recent literature, there seems to be some consensus on the definition and components of BM. Several distinct components have been mapped out to illustrate the multifaceted character of a BM. For example, studies by Amit and Zott (2001), Magretta (2002), Morris et al. (2005), Osterwalder et al. (2005), Teece (2010) and others have provided component breakdowns of BMs (see Table 6.1).

Table 6.1 Business model components defined in various literature

Source	Focus area	Broader BM components
Alt and Zimmermann (2001)	Electronics market	Mission; structure; processes; revenues; legal issues; technology
Amit and Zott (2001)	e-Business	Transaction content; transaction structure; transaction governance
Magretta (2002)	Retail	Customer; customer value proposition; value delivery method; economic logic for value delivery; appropriate cost
Osterwalder et al. (2005)	Information systems	Value proposition; core competency; target customer; distribution channel; relationship configuration; partner network; cost structure; revenue model
Morris et al. (2005)	Entrepreneurship, airlines	Offerings; market factors; internal capability; competitive strategy; economic factors; growth/exit factors
Teece (2010)	General business strategy	Market segment; value proposition; value capture mechanism; isolating mechanism
Mason and Spring (2011)	Recorded music market	Technology; market offering; network architecture
Dmitriev et al. (2014)	Technology innovation in various industries	Value proposition; market segmentation; revenue model; organisational block (partners network, facilities and resources, complementary assets); cost and profit estimation
Wirtz (2020)	General business management	Strategy model; resources model; network model; customer model; market offer model; revenue model; value creation model; procurement model; finance model

Zott et al. (2011) identified some common themes in the BM literature, specifically the value creation and value capture components. Moreover, even with the variations concerning some specific BM aspects, the BM construct can generally be summarised as including the following key elements: value proposition, target market, revenue model, partner network, internal infrastructure and processes (Dmitriev et al., 2014). Among others, Morris et al. (2005) offered a comprehensive explanation of the six components that make up a BM. In the following, the description of the components are explained to illustrate how they are employed in the BM design.

1. The first component focuses on how the company can create value. It involves analysing the company's offerings, which include the nature of offerings, customisation, product line, customer access, manufacturing and delivery approaches and distribution methods.
2. The second component focuses on the question *For whom is the value being created?* It considers market factors, particularly target customers. It includes the organisation, geographic scope, customer position, market nature and relationship approach.
3. The third component focuses on the question *What are the company's internal sources of advantage?* It explores the internal capabilities, including the strengths in production, operations, sales, marketing, information management, technological innovation, financial transactions, supply chain oversight and resource networking.
4. The fourth component focuses on how the company can position itself in the market. It involves establishing a strong market position, providing high-quality products or services, leading innovation, offering low-cost products and building personalised relationships.
5. The fifth component focuses on how the company can make money. It involves analysing economic factors to determine pricing, revenue sources, operating leverage, sales volume and profit margins.
6. The sixth component focuses on analysing the company's growth and exit strategies. It includes exploring different models, such as the subsistence model, income model, growth model and speculative model.

BMs are categorised mainly into two types: efficiency-centred business models and novelty-centred business models. A company's strategy may be driven by either of these models based on the industry, market trends and specific organisational goals. An efficiency-centred BM focuses on optimising operations, processes and resources to achieve maximum productivity, cost-effectiveness and streamlined performance. This model is particularly relevant, where competition is based on price and operational performance. At the core of efficiency-centred BMs lies the objective of reducing transaction costs and attaining better efficiency costs (Chatterjee, 2013).

On the other hand, novelty-centred BMs involve embracing novel undertakings and finding fresh approaches to managing these undertakings. Thus, a novelty-centred BM emphasises innovation, differentiation and unique value propositions as the main drivers of success (Amit & Zott, 2001). This model is particularly suitable in industries where customer preferences change rapidly and the competitive landscape requires constant adaptation. Some companies may find ways to integrate elements from both models to create a hybrid system that balances efficiency and innovation. Table 6.2 lists the key characteristics of the two types of BMs.

Both forms of BM have their own advantages. Efficiency-centred BMs can be more effective in delivering on the established value proposition by making transactions more efficient. Amit and Zott (2001) proposed that transaction efficiency is the primary driver of value. With greater transaction efficiency, the cost of each transaction would decrease, leading to an increase in each transaction's profit. It fits with the complex and multi-project environment of construction companies, which use a unified routine to deal with complex transactions (Zott & Amit, 2007).

The advantage of novelty-centred BMs is that they can lead to differentiation and innovative propositions through which value is created. They provide a competitive advantage for the company. By focusing on novel approaches, companies can position themselves in their industry or market segment with first-mover advantage that enables them to capture a larger market share and build brand loyalty before competitors catch up.

In reality, many businesses fail. Several factors can lead to BM failures, but perhaps one of the most common reasons is the lack of alignment with market needs and customer preferences, thus

Table 6.2 Key characteristics of efficiency-centred and novelty-centred business models

Efficiency-centred business models	Novelty-centred business models
• Cost Optimisation: reducing costs without compromising on quality by improving processes, minimising waste and working with cost-efficient suppliers.	• Innovation and Creativity: developing innovative goods, services or solutions.
• Lean Operations: utilising lean concepts to remove unnecessary processes, minimise cycle times and improve overall operational efficiency.	• Differentiation: creating a strong brand identity and value proposition that sets it apart in the market.
• Scalability: increasing production, distribution and services without incurring large cost increases.	• Agility: responding to changing market trends and customer demands is essential to remain competitive.
	• Risks: embracing a higher level of risk to explore opportunities in uncharted potential markets.
• Consistency: offering good quality and consistent products and services and catering to the needs of specific customer segments.	• Customer Experience: prioritising customer engagement and satisfaction through innovative experiences and solutions that meet rising consumer expectations.

providing poor value propositions. It is also related to the BM components not being carefully planned and not aligned with the company's core strategies (Casadesus & Ricart, 2011). Moreover, for the BM to work well, the various BM components need to complement one another, and there needs to be feedback loops that inform and improve the various BM components. A well-designed BM, according to Shafer et al. (2005), should aid in articulating and analysing cause-and-effect linkages and the coherence of strategic decisions. Furthermore, it is crucial to acknowledge and embrace the possibility that the BM will become outdated and irrelevant in the future. Thus, the BM must be monitored, updated and improved upon until it becomes obsolete, discarded and replaced by a new and more up-to-date BM (Wirtz, 2020; see Figure 6.2).

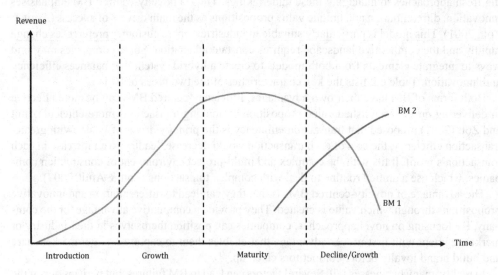

Figure 6.2 Business model stages.

Source: Adapted from Wirtz (2020)

6.5 Business Models in Construction Companies

BM knowledge and practice are evolving rapidly in the fields of strategy, technological innovation and environmental sustainability (Massa et al., 2017). While the construction sector has lagged behind other industries, the interest in BMs has increased recently. Comparatively little research has been undertaken to understand BMs for construction organisations. To some extent, this is because the construction industry has placed a low priority on innovation in past decades (Pan & Goodier, 2012). However, those who have examined BMs in the context of the industry have strongly highlighted the significant impact BMs have on the performance of construction enterprises (Abeynayake et al., 2022).

According to Pekuri et al. (2013), the utilisation of the BM concept in the construction industry differs from its application in other sectors. This discrepancy arises from the distinct characteristics of the construction industry. In the construction industry, companies are inherently project centric, meaning their organizational culture, processes and overarching strategies are tailored to accommodate the unique demands of project management. Furthermore, the ever-evolving customer landscape in construction further complicates matters. Bygballe and Jahre (2009) highlighted that construction companies often serve diverse customer groups, each necessitating distinct operational approaches to cater to their specific needs. Moreover, even the company itself takes on different roles – sometimes being the supplier (subcontractor) and sometimes being a customer (head contractor). Consequently, it becomes imperative to customise BMs to cater to individual customer segments or types of activity.

However, Pekuri et al. (2013) argued that, presently, the construction industry tends to apply uniform BMs across various project types despite significant disparities in revenue models and principles of value creation. This underscores a pressing requirement to develop BMs that align with the unique characteristics of each business segment, enabling a deeper understanding of the intricacies involved in value creation. As Wikström et al. (2010) articulated, BMs can serve as a crucial bridge between strategy and operational execution for companies operating in such contexts.

6.5.1 Types of Business Models in Construction

BMs in the construction industry can take many forms, including new ways of financing, designing, building and operating construction projects. Traditionally, construction companies have relied more on efficiency-centred BMs. The construction industry inherently presents companies with different models to choose from. In essence, various forms of contracting, procurement and project delivery methods can be viewed as forms of BM that the company can utilise. For example, a contractor may choose to participate in a fixed-price contract, exploiting their risk-bearing capabilities to create value for the clients while making a profit for the company.

The *Calcutt Review* (Calcutt, 2007) was one of the first reports that comprehensively investigated BMs in the UK house building industry. The report identified four types of models that companies utilised. They included the 'current trader' model, the 'investor' model, the 'self-build model' and the 'registered social landlord' model. The current trader model, which involves land purchase, development and outright sale, was found to be the most prevalent, with no long-term stake kept by the house builder. In contrast, the investor model involves developers maintaining a long-term interest in a developed site and trading a portion of the initial development profit for future revenues and capital growth, resulting in potentially smaller but more secure returns.

The self-build approach refers to individuals constructing their own dwellings or overseeing the process with architects and builders, while the registered social landlord model focuses on integrating communities with similar social and market sale properties, emphasising quality, sustainability and innovation but often lacking the delivery expertise of big house builders owing to financing limitations. According to Pan and Goodier (2012), the current trader model described above is a common BM property developers adopt in many parts of the world. However, the house builder model in Australia differs from the common developer model. In the Australian model, land purchase is generally separate from the house building process, and most house builders focus primarily on constructing and selling houses.

Construction companies, particularly larger ones, may also develop their BMs around horizontal and vertical integration. *Horizontal integration* is where a company seeks to expand through acquisitions of other companies that are at the same level in the value chain for various reasons, including reducing competition, entering new markets or diversification. *Vertical integration* is where a company seeks to take control of multiple stages of the construction process, including material production, design and facilities management, mainly to increase efficiency and reduce costs.

In general, the construction industry has been very receptive to supply chain management, where companies have developed their BMs around recognising and managing the roles and relationships within the supply chain. Many companies, particularly larger ones, have invested in the management and control of the supply chain to create and capture value. As Bygballe and Jahre (2009) argued, successful BM development for construction companies necessitates considering both project management and supply chain management perspectives, emphasizing the importance of intra- and inter-organizational interactions.

In a recent chapter by Murray and Kulakov (2022), they discussed a BM utilised by larger construction companies in the United Kingdom, where, companies increase their turnover by taking on a high volume of contracts, which they then predominantly subcontract to other companies. Rather than growing by increasing inputs or focusing on efficiency like other models, this model achieves growth by increasing the purchase of intermediate output from other companies in the supply chain. Murray and Kulakov (2022) also discussed the BMs of specialist contractors that involve working on numerous projects simultaneously, reducing the dependency on each project.

Recent literature has also drawn from theories and experience from other fields to study and recommend BMs specifically suited to the changing environment of the construction sector. There has been a new focus on studying BMs related to technology and the environment. Table 6.3 highlights two examples of the BMs identified within some specific contexts in recent years.

Table 6.3 Examples of BMs identified in recent construction literature

Source	Study area	Broader BM components	Specific BM components	Identified BMs
Brege et al. (2014)	Industrialised buildings in Sweden	• Offering • Market position • Operational platform	Prefabrication modes Building process positioning End-user segments System augmentation Complementary resource base	Systems supplier: turnkey Systems supplier: free factory In-house developer and system supplier Frame system supplier Frame system supplier: free factory Component system supplier: tech support Component system supplier

(*Continued*)

Table 6.3 (Continued)

Source	Study area	Broader BM components	Specific BM components	Identified BMs
Liu et al. (2018)	Chinese real estate companies	• Strategy • Value offer • Value creation • Value capture	Clients Products offered Market locale Financial structure Value chain embeddedness Core competency Revenue source	Asset management model Government servicing model High turnover model Cost efficiency model Commercial property model

Specifically, looking at the study by Liu et al. (2018), their detailed analysis of seven specific BM components for 117 real estate companies in China identified five distinct clusters of BMs. The first cluster of companies employ the asset management model, where the emphasis is on managing real estate assets effectively, often involving close cooperation with financial institutions for investment. The second group of companies utilise the government servicing model, primarily engaging in projects and services catering to government needs, often including infrastructure projects and public service facilities. The third cluster of companies employs a high turnover model that is characterised by a focus on rapid turnover of properties, aiming for quick sales and high volumes of transactions. The fourth cluster uses the cost efficiency model, which emphasises minimising costs and maximising efficiency in operations, often through systematic cost management and lean operations strategies. Finally, the fifth cluster utilises the commercial property model, which focuses on developing and managing commercial properties such as office spaces, retail outlets and other commercial buildings.

However, it may need to be noted here that the properties of the Chinese real estate market and companies are different from those of other parts of the world, particularly concerning the involvement of state-owned enterprises as well as real estate companies' access to financing. So, the above-identified BMs may not be directly relevant in a global context; nonetheless, the study offers a good illustration of how different BM components can be utilised and to understand how organisations develop their BMs.

6.5.2 Challenges and Opportunities of Moving to a New Business Model

Traditional BMs will need to evolve with time. As Flanagan (2022) highlighted, the industry is now changing, with more focus on technology and the circular economy, and the old model with the build-and-throwaway system is being replaced by restoration and innovative BMs. The traditional models were heavily impacted during the recent COVID-19 pandemic. Disruptions in global supply chains, labour shortages and increased compliance requirements led to additional costs and delays. Particularly, the fixed price BM in the building industry encountered significant challenges because of economic volatility, which made it difficult to adhere to fixed pricing structures. For example, in Australia, the high rate of insolvencies of construction businesses post-COVID has been strongly tied to this issue. Thus, the pandemic highlighted the need for more flexible contract arrangements and risk-sharing mechanisms in construction to better manage unexpected crises.

There have been challenges for construction companies in transitioning to newer BMs. Beyond the traditional forms of economic models utilised in the construction industry, the sector has been notoriously slow to adopt change and innovation. Many companies and professionals have been accustomed to traditional practices and are hesitant to embrace new business models. According to Jones et al. (2019), one of the biggest issues for large and established construction companies is how to transition and adapt to BMs that can generate disruptive innovation because these companies have generally depended on their dominant logic for some time, and disruptive ideas and behaviours that do not conform to this logic may require significant adjustments. Construction companies are faced with significant risks associated with construction projects, and the companies are faced with low margins. The introduction of a new BM might further create uncertainties, and risk-averse companies may reject changes they see as posing a threat to business objectives.

But companies also realise that they need to keep up with the times to remain competitive. When building information modelling was first introduced to the industry in the early 2000s, there was a mixed reaction, with some companies seeing it as an opportunity and others resisting the idea and not seeing value in it. Companies such as Turner Construction and AECOM, which were the first to adopt building information modelling, were generally larger or more technologically inclined companies. Through early adoption, they positioned themselves as leaders in the field. Similar lessons can be learned from other disruptive technologies, such as prefabrication and modular construction. For example, Liu et al. (2017) discussed the Vanke Group in China investing considerable effort and money to develop modular prefabrication and construction technologies. Building long-term partnerships with manufacturers allowed them to successfully implement the new technology, resulting in increased commercial value. This led to them becoming leaders in delivering high-rise residential buildings in China.

6.6 Business Model Innovation

Businesses, no matter their size, need to explore forms of innovation at some point if they want to be successful. Organisations may have different approaches to innovation. These approaches can take the form of incremental innovation, breakthrough innovation or radical innovation. The most common forms of innovation are more likely to be incremental in business practice as they are relatively easy to execute and carry fewer risks. Incremental innovation may entail making adjustments to increase the value of goods, services or processes. Examples of incremental innovation may include the improvement of building materials and processes and the use of digital tools to enhance the performance of construction projects. According to Jones et al. (2019), when a company's BM changes in a way that builds on its current resources and infrastructure to better serve its current customers, that change is likely to be in line with the company's prevailing logic. In this situation, small changes are likely to be accepted and may even be encouraged.

Breakthrough innovations are those that lead to significant advancements in performance or capabilities, while radical innovations result in fundamental changes that can completely disrupt industries. Breakthrough innovation and radical innovation are generally both regarded as disruptive innovations. Examples of breakthrough innovations may include the use of prefabrication and modular construction, 3D printing and advanced robotics. There may be fewer instances of radical innovation in construction, which may include concepts that revolutionise the whole sector, such as 'smart cities'.

Business model innovation (BMI) is an essential medium for innovation. BMI is seen as a form of organisational innovation that is crucial to the success of a company. According to

Andreini et al. (2022), most organisations today feel that BMI will have a significant effect on their performance and efficiency. As a result, they are progressively adopting the notion of BMI to explain and analyse their strategic problems. So, BMI refers to the alteration of one or more of the BM's components. It is not limited to the innovation of products and processes but also involves reinventing the BM itself (Chesbrough, 2010; Roos, 2014). According to Zott and Amit (2007), because organizational innovation is difficult to imitate, it leads to lasting advantages for the company. BMI accelerates the transformation of innovations into commercial value and supports value creation and capture for companies (Chesbrough, 2010). Several major, well-established companies have successfully adopted BMI by capitalising on their already-established consumer bases and extensive resources. IBM conducted a worldwide poll in 2006 and found that financially successful companies were twice as focused on BMI compared to those that were unsuccessful (Giesen et al., 2007).

In the literature, BMI is often held to be inherently disruptive (Wirtz, 2020). Typically, incremental changes are considered part of a comprehensive transformation, and BMIs are generally seen as breakthrough innovations. Winterhalter et al. (2017) identify three main types of BMI processes larger companies utilise. They include (a) back-end BMI, which are pure business innovations where the logic of an existing business is altered without a new product or technology; (b) front-end BMI, which is for technology or product innovations that are in the final development stage with a clear target market; and (c) fuzzy front-end BMI, which is for radical early-stage technology developments with an undefined target market.

Regardless of the approach, the goal of BMI is to determine and transform generic value capture logic into company-specific activities. Andreini et al. (2022) explained that the BMI process includes four stages: (a) the cognition process, where the company senses problems and considers adaptations; (b) the strategising process, where the company defines and selects alternative BMI strategies; (c) the value creation process, where the company defines value priorities and selects ways to create value; and (d) knowledge-shaping process, where companies perform systematic experimentation and create tacit knowledge based on trial and error. The end result is generally a comprehensive activity system map that the company can utilise.

Research on the topic of innovation in the construction industry is not new. The industry is now under growing pressure to innovate to meet the expectations of customers and the demands of an increasingly competitive market (Barrett et al., 2008). BMI can play an important role as the construction industry is undergoing strategic transformation and industrial upgrading. Moreover, the construction sector is seeing shorter business model life cycles due to the increased frequency of disruptions and dislocations (Pan & Li, 2016). So, a construction company's ability to achieve its goals across a wide variety of projects may depend on its ability to innovate its BM, and this innovation may lead to the company's strategic advantage. There has been a growing body of research demonstrating how construction organisations benefit from an emphasis on innovation. Most studies have identified both internal and external drivers of innovation and the importance of feedback loops between the phases of the innovation process (Manley et al., 2009). Jones et al. (2019) emphasised the need for companies to focus on both 'exploration' and 'exploitation' to capitalise on emerging value possibilities while continuing to serve their current clientele. *Exploration* refers to seeking out and trying new options and possibilities, whereas *exploitation* refers to making the most of known, reliable options or resources.

6.6.1 Technological Innovation and BMI

In the past 2 decades, BMs have emerged as a central concept in technology and innovation management. Often associated with novelty-centred BMs, technological innovation is seen as a vital component in creating and capturing value. However, according to Teece (2010), business success is not guaranteed by technical innovation alone. The success requires a BM that specifies how the company will go to market and how it will capture value. In fact, the mediatory role of BMs in the relationship between innovation and construction company performance has been demonstrated by Li et al. (2023).

Technological changes within an organization are also regarded as one of the major drivers of BMI. Changes include the implementation of new technologies, the refinement of current technologies and the combining of different technologies. The term *technology-driven BMI* is used to describe the evolution of BMI as a direct consequence of technological advancements (Amit & Zott, 2001; Willemstein et al., 2007). Companies that adopt new technologies will alter different aspects of their current BM. For example, product innovation may lead to the emergence of new market needs, new production methods and new distribution channels, resulting in BMI. In addition, the use of emerging technologies may change the delivery of goods and services, the current cost–benefit structure and the traditional industrial supply chain. So, the original BM will also need to change in line with these changes. Moreover, these BMIs will require companies to develop new resources and competencies at the organisational level (Andreini et al., 2022). When BMI does not occur along with technological innovations, it could likely lead to business failure (Johnson et al., 2008).

BMI, or particularly the creation of new BMs, is closely tied to the exploration and exploitation of new technologies (Andreini et al., 2022). BMI is believed to be impacted not just by technological advancements within the industry but also by technological advancements in other industries. Thus, technology convergence is a topic that is gaining interest in construction practice and research. Technology convergence is a phenomenon in which new technologies are formed by combining the knowledge of two or more other fields (Kim et al., 2013). The convergence of various technologies is seen as a primary driver of technological innovation (Han & Sohn, 2016). This trend has recently been more visible in the construction industry, where several technologies have been steadily used to improve building components. Examples include solar technologies, photovoltaic windows and 3D printing technologies.

Companies with more capital and resources may afford investments in technological innovations. However, smaller businesses may not always have that capacity. In fact, Barrett et al. (2008) found that smaller construction companies prioritise survival and stability. When they innovate, they do so in reaction to external environments and focus on areas of client relationships, project management and, in some cases, technology. In essence, many smaller companies may not want to grow, and much of the BMI may occur for the purpose of holding on to their clients and customers and improving their efficiency and productivity.

6.6.2 Sustainability Innovation and BMI

Over the last century, a company's success has been measured in large part by its ability to generate profits for its stockholders. This overemphasis on economic success has led to a host of environmental issues, but a growing number of established businesses are now putting sustainability at the top of their corporate agenda. According to Nidumolu et al. (2009), only businesses that prioritise sustainability will thrive in the years to come. So, this requires many construction companies to re-evaluate their BMs to incorporate sustainable products, technologies and practices.

The construction industry is a major source of environmental degradation. In recent years, there has been a focus on how the industry can transform to become more sustainable. A large number of research and policy papers have been published that highlight the need for sustainable products, processes, policies and systems. There is now a consensus that sustainability presents many opportunities for construction companies. However, the transition has been rather slow. This is mostly owing to greater upfront costs and longer payback periods of sustainable features, as well as a lack of client demand resulting from a lack of understanding regarding sustainable buildings (Zhao & Pan, 2022).

Businesses are aware of these challenges as well as the opportunities that the sustainability focus presents. The main difficulty that businesses confront is how to create a BM that uses sustainability features to generate revenue. Recent research has shown that sustainable BMIs do, in fact, lead to various advantages for the company. For example, Treptow et al. (2022) identified the values that are created and captured through sustainable BMI. They highlighted various elements necessary for BMs to have a genuine commitment to sustainability. These include not only product and project energy efficiency but also the 'stewardship' role of businesses and their community involvement. They found that integrating these elements could lead construction companies to offer customers value associated with long-term cost savings while capturing value through improved company reputation and market positioning.

Construction companies in many countries have begun the transformation to become more sustainable in response to government pressure and rising customer demand. Bocken et al. (2014) identified three main higher-order categorizations under which sustainable BMIs fall, namely, technological, social and organisational categories. Technological BMIs prioritise material and energy efficiency, waste reduction and renewable alternatives. Social BMIs emphasise functionality and stewardship. Organisational BMIs focus on environmental repurposing and scaling up.

Similarly, Zhao, Chen et al. (2017) identified four types of innovative BMs focusing on sustainable buildings. They include (a) extended operation and maintenance service models, which offer value by prioritizing long-term, relationship-based services to enhance the sustainable building experience; (b) customer-oriented service models, which span multiple stages of the building life cycle to maximise flexibility and value extraction; (c) collaborative design and construction models, which emphasise teamwork in decision-making to fostering cost control and effective stakeholder collaboration; and (d) energy performance contracting models, which involve partnership with energy service companies to outsource energy-related risks. Their analysis found that extended operation and maintenance service models and collaborative design and construction models are the most effective in value creation and capture.

Sustainability-focused BMs may also emerge in smaller companies. In the building industry, many larger companies focus more on efficiency and productivity. For smaller companies, competing on efficiency with bigger companies is not possible, so they may seek to rely on specialised products or focus markets. For example, in Australia, where volume home builders dominate the market for the construction of new residential houses, Warren-Myers and Heywood (2016) found that smaller companies focus on sustainable homes to stay in business and offer value to the customers. Since sustainability is not always a major priority for volume builders, smaller companies find opportunities to differentiate themselves by providing this unique value proposition.

6.7 Case Studies

6.7.1 Metricon Homes, Australia

Metricon is the largest and most well-known volume home builder in Australia. Volume home builders in Australia sell a large number of homes, and their BM is generally designed to optimise processes and economies of scale to deliver homes at competitive prices. While many volume builders, both large and small, went into insolvency as a direct result of the COVID-19 pandemic, Metricon has not only survived but is also performing well post-pandemic.

Metricon's BM is slightly different from that of its competitors, and that makes it one of the most successful companies in Australia. Like all volume builders, they typically offer a range of standardised home designs that cater to various market segments. These designs are pre-developed, tested and optimised for efficiency and cost-effectiveness. Metricon stands apart from the competition in part because it serves such a diverse clientele, from first-time homeowners to those looking to upscale or downscale and from luxury property owners to government agencies. They offer innovative designs with more options to choose from and more flexibility in terms of customisation, which attracts homebuyers who want to personalise their homes while benefitting from the efficiency of a volume builder. They have built a strong brand reputation for quality construction and customer service. All these present strong value propositions for customers. Moreover, they have recently also established ABC Homes, a sister company that provides cheaper options and whose main purpose is to compete with smaller home builders in the market.

Metricon has also built strong relationships with suppliers and trades over the years, many of whom have been working with Metricon for over 20 years and have found security knowing they will be treated fairly, paid on time and have a reliable stream of business going forward. This strong relationship was on display in 2022 when speculation circulated that Metricon was going bankrupt. The suppliers and trades at the time came together to support them and dispel these concerns.

The company captures value in a range of ways. Apart from the economies of scale, they have invested in technology to streamline various processes, from design to construction. For example, customers may modify a standard design simply by clicking their mouse and picking from a number of available options; the resulting layout and any associated costs are shown in real time. They have also invested in technology to streamline the construction process. The use of cutting-edge project management software, such as ClickHome, enables them to maintain operational efficiency. Moreover, Metricon has focused on continuous improvement, learning from past projects and adapting to changing market dynamics. They are currently invested in delivering environmentally sustainable homes with a 7-star green star rating and have also begun working for the government's low-income housing programme. Some of these endeavours may not be as lucrative right from the start, but they provide growth and entry into new areas.

6.7.2 *Orascom Construction, Egypt*

Orascom Construction PLC is a leading global engineering and construction company with activities across the infrastructure, industrial and commercial sectors in the Middle East, Africa, and the United States. Their core strategies include strengthening their market leadership through growth. Originally established in 1976 in Egypt as a construction company, it has now grown to be a major multinational company with more than 20 subsidiaries active in 30 countries. The company has a large presence in the Middle East and North Africa region and the United States. Orascom implemented a horizontal integration approach to expand. After the acquisition of a 45% stake in Contrack Construction, a U.S.-based company with a presence in the Middle East, the company's foothold in the U.S. market expanded further as it capitalised on opportunities arising from the Iraqi reconstruction effort. Additionally, Orascom entered the European construction market in 2003, participating in a 50% buyout of Besix, one of Belgium's largest construction groups, which had a strong presence in the Middle East.

Orascom also effectively implemented a vertical integration strategy to control its supply chain and gain a market advantage. It began with the creation of the National Steel Fabrication Company, producing steel components for construction, followed by the Egyptian Cement Company, becoming a significant player in the cement industry. Subsequently, the company's vertical integration extended to other sectors, including logistics, which provided materials handling and transport services as well as paint and chemicals. These initiatives provided Orascom with a distinct advantage by controlling critical aspects of its supply chain, from raw material production to logistics, ensuring a steady supply of materials and services essential for its construction projects. This comprehensive control over the supply chain has enabled OCI to reduce dependencies, manage costs more effectively and maintain a strong market position in the construction and infrastructure sectors in the Middle East and North Africa region.

6.7.3 Wanda Group, China

The Dalian Wanda Group was established in 1988 as a real estate company and operates in commercial property, luxury hotels, culture and tourism and department stores. In recent years, they have expanded to many countries around the world, including the United States, Australia and Europe, and have even acquired major cinema chains and movie studios.

The BM of Wanda Group is based on diversification across many sectors, synergy between its numerous companies and brand expansion into new international markets. The company leverages its diversified portfolio to tap into China's growing middle-class consumer market and to expand its global footprint. In doing so, the company has formed strategic partnerships and collaborations with other businesses, both domestically and internationally, to access new markets and resources. Their goal is to transition from a company largely focused on real estate to a worldwide entertainment and leisure business.

As reported by Liu et al. (2018), the company's profit formula in China mainly includes revenue generation from property sales and rentals. But their model also involves negotiating with local government to purchase land close to its Wanda Plazas at a low price, on which they can then construct residential and/or commercial properties. They are able to negotiate a low price on the basis that the long-term operation of their plazas provides several benefits. And, when these properties around the plazas appreciate in value (mostly as a result of their brand influence), they share the profits with the government and financiers. This model helps Wanda reap a high level of profits.

6.8 Conclusion

The future of construction will look quite different due to technological advances, industrialisation and globalisation. Construction companies need to prepare for these changes. To survive and prosper in the new economic climate, businesses need to adapt their business logic and prioritise business model innovations that serve both the company and its customers well. For this to occur, companies need to understand the core ideas behind BMs and how they function. Moving towards sustainable building practices and innovative technologies, construction companies must also adapt their strategies and models to be successful. Drawing from a wide range of literature and case studies, this chapter highlights the important role business models play for construction businesses and presents some important insights into business model innovation that can drive growth, efficiency and success for construction companies. As the industry continues to evolve, the adoption of these models, coupled with innovation and adaptability, will be pivotal in shaping the future of construction companies.

References

Abeynayake, D. N., Perera, B., & Hadiwattege, C. (2022). A roadmap for business model adaptation in the construction industry: A structured review of business model research. *Construction Innovation, 22*(4), 1122–1137.

Alt, R., & Zimmermann, H.-D. (2001). Preface: Introduction to special section – Business models. *Electronic Markets, 11*(1), 3–9.

Amit, R., & Zott, C. (2001). Value creation in e-business. *Strategic Management Journal, 22*(6–7), 493–520. https://doi.org/10.1002/smj.187

Andreini, D., Bettinelli, C., Foss, N. J., & Mismetti, M. (2022). Business model innovation: A review of the process-based literature. *Journal of Management and Governance, 26*(4), 1089–1121.

Barrett, P., Sexton, M., & Lee, A. (2008). *Innovation in small construction firms*. Routledge.

Bocken, N. M., Short, S. W., Rana, P., & Evans, S. (2014). A literature and practice review to develop sustainable business model archetypes. *Journal of Cleaner Production, 65*, 42–56.

Brege, S., Stehn, L., & Nord, T. (2014). Business models in industrialized building of multi-storey houses. *Construction Management and Economics, 32*(1–2), 208–226.

Bygballe, L. E., & Jahre, M. (2009). Balancing value creating logics in construction. *Construction Management and Economics, 27*(7), 695–704.

Calcutt, J. (2007). *The Calcutt review of housebuilding delivery*. Department for Communities and Local Government. https://housingforum.org.uk/wp-content/uploads/2020/05/callcuttreview_221107.pdf.

Casadesus, R., & Ricart, J. E. (2011). How to design a winning business model. *Harvard Business Review, 89*(1/2), 100–107.

Chatterjee, S. (2013). Simple rules for designing business models. *California Management Review, 55*(2), 97–124.

Chesbrough, H. (2010). Business model innovation: Opportunities and barriers. *Long Range Planning, 43*(2–3), 354–363.

DaSilva, C. M., & Trkman, P. (2014). Business model: What it is and what it is not. *Long Range Planning, 47*(6), 379–389.

Dmitriev, V., Simmons, G., Truong, Y., Palmer, M., & Schneckenberg, D. (2014). An exploration of business model development in the commercialization of technology innovations. *R&D Management, 44*(3), 306–321.

Ferrero, F., Perboli, G., Vesco, A., Musso, S., & Pacifici, A. (2015). *Car-sharing services – Part B business and service models*. Interuniversity Research Centre on Enterprise Networks, Logistics and Transportation. https://www.cirrelt.ca/documentstravail/cirrelt-2015-48.pdf

Flanagan, R. (2022). The race to the future for the construction sector. In R. Best & J. Meikle (Eds.), *Describing construction* (pp. 276–293). Routledge.

Giesen, E., Berman, S. J., Bell, R., & Blitz, A. (2007). Paths to success: Three ways to innovate your business model. *IBM Institute for Business Value, 23*(7), 436–438.

Han, E. J., & Sohn, S. Y. (2016). Technological convergence in standards for information and communication technologies. *Technological Forecasting and Social Change, 106*, 1–10.

Hedman, J., & Kalling, T. (2003). The business model concept: Theoretical underpinnings and empirical illustrations. *European Journal of Information Systems, 12*(1), 49–59.

Johnson, M. W., Christensen, C. M., & Kagermann, H. (2008). Reinventing your business model. *Harvard Business Review, 86*(12), 50–59.

Jones, K., Davies, A., Mosca, L., Whyte, J., & Glass, J. (2019). Changing business models: Implications for construction. *Transforming Construction Network Plus, Digest Series, No. 1.*

Karam, E. (2022). *General contractor business model for smart cities: Fundamentals and techniques.* John Wiley & Sons.

Kim, E., Cho, Y., & Kim, W. (2013). Dynamic patterns of technological convergence in printed electronics technologies: Patent citation network. *Scientometrics, 98*, 975–998.

Li, K., Yang, Q., Shrestha, A., & Wang, D. (2023). Impact of technology recombination on construction firm performance: Evidence from Chinese construction sector. *Journal of Management in Engineering, 39*(3), 04023014.

Liu, G., Li, K., Shrestha, A., Martek, I., & Zhou, Y. (2018). Strategic business model typologies evident in the Chinese real-estate industry. *International Journal of Strategic Property Management, 22*(6), 501–515.

Liu, G., Li, K., Zhao, D., & Mao, C. (2017). Business model innovation and its drivers in the Chinese construction industry during the shift to modular prefabrication. *Journal of Management in Engineering, 33*(3), 04016051.

Magretta, J. (2002). Why business models matter. *Harvard Business Review, 80*(5), 86–92, 133.

Mangematin, V., Ravarini, A. M., & Scott, P. S. (2017). Practitioner insights on business models and future directions. *Journal of Business Strategy, 38*(2), 3–5.

Manley, K., McFallan, S., & Kajewski, S. (2009). Relationship between construction firm strategies and innovation outcomes. *Journal of Construction Engineering and Management, 135*(8), 764–771.

Mason, K., & Spring, M. (2011). The sites and practices of business models. *Industrial Marketing Management, 40*(6), 1032–1041.

Massa, L., Tucci, C. L., & Afuah, A. (2017). A critical assessment of business model research. *Academy of Management Annals, 11*(1), 73–104.

Morris, M., Schindehutte, M., & Allen, J. (2005). The entrepreneur's business model: Toward a unified perspective. *Journal of Business Research, 58*(6), 726–735.

Murray, A., & Kulakov, A. (2022). Using financial concepts to understand failing construction contractors. In *Describing construction* (pp. 249–275). Routledge.

Nidumolu, R., Prahalad, C. K., & Rangaswami, M. R. (2009). Why sustainability is now the key driver of innovation. *Harvard Business Review, 87*(9), 56–64.

Osterwalder, A., & Pigneur, Y. (2010). *Business model generation: A handbook for visionaries, game changers, and challengers* (Vol. 1). John Wiley & Sons.

Osterwalder, A., Pigneur, Y., & Tucci, C. L. (2005). Clarifying business models: Origins, present, and future of the concept. *Communications of the Association for Information Systems, 16*(1), 1.

Pan, W., & Goodier, C. (2012). House-building business models and off-site construction take-up. *Journal of Architectural Engineering, 18*(2), 84–93.

Pan, W., & Li, K. (2016). Clusters and exemplars of buildings towards zero carbon. *Building and Environment, 104*, 92–101.

Pekuri, A., Pekuri, L., & Haapasalo, H. (2013, May 29–31). *Business models in construction companies – Construction managers' viewpoint* [Paper presentation]. International Conference on Technology Innovation and Industrial Management, Phuket, Thailand.

Porter, M. E. (1980). *Competitive strategy: Techniques for analyzing industries and competitors.* The Free Press.

Renz, A., Solas, M. Z., de Almeida, P. R., Bühler, M., Gerbert, P., Castagnino, S., & Rothballer, C. (2016). *Shaping the future of construction: A breakthrough in mindset and technology.* World Economic Forum. https://www3.weforum.org/docs/WEF_Shaping_the_Future_of_Construction_report_020516.pdf

Ribeirinho, M. J., Mischke, J., Strube, G., Sjödin, E., Blanco, J. L., Palter, R., Biörck, J., Rockhill, D., & Andersson, T. (2020). *The next normal in construction: How disruption is reshaping the world's largest ecosystem.* McKinsey & Company.

Roos, G. (2014). Business model innovation to create and capture resource value in future circular material chains. *Resources, 3*(1), 248–274.

Shafer, S. M., Smith, H. J., & Linder, J. C. (2005). The power of business models. *Business Horizons, 48*(3), 199–207.

Steininger, D. M., Huntgeburth, J. C., & Veit, D. J. (2011, August 4–7). *Conceptualizing business models for competitive advantage research by integrating the resource and market-based views* [Paper presentation]. Seventeenth Americas Conference on Information Systems, Detroit, MI, United States.

Teece, D. J. (2010). Business models, business strategy and innovation. *Long Range Planning, 43*(2–3), 172–194.

Treptow, I. C., Kneipp, J. M., Gomes, C. M., Kruglianskas, I., Favarin, R. R., & Fernandez-Jardón, C. M. (2022). Business model innovation for sustainable value creation in construction companies. *Sustainability, 14*(16), 10101.

Warren-Myers, G., & Heywood, C. (2016, May 30–June 3). *Identifying client roles in mainstreaming innovation in Australian residential construction* [Paper presentation]. CIB World Building Congress, Tampere, Finland.

Wikström, K., Artto, K., Kujala, J., & Söderlund, J. (2010). Business models in project business. *International Journal of Project Management, 28*(8), 832–841.

Willemstein, L., van der Valk, T., & Meeus, M. T. H. (2007). Dynamics in business models: An empirical analysis of medical biotechnology firm in the Netherlands. *Technovation, 27*(2), 221–232.

Winterhalter, S., Weiblen, T., Wecht, C. H., & Gassmann, O. (2017). Business model innovation processes in large corporations: Insights from BASF. *Journal of Business Strategy, 38*(2), 62–75.

Wirtz, B. W. (2020). *Business model management: Design – Process – Instruments*. Springer.

Zhao, X., Chang, T., Hwang, B.-G., & Deng, X. (2017). Critical factors influencing business model innovation for sustainable buildings. *Sustainability, 10*(1), 33.

Zhao, X., Chen, L., Pan, W., & Lu, Q. (2017). AHP-ANP–fuzzy integral integrated network for evaluating performance of innovative business models for sustainable building. *Journal of Construction Engineering and Management, 143*(8), 04017054.

Zhao, X., & Pan, W. (2022). The characteristics and evolution of business model for green buildings: A bibliometric approach. *Engineering, Construction and Architectural Management, 29*(10), 4241–4266.

Zott, C., & Amit, R. (2007). Business model design and the performance of entrepreneurial firms. *Organization Science, 18*(2), 181–199.

Zott, C., Amit, R., & Massa, L. (2011). The business model: Recent developments and future research. *Journal of Management, 37*(4), 1019–1042.

7 Resource Management

Kumar Neeraj Jha

7.1 Introduction

Barney (1991, p. 101) defined *resources* as 'all assets, capabilities, organisational processes, firm attributes, information, knowledge, etc., controlled by a firm that enables the firm to conceive of and implement strategies that improve its efficiency and effectiveness'. Resource management can be defined as the efficient and effective deployment of organisational resources to maximise their value and achieve the required objectives (Karaa & Nasr, 1986).

Resource management in the context of construction company management involves policies, processes and systems to enable planning, acquiring, allocating and optimising the use of resources to help the company achieve its business goals and objectives. Proper resource management ensures that resource requirements are satisfied and that the right resources are available on time to offer the maximum benefit to the company. Fundamental to effective resource management is a clear understanding of what resources are needed, what resources are available, when they are required, where they are located and the ability to acquire them to meet organisational needs.

The ability to successfully leverage various essential resources and assets has become increasingly important as construction companies compete and try to succeed in a fierce business environment characterised by high competition, low profit margins and resource shortages. The challenge for construction companies is to make necessary resources available in a timely and cost-effective manner in a resource-constrained environment. How efficiently a construction company can manage resources that are in short supply to address competing resource requirements across its projects determines the effectiveness of its resource management practices and, thus, its success. Construction companies that adopt effective resource management policies, systems and practices improve not only time and cost performance in their projects but also key business processes, leading to overall operational efficiency and enhanced productivity.

The chapter discusses various resources and the importance of resource management in construction companies. It highlights the key differences in resource management at the project and company levels. It discusses sustainable resource management in construction companies. The chapter then provides insights into the management of three critical resources, viz. materials, plant and equipment (P&M) and human resources (HR), in the context of construction companies. The financial resources are discussed as a separate chapter in the book. Finally, the chapter examines some resource management challenges construction companies face, along with potential solutions.

DOI: 10.1201/9781003223092-7

7.2 Examples of Key Resources

Construction is a resource-intensive industry. Construction projects consume significant amounts of material resources such as concrete, steel and timber and require many different types of P&M or equipment, such as excavators, tower cranes, mobile cranes, concrete trucks, forklifts, etc. Similarly, a large number of workers, tradespeople and professionals with different backgrounds and skills are needed to execute a typical medium- or large-size construction project.

Moreover, financial resources play a central role in ensuring the success of a construction company. Construction companies make significant capital investments in construction projects, necessitating sound financial resource management to stay profitable in the highly competitive construction industry. Similarly, maintaining a positive working capital is crucial for meeting the significant upfront investments in construction projects. A healthy financial buffer not only facilitates the financing of initial project stages but also sustains ongoing operations and helps companies meet their diverse financial obligations. Furthermore, investment capital, earmarked for equipment acquisition, technology integration and infrastructure development, requires strategic financial planning to align with the long-term goals of the company. Therefore, financial resources are critical for construction companies to achieve efficiency, competitiveness and sustained growth in the dynamic construction industry.

Physical spaces such as office and storage facilities, as well as appropriate infrastructure and utilities, provide a strong foundation for the seamless operation of a construction company. Medium or large construction companies often require separate physical spaces or head offices for administrative functions. Moreover, material and equipment storage facilities or warehouses facilitate the effective coordination of resources across multiple project sites. Similarly, the availability of transportation and vehicle resources could ensure timely and cost-effective mobilisation of resources, materials and P&M deliveries and streamline personnel movement on and across construction sites.

Construction is an information-intensive industry where knowledge is continuously produced in construction projects. Data, information and knowledge are important resources for construction companies that need to be carefully managed. The implementation of effective document and information management systems and procedures ensures accurate access to information and project documentation, such as drawings and specifications, and provides essential information for construction activities, promoting streamlined communication across project teams. Complementary to this, real-time site data offer invaluable insights into the ongoing project progress, resource utilisation and potential issues for their proactive management. Staying abreast of dynamic industry trends, regulations and best practices is also essential for construction companies to adapt to changes and continually enhance construction processes. Therefore, information resources are fundamental components of resource management, fostering informed decision-making and contributing to the overall success of construction companies (Ma et al., 2011). Knowledge management is discussed in a separate chapter in the book.

7.3 Importance of Resource Management

The success of a construction company depends on the successful completion of projects within the timelines and budgeted cost according to the defined scope and quality specifications. Proper resource management in construction companies ensures the availability of resources as and when needed and could help minimise under- and over-utilisation of resources. It also leads to operational efficiency, profitability and long-term sustainability. Moreover, it can help companies improve time and cost performances, improve profit margins, deliver better quality and improve risk management in construction projects. Companies can significantly reduce waste and redundancy by optimising resource usage and ensuring that money, materials, equipment and labour are utilised efficiently.

Proper resource allocation means construction companies can maximise the value derived from their investments by reducing idle time resulting from the unavailability of resources. Additionally, the data-driven insights gained from resource management could play a pivotal role in enabling companies to make informed decisions regarding project selection, bidding, cost estimations and operational planning. For instance, the ability to predict resource demand and identify and establish reliable supply sources could help construction companies make accurate cost estimates and minimise delay and cost overrun risks in projects.

The disruptions in the supply chain during the COVID-19 pandemic and their impact on construction projects and companies have further highlighted the importance of robust resource management practices. Proper resource management processes can aid in the proactive identification of potential risks associated with resources, such as resource availability constraints, skill gaps and potential supply chain disruptions. The early identification of problems will allow construction companies to develop comprehensive contingency plans and take steps to mitigate risks effectively, minimising potential disruptions and their impact. It will help them improve their resilience and ensure business continuity even in the face of unexpected challenges and events. Therefore, effective resource management could help construction companies anticipate, plan for and mitigate resource-related risks in construction projects and is fundamental to achieving efficiency, cost-effectiveness and overall success in the construction industry.

Finally, by minimising waste and optimising resource utilisation, construction companies can reduce their negative environmental impact and help achieve sustainability in the built environment. Sustainable resource management (SRM) practices could enable the procurement of sustainable materials, responsible resource use and resource optimisation and reuse of construction and demolition waste. Construction companies can enable a circular economy by reducing resource wastage and promoting the recycling and reuse of waste or used resources.

7.4 Difference between Company and Project-Level Resource Management

Company and project-level resource management are integral to effective resource management in construction companies. At the company level, organisational policies and strategic decisions shape the company's overall procurement strategies, leveraging economies of scale and fostering long-term partnerships with suppliers and vendors. On the other hand, project-level resource management is more hands-on, tailoring resources to meet specific project needs and adapting to immediate on-site requirements. It also ensures that resources are being used in a sustainable and optimal way on project sites. Some differences are outlined in Table 7.1.

At the company level, the procurement team engages in strategic decision-making and bulk procurement of major construction materials such as steel, cement and fuel. Additionally, the central resource management team oversees the initial site mobilisation, strategically selects subcontractors and manages international procurement through imports and exports. The strategic and organisational approach to procurement or centralised procurement capitalises on economies of scale and establishes rate contracts to maintain consistency across various projects.

On the other hand, the project-level resource management team operates with a more hands-on approach. They are responsible for tailoring procurement to meet the specific needs of individual projects. They handle project-specific procurement, ensure day-to-day inventory management, coordinate logistics for timely material and P&M delivery, adapt rate contracts to

Table 7.1 Example of differences in company and project-level resource management

S. No.	Aspect	Company-level resource management	Project-level resource management
1	Strategic oversight	Focuses on high-level, strategic decisions impacting the entire company	Focuses on the unique resource requirements and management aspects in individual projects
2	Economies of scale	Capitalises on bulk purchasing and centralised negotiations for cost efficiency due to larger quantities	Manages inventory and logistics specific to the project site, addressing immediate needs
3	Procurement reach	Manages international procurement, imports and exports	Adapts to regional factors and adjusts procurement strategies accordingly
4	Strategic partnerships	Establishes long-term partnerships, rate contracts and supply chain agreements	Adopts more hands-on, day-to-day management of project-specific activities and short-term partnerships

regional variations, find local human resources such as labour and optimise procurement strategies based on project characteristics (Ma et al., 2011).

Resource management at company and project levels must be done tactfully so that projects can be completed successfully within the constraints of cost, quality, time and scope. A coordinated approach requires various strategies to be adopted at the corporate and project site levels in construction companies. A clear division of roles and responsibilities ensures that resource management practices are strategically aligned with overarching organisational goals while remaining responsive to the specific and dynamic resource requirements of individual construction projects.

7.5 Sustainable Resource Management

The growing emphasis on sustainable practices has necessitated that construction companies proactively manage resources with environmental impact in mind. A strategic solution involves a holistic approach to sustainability, incorporating practices that extend beyond immediate project needs. This includes the implementation of sustainable construction practices, comprehensive life cycle assessments and seamlessly integrating environmental considerations into resource planning and use processes to enable the circular economy principles in the built environment. Business practices such as eco-efficiency, eco-design, green purchasing and reverse logistics could decrease the harmful environmental impact of construction companies and improve their environmental performance (Payán-Sánchez et al., 2021).

Resource management decisions and activities in construction companies should focus on sustainable resource management. Sustainability considerations should be integrated into managing the supply chain and also at the operational level into the different departments or functions of the firm (e.g., marketing, accountancy, operations, human resource management; Payán-Sánchez et al., 2021). The integration of environmental thinking into supply chain management, including product design, material sourcing and selection, manufacturing processes and delivery of the final product, as well as end-of-life management of the product, is important for achieving SRM (Srivastava, 2007).

SRM can be achieved by reducing the need for virgin materials, energy and consumables; substituting scarce natural materials with recycled materials; and reusing and recycling construction and demolition waste. The integration of renewable energy sources, such as solar panels, could reduce operational costs and the negative environmental impact of energy-intensive construction activities. Efficient waste reduction and recycling practices play a critical role in reducing waste, minimising the environmental footprint and promoting resource conservation. Adopting circular economy principles and systematic waste reduction strategies ensures that materials are repurposed and reused whenever possible. Moreover, embracing sustainable materials and construction technologies, such as prefabrication and modularisation, contributes to environmental sustainability by reducing resource requirements and wastage and aligning construction practices with broader eco-friendly goals.

It is important to manage resources effectively by allocating and efficiently utilising them. Resource optimisation focuses on all aspects of resource management, starting from planning and allocation of resources to their usage, monitoring and control. The first step is to develop a detailed plan of resource requirements, considering all possible risks and uncertainties that may be encountered. The quantity and type of all resources and their variable requirement must be identified. During allocation, resources are optimised based on the availability of resources and project requirements at multiple locations. Next, resource utilisation must be monitored so that any deviations from the plan can be identified and corrected.

Together, these sustainable actions can form a holistic approach, fostering responsible and environmentally conscious resource management in the construction industry.

7.6 Material Resource Management

Construction material costs comprise approximately two-thirds of the overall construction project costs (Omotayo et al., 2023). Even though materials account for a large portion of construction expenditures, a rigorous approach to the management of materials is lacking in many construction companies.

Materials resource management is a comprehensive process encompassing a wide range of activities, including identifying materials requirements, acquisition strategies, procurement and subcontracting, supplier quality management, transportation and site material management. It involves planning, organising, directing and controlling the flow of materials within a construction company across its multiple project sites. Efficient material procurement and inventory management processes are paramount in ensuring the timely availability of various construction materials, reducing the project cost and minimising waste.

The organisation of the material procurement team in construction companies can be done in many ways depending on the company size and type. Large companies often have separate teams for managing material procurement, starting from the vendor selection process and placing orders for various materials to supply at various site locations. The site management takes responsibility for raising the requirements of different materials with the head office and managing them once they have been delivered. Generally, the project planning team prepares material procurement plans, which are sent to the central procurement team. A monthly procurement schedule of the materials, as shown in Table 7.2, could facilitate the timely procurement of the right materials in the right quantities.

Construction companies generally adopt centralised procurement methods for materials like steel, cement, brick, fine and coarse aggregates and timber to benefit from bulk orders due to their huge quantity requirements in construction projects. They can receive competitive rates

Table 7.2 A typical schedule of material procurement

Description	Unit	Unit cost	Vendor details	Quantity			Remarks
				First month	Second month	nth month	

from major suppliers. The negotiating power allows them to save money on bulk orders. Similarly, for consumables such as fuel and safety gear required at construction sites, rate contracts can be entered at the company level for the duration the item is needed at project sites. The pre-agreed rate protects the company from market price fluctuations.

Even if bulk orders are placed at the company level, small purchase orders at the site level are sometimes needed to meet the gaps in quantity supplied and required on project sites resulting from estimation errors, material wastage and rework. Furthermore, construction materials are more varied and less standardised. There can be several types and specifications for the same material. Moreover, the nature, type and quantity of a particular material required in projects of the same type and size could be very different depending on the exact material specifications per the client's requirements. Since every project is unique, some different materials may need to be procured. For instance, in two building projects, the type of cladding and fit-out can be different based on the aesthetics and design requirements of the client. Therefore, project-specific items are sometimes procured locally, especially when in small quantities. Construction companies can also avoid the procurement and management of certain materials through subcontracting. They can make subcontractors responsible for material management by including it in the scope of their work.

It is important to consider the equipment and additional workers that may be required to handle, transport and mix or assemble materials on construction sites. For instance, tower cranes are often required on large construction projects to lift the materials from one point (e.g., point of storage) and deliver them to another (e.g., point of use), whereas scissor lifts and other equipment are used to reduce manual handling by transporting materials on building floors. Similarly, unskilled workers are frequently hired by construction companies to manually lift and carry materials on construction sites, especially in developing countries.

7.6.1 Material Purchase Process Management

The request for quotation mainly consists of specifications, scope matrix, drawings, bill of quantities, technical and commercial terms, supply schedule and additional information, if any. The specifications are as per the contract document or design requirement. Next, various suppliers or vendors are consulted and the request for quotation is shared with them. The clarifications on the work are made to the vendors, and their pricing for the work is collected. The supplier clearly describes the work or product, their specification, price, supply schedule, place of delivery, freight charges, tax and additional terms and conditions in their quotation.

The procurement team may decide to negotiate the price to bring down the total cost. It is important to ensure that there is no hidden price and that the technical and commercial terms

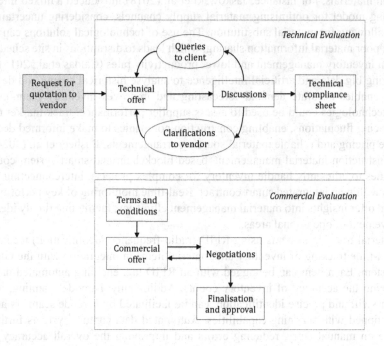

Figure 7.1 Technical and commercial evaluations in the purchase process.

provided by the supplier meet the requirements. Usually, a number of quotations (preferably three or more) are collected, and a comparative analysis is done based on selected parameters such as cost, supply terms and conditions, reputation, etc., before finalising the preferred supplier. Therefore, the purchase order process can be divided into two stages: technical evaluation and commercial evaluation, as shown in Figure 7.1.

7.6.2 Supplier Management

Suppliers in the construction industry face many issues due to various reasons such as scale of business, supply on credit, dependence of cash flow on timely payments, relation with contractors and so on. Therefore, it is important that supplier relation management is adequately managed by the company for their continued services to the company. The construction company can engage in the development of the supplier as there is mutual benefit resulting from sharing technology, information and planning. The company should create and maintain a vendor database that consists of the contact details, address, previous orders, specialisation, performance ratings and other details. Maintaining a healthy supplier database is an important step in material management. Moreover, supplier mapping with respect to ongoing projects helps in locating nearby sources of supply and may save time and cost.

7.6.3 Material Management Technology

Material management modelling and technology can play an important role in material management and help construction companies reduce costs and improve efficiency in the supply of construction materials. For instance, Jaśkowski et al. (2018) introduced a mixed integer linear programming model for optimising material supply channels, considering uncertain material prices and allowing for material substitution. The use of technological solutions can overcome the issue of poor material information sharing, which leads to disruptions in site schedules, inefficiencies in inventory management and lower productivity rates (Caldas et al., 2015).

Leveraging big data and artificial intelligence to analyse historical supply and demand data and prices enables accurate demand forecasting and improves inventory planning. Moreover, these technologies could be used to assess supplier performance, track market trends and evaluate pricing fluctuations, enabling construction companies to make informed decisions for competitive pricing and reliable material sourcing arrangements. Basheer et al. (2024) showed that the construction material management–based blockchain and smart system could enable project parties to efficiently handle inventory for complex projects, interconnecting all supply chain tasks within an integrated smart contract. Real-time monitoring of key performance indicators could offer insights into material management efficiency, facilitating timely identification and improvement of operational areas.

The material tracking systems using RFID (radio-frequency identification) technology can provide real-time tracking of inventory and material items in integration with the Global Positioning System. Each item can be tagged with an RFID tag, enabling automated data capture and enhancing the accuracy of inventory counts. Additionally, barcode scanning, a common practice for swift and precise identification, can be facilitated by barcode scanners and mobile devices equipped with scanning capabilities. Automated data capture systems further reduce dependence on manual entry, reducing errors and improving the overall accuracy of inventory records. Integration with blockchain technology can create transparent and traceable supply chains, reducing the risk of counterfeit goods and ensuring the authenticity of materials (Elghaish et al., 2023).

Sharing specific project requirements, updated delivery schedules and other relevant data with the help of technology can help construction companies and supply chain partners collaborate and make informed decisions (Dallasega et al., 2018). Material passports are increasingly being used in this context in integration with building information modelling. Material passports are comprehensive digital records of materials, which define characteristics of materials on the composition, origin and life cycle of materials, facilitating traceability, recovery and reuse (Honic et al., 2019).

Digital material procurement systems can facilitate the online ordering and management of material supplies, reducing paperwork and enhancing the speed of procurement processes. Online portals such as Enterprise Information Portal could manage and store material-related data and documents electronically, integrating data from diverse sources and providing a holistic view of the procurement landscape. Additionally, e-approval systems optimise and streamline the approvals workflow, involving the review of purchase requests, purchase orders and other documents with the help of digital signatures and in-built authorisation controls based on predefined rules and hierarchies (He et al., 2018).

Document management systems facilitate the organised handling of various material management documents. They involve maintaining a centralised repository for material specifications, standards, supplier information, purchase orders, quality control documentation, inventory records, change orders, safety data sheets and communication logs. Collaborative tools like common data environments and electronic document management systems enhance transparency and effectiveness, improving communication and coordination among material management team members. Such systems and processes are crucial for ensuring compliance, minimising errors and providing accurate and up-to-date information to all construction project stakeholders (Guo et al., 2021).

7.7 Plant and Machinery Resource Management

A wide variety of P&M or equipment, ranging from cranes and excavators to bulldozers and forklifts, is used on construction sites to execute tasks like excavation, material lifting, site preparation and other construction activities. Effective management of P&M resources is paramount for project success. Medium and large construction companies often have a separate P&M department or team of staff to perform functions such as P&M procurement, purchase of spares, inventory management, predictive and preventive maintenance, cost management, refurbishment and asset disposal.

Effective P&M management requires categorising them to streamline operations, plan maintenance and optimise resources. The categorisation can be done based on the aspects such as functionality, size and capacity and criticality. Functionality-wise categorisation of construction P&M is shown in Table 7.3. Based on size and capacity, the equipment can be categorised as small, medium and large equipment. It is also helpful to classify them as critical and non-critical equipment, depending on the cost, productivity and schedule, for improved planning and maintenance activities.

7.7.1 Purchase versus Hire Decision

The equipment purchase versus hire or lease decision demands careful consideration. In some cases, purchasing a piece of equipment could be advantageous if the company is going to use it for the long term or across several projects, while in others, hiring equipment could offer flexibility and cost savings, and operational and maintenance risks can be offloaded to the supplier. Some of the factors influencing the purchase versus hire decisions are project duration, equipment utilisation rate, cost, availability, tax benefits, future needs and maintenance support.

If the project requires scaling up of equipment, hiring is a better option, offering the flexibility to expand or reduce the fleet as per the varying project needs. Also, rental companies usually offer upgraded equipment and the latest models, which lessens the need for replacement to access the new technology. In contrast, owned equipment becomes outdated and is difficult to replace. Sometimes, even if the company owns a piece of equipment, it may decide to hire it locally in a project located far away to save the transportation cost of equipment. The availability of skilled operators and maintenance crews could also be a major deciding factor.

Table 7.3 Construction P&M classification based on functionality

Type	Examples
Earth-moving	Excavators, scrapers, graders, bulldozers, road rollers, compactors, trenchers, dredgers, backhoes, rippers
Lifting machinery and equipment	Tower cranes, mobile cranes, forklifts, hoists, conveyors
Concreting equipment	Batching plant, concrete mixers, concrete pumps
Foundation and piling equipment	Piling rigs, pile drivers, reverse circulation drills
Transportation vehicles	Dumpers, trailers, trucks, tankers, tippers
General equipment	Generators, compressors, pumps
Specialised equipment	Tunnel boring machines

Capital expenditure or CAPEX budget in companies allows for expenditures related to the purchase, upgrade and maintenance of physical assets such as equipment, buildings, etc. Some of the costs associated with equipment are as follows.

1. Capital investment: The cost of purchasing the equipment, including purchase price, financing, insurance and tax.
2. Depreciation: Purchased equipment depreciates over time, lowering resale value. The final resale value shall be considered in the cost of ownership.
3. Maintenance and repairs: In the case of purchased equipment, the cost of maintenance and repair is to be borne by the company.
4. Storage and transportation: There are costs for transportation of equipment to the project location and storing them on-site or in a storage yard or warehouse.
5. Other costs: Costs such as road tax, operation costs (for operators) and overhead costs also need to be considered.

7.7.2 Equipment Maintenance

It is also important to predict equipment failure and monitor the potential causes of breakdown. Equipment maintenance records should be maintained based on the manufacturer's recommendations and guidelines.

Preventive maintenance involves scheduled or regular maintenance activities such as cleaning, repair and inspections to prevent failures and unplanned downtime. Based on the maintenance records and the manufacturer's recommendations, preventive maintenance activities can be planned. The costs associated with preventive maintenance activities are inspections, lubrication, calibration, change of parts, routine servicing, service charges, etc. Preventive maintenance should be optimally scheduled; otherwise, it may create unnecessary maintenance and cost.

On the other hand, predictive maintenance uses advanced analytical tools applied to equipment data to predict failure and costly breakdowns. It increases overall productivity and reduces overall cost due to improved efficiency and reduced downtime. However, without the availability of quality data and analytical skills, predictive maintenance is challenging to implement in practice.

Companies should regularly review breakdown records and maintenance history to determine the frequency and costs of repairs and identify recurring issues. The cost-effectiveness of repair versus replacement decisions for major equipment components must be evaluated based on factors such as the age of the equipment, the frequency of breakdowns and the potential for improved performance and reduced maintenance costs. Additionally, the costs and turnaround time of outsourcing repairs versus in-house repairs should be compared. Operator skills and training can improve the performance of equipment while reducing operating costs, wear and tear and breakdowns.

Equipment maintenance costs can be compared with manufacturer and industry benchmarks to identify areas of improvement and potential cost-saving opportunities. Performance metrics can track and measure key cost indicators, such as cost per hour of operation or cost per unit of production, to identify areas for improvement and monitor performance. After the equipment's life span, refurbishment is not cost-effective, and asset disposal becomes necessary. The disposal method is based on equipment conditions, market demand and regulations. Used equipment can be scrapped, exchanged or put for resale.

7.7.3 *Spare Parts Inventory and Cost Management*

Inventory management is another important function of the P&M department, as it reduces the risk of running out of stock of critical equipment spare parts. Making the important spare parts and components of equipment available in time is critical for reducing downtime and ensuring the continuous smooth operation of the equipment. The spare part inventory analysis considers spare part type, consumption patterns and rate, criticality and lead time, and requirement is identified as per equipment owned by the company. Standardising the equipment fleet helps maintain minimum spare inventory and reduce costs.

Statistical methods such as moving averages or exponential smoothing on consumption patterns could help in making demand predictions for spare parts. Based on the forecast, the spare procurement schedule can be prepared to ensure the availability of spare parts for regular maintenance as well as breakdown. The reorder point – that is, the level of inventory at which a request for procurement is triggered – could differ for different spares based on the lead time, usage patterns and maintenance schedules. It is recommended that the reorder point be identified for different spare parts and a buffer stock be maintained for critical spare parts to handle any unexpected breakdowns and delays in procurement or supply.

Implementing a P&M maintenance management software system can help track equipment maintenance schedules, record breakdowns and manage spare part inventory. Many commercially available software packages can also help improve the forecasting and procurement process by generating alerts and reports based on predefined parameters. By effectively forecasting spare part requirements and preparing procurement schedules for long lead spares, construction companies can proactively address equipment maintenance needs, minimise downtime and optimise the overall productivity of the equipment fleet.

The relation between working capital and spare parts stock entails striking a balance between guaranteeing appropriate spare parts availability and maximising working capital. To support efficient maintenance operations while maintaining healthy working capital levels, appropriate inventory management techniques must be implemented, taking into account demand patterns, lead times, operational needs and financial concerns. The costs associated with acquiring and holding spare parts inventory, including purchase or procurement costs, storage expenses, insurance and potential obsolescence costs, must be assessed. Reorder points should be adjusted to avoid both the lack of spare parts and excess inventory costs.

7.8 Human Resource Management

Human resources are one of the most important or critical resources that decide the success and growth of an organisation (Becker & Gerhart, 1996; Pfeffer, 1994). The resource-based view of the firm views resources available to an organisation as a source of competitive advantage if they are unique, add positive value to the firm and cannot be substituted with other resources by competitors (Barney, 1991). While the finance, material and equipment resources are extremely important for construction companies, they can be easily acquired or replicated by competitors. In contrast, human capital is a unique resource that can provide a construction company with a competitive advantage. Therefore, high-performance human resources are an essential asset for an enterprise to create competitive advantages in the market (Becker & Gerhart, 1996).

Human resources play a pivotal role in the success of construction projects and companies. The skilled workforce consists of trades such as carpenters, electricians, plumbers, etc., as well as management professionals such as contract administrators, site engineers, and project managers. The trades, along with semi- and unskilled workers or labourers, often physically perform the construction tasks in projects. In comparison, the management team assumes the responsibility for planning, coordinating and overseeing construction activities. Additionally, support or overhead staff, including administrative personnel, accountants and marketing professionals, play indispensable roles in providing essential support functions. The workforce may include a combination of full-time and part-time roles, permanent and contract staff and experienced workers and trainees or apprentices.

Operating in a labour-intensive industry, construction companies depend primarily on human capital for their effective functioning and successful completion of construction projects (Sinesilassie et al., 2019). If human resources are unavailable in the local market, construction companies have to rely heavily on migrant workers. In Gulf Cooperation Council member countries, approximately 90% of workers in the construction industry are migrant workers (Umar, 2022), while international construction projects by Chinese construction companies have an approximately 50% Chinese workforce (Gao et al., 2024). In construction markets experiencing acute skill shortages, ensuring the availability of a skilled and qualified workforce by implementing sound human resource management policies and practices not only improves project outcomes and company performance but also provides companies with a competitive advantage. The collaboration of diverse human resources within and across construction companies is integral to the seamless and successful execution of construction projects.

Human resource management (HRM) is of strategic importance in all organisations (Huemann et al., 2007). Proper HRM in construction companies can help them attract, cultivate, inspire and retain a bright and motivated team to create a competitive advantage. HRM covers all aspects of how employees are managed, including employment and work conditions (Wilkinson, 2022). HRM practices aim to ensure that the organisation's human resources are optimally utilised and linked with its strategic goals. Generally, medium- and large-size companies have a separate formal department that develops and implements policies, programs and procedures for the acquisition and development of the company's human capital, consistent with its business needs. In contrast, small companies generally do not have an independent HRM department (Čech et al., 2016).

7.8.1 Organisational Structure

The employees of an organisation are arranged in a particular structure known as organisational structure, defining various roles, responsibilities, accountability and decision-making authority. A well-defined structure and hierarchy increases organisational capacity for information processing and coordination by empowering managers to allocate tasks, fix accountability and manage subordinates (Puranam, 2018).

Selecting the right organisational structure is crucial for any company. There are various types of organisational structures, such as functional organisational structure, divisional organisational structure and matrix organisational structure. A functional or role-based organisational structure is characterised by a centralised leadership team at the top and a vertical hierarchy below. This structure enhances specialisation, scalability and accountability, establishing clear expectations with a well-defined chain of command. However, drawbacks include the risk of being overly confining, hindering employee growth and potentially impeding cross-department communication and collaboration due to silos within the organisation. Many construction companies follow a functional organisational structure with separate functional teams responsible for procurement, contracts, quality and safety, etc.

A divisional organisational structure is common in large construction companies where dedicated resources are allocated to specific divisions rather than centralised departments looking over a portfolio of projects. For instance, a large construction company could have different divisions responsible for different sectors, such as residential, commercial and civil engineering projects. Similarly, there could be different divisions looking after the company's business in different states or countries. This promotes independence and a customised approach to suit the relevant client base and market. However, it can lead to an increase in overhead costs due to duplicate resources and may lead to unhealthy competition between different divisions.

In a matrix organisational structure, individuals have reporting obligations to multiple teams. This provides flexibility, encourages resource sharing and facilitates internal collaboration. However, its complexity may lead to challenges in defining accountability and communication, particularly for new employees. There are weak, balanced or strong matrix organisational structures, depending upon the distribution of authority between the project manager and the functional manager.

7.8.2 Human Resource Management Functions

Workforce Planning and Recruitment

The HRM team analyses the company's workforce requirements, identifying the need for additional personnel based on current and upcoming project demands. The roles are defined on the basis of the skill sets, experience levels, license and certification requirements, etc., needed for specific positions within the company. Once the roles and mandatory and desired qualifications or skills are identified, the recruitment and hiring process begins. The HRM policy often dictates the decisions about hiring outside candidates or promoting from within, the search criteria and selection process and the onboarding process. Sometimes, critical positions are initially filled by reassigning the experienced existing personnel to ensure the continuity of work until a new person is appointed. Some companies outsource the recruitment tasks to independent recruitment agencies that have a better reach to the talent pool, whereas in others, dedicated HR departments or senior leaders such as project managers are responsible for recruitment processes.

Training, Development and Performance Management

Workforce training and development is a critical component of HRM in construction companies. To enable staff to execute their responsibilities, management should make sure that all staff members receive the necessary training. The training needs should be identified based on job analysis, performance evaluation, knowledge gaps, organisational need analysis and employee requests and feedback. The implementation of robust training and development initiatives not only focuses on skill enhancement but also aims to develop a workforce that is resilient and ready for the future.

Training can be measured in four main ways: (a) absolute measures (e.g., amount of training), (b) proportional measures (e.g., percentage of workers trained), (c) content measures (e.g., type of training) and (d) emphasis measures (e.g., perceived importance of training; Úbeda-García et al., 2013). Off-the-job and on-the-job (OJT) training are two important approaches to training construction workers (Tabassi & Abu Bakar, 2009). The most widely used techniques of off-the-job training include simulation exercises, case studies and lectures in the classroom. In the conventional off-the-job training model, employees are expected to apply abstracted knowledge they had acquired through a pre-arranged course on the new rules, procedures or processes, frequently at a location other than their place of employment, to promote new practices. A unique kind of OJT, known as apprenticeship training, is used by construction companies to train and employ 'skilled' labourers, such as carpenters, plumbers and welders. Most of this training is conducted in accordance with standards set by government entities, including the curriculum, the number of hours and the aims of affirmative action.

Developing required competencies should be the focus of training programs. Supervisors and managers can positively influence the transfer of tacit knowledge and skills in a significant way. Since construction is a people industry, training in soft skills development, such as communication, problem-solving, leadership and team management, should also be provided in addition to technical training. In addition to need-based training, learning and development as a continuous process of professional development should be promoted by providing relevant opportunities and encouraging employees to participate. Learning and development opportunities improve workplace culture, increase employee engagement and improve retention rates. Every training and learning accomplishment should be documented to update records and identify and fill any training and competency gaps (Howarth & Watson, 2012).

One of the HRM functions is to establish metrics and evaluation processes for employee assessment. Performance management includes regular performance reviews, feedback sessions and recognition programs to motivate and retain skilled workers. The performance evaluation encourages employees to perform better by acknowledging and rewarding their achievements. Wilkinson (2022, p. 59) stated that 'performance management should link together strategy, performance objectives, and standards by measuring and developing individuals, and performance appraisal is part of this process'.

Remuneration Management

Conducting salary benchmarking to ensure competitiveness in the industry is crucial for retaining talent. Other than direct pay, employees can also be offered several other benefits such as health care, insurance, paid leave, etc. By effectively managing compensation structures and implementing fair and transparent processes, the HRM team can improve employee satisfaction, productivity and retention. The gender pay gap and pay inequity need to be addressed in construction companies. Fairness in remuneration and reward is essential for improving employee well-being and retention in an organisation (Tetteh et al., 2019). Employees can perceive fairness by comparing their pay, rewards and benefits, recognition, promotion opportunities, etc. (Hazeen Fathima & Umarani, 2023). If employees perceive fairness in HRM practices, they exhibit positive work attitudes such as satisfaction, loyalty, commitment and organisational citizenship behaviour, whereas lack of fairness may result in absenteeism, counterproductive work behaviour and turnover (Addai et al., 2018; Perreira et al., 2018).

Compliance with Laws and Regulations

Ranging from power tool handling and emergency response to minimum employment standards, construction companies need to comply with a myriad of labour laws and regulations. HR compliance is the process of bringing an organisation's HR policies in sync with the employment laws prevalent at the local, state, country and international levels. For instance, the National Employment Standards are the minimum employment entitlements that have to be provided to all employees in Australia. The minimum entitlements of the National Employment Standards are maximum weekly hours, requests for flexible working arrangements, parental leave and related entitlements, superannuation contributions, notice of termination, redundancy pay, etc. (Australian Government, 2024). Therefore, construction employers operating in Australia must meet these minimum standards. Similarly, other countries have laws and regulations that apply to workers and their work in organisational contexts. Compliance with applicable laws and regulations not only fulfils legal obligations but also helps companies maintain ethical conduct and a positive corporate image. Any non-compliance with laws and regulations can lead to temporary or permanent business closure, fines and reputational damage. Therefore, companies often have a dedicated compliance team or an HR employee or team responsible for navigating various laws and regulations to mitigate legal risks.

Health and Well-being

Ensuring employee well-being and ethical treatment of workers in project-oriented companies has been identified as an issue that is important for HRM (Huemann et al., 2007). HRM practices in construction companies must pay particular attention to the occupational health and safety management of workers, given the high rate of physical injuries and mental health disorders in the construction industry. In many countries, the presence of a predominantly informal and unorganised labour market demands proactive measures to ensure that the rights and well-being of workers are protected. Similarly, HRM needs to pay attention to women and expatriates or migrant workers who experience higher levels of mental health issues such as anxiety, depression and burnout (Gao et al., 2024; Hasan & Kamardeen, 2022).

7.8.3 Strategic Human Resource Management

A construction company must think strategically about HRM, focusing on the long-term and external issues that could impact its business and operations in the coming years. The following six traits characterise the strategic HRM strategy used in the construction industry (Loosemore et al., 2003).

1. Acknowledging and reacting to the environment: Construction companies are not isolated entities; external contexts offer both possibilities and challenges for the company's continued growth. The role of the HRM strategy is to capitalise on the opportunities and to mitigate the threats through its people management policies.
2. Identifying and addressing the characteristics of the labour market: Since the labour market is just as competitive as the business sector, the management strategy should prioritise acquiring, rewarding, deploying and maintaining human capital. This is a particularly complicated problem in the construction sector because of the mobile nature of the workforce and the need for a wide range of skills. The organisation needs to create a strategy that adapts to the labour market conditions in which it works if it is to be successful.
3. Considering every employee in the organisation: Regardless of gender, race, physical ability or seniority, every employee of the company must be seen as a component of the larger strategy. Diversity in the workforce is a strength that should be properly planned for and managed to maximise its potential for productivity.
4. Taking a broad perspective: It is imperative for an organisation to adopt a 3- to 5-year long-term perspective to navigate the external environment and labour market effectively. The challenge facing the construction industry is that practically all company tasks are viewed from a short-term perspective due to the market's cyclical nature.
5. Putting the emphasis on making decisions and choices: Considering the future and selecting the best course of action from a variety of possibilities is known as *strategising*. Choosing an HRM strategy requires deciding on an organisation's future course and how its personnel will contribute to achieving it.

6. Including strategic human resource management in the broader business plan: Alignment with the company's business strategy is a prerequisite for strategic HRM formulation. A company that wants to establish a reputation for excellence must start modifying the working culture.

Moreover, the following questions must be taken into consideration when developing the strategies (Loosemore et al., 2003):

1. Does the organisation's corporate culture align with the overarching business direction and performance standards of the contemporary industrial context?
2. Do the current employees of the organisation have the necessary skills and adaptability to handle any anticipated changes in the organisation's external position?
3. Has the company considered outside influences that might affect its capacity to attract and retain top talent?

After formulating strategies based on the above considerations, assigning accountability for each component of the HRM strategy to those best equipped to implement it is the next step in the strategy design process. The last step in developing a strategy is to keep an eye on how the tactics being used are helping the company achieve its goals. To create, review and modify strategies as needed, this knowledge must be included in the strategic planning process. The organisational HRM strategy may need to adapt continuously because erratic workloads, workforces and projects would likely produce drastically variable HR requirements.

7.9 Resource Management Challenges

Resource management in construction companies poses various challenges that require careful consideration and strategic solutions. Unlike manufacturing, where supply networks may be strengthened and made more efficient via the establishment of long-term connections, construction supply chains are typically built on adversarial methods and transient ties. Many construction companies lack robust resource planning and contingency plans.

The need for sustainable and efficient management of materials resources in construction projects is aggravated by the unsustainable use of these resources and their adverse effects on the environment. Therefore, sustainable management and governance of resources is an important challenge. For instance, Chen et al. (2017) discussed two obstacles to the implementation of sustainable materials management. First, it takes effort and expertise to collect information and synthesise knowledge regarding a material's complex life cycle to formulate integrated sustainable materials management actions. However, such comprehensive life cycle information of material flows and expertise to utilise it to make informed decisions are unavailable in many companies. Second, poor inter-agency collaboration means a lack of understanding of various aspects of the whole life cycle of materials and its linkage with economic activities, resulting in inadequate policies, decision support systems and governance. Construction companies and stakeholders must work together to develop processes and tools that support informed decisions concerning SRM. SRM can only be achieved if an integrated or end-to-end approach to sustainability is adopted by all supply chain partners.

Since construction is a project-based industry, resource management at the company level often involves resource management for multiple projects, ensuring that the use of resources is optimised between projects. However, this may result in conflicts when the resources required to optimise the delivery of a single project are not the same as the resource allocation required to optimise performance across a portfolio of projects (Karaa & Nasr, 1986). Many resources used by construction companies are often in short supply, affecting project and company performance in the construction industry. For instance, Hasan et al. (2018) found material shortage, equipment unavailability and skill shortages among the top 10 factors affecting construction productivity globally.

A major problem that construction companies face while managing their resources is uncertainty due to dynamic demands, changing priorities and unforeseen changes. Implementing a dynamic approach by regularly updating resource allocation based on real-time project data could allow construction companies to navigate unforeseen circumstances efficiently, ensuring adaptability and effective resource utilisation. Advanced resource management tools and software can provide real-time resource tracking, allowing companies to adapt to evolving project timelines and priorities swiftly, ensuring optimal resource utilisation and project success. The seamless incorporation of advanced technologies based on big data and artificial intelligence could change resource management from a reactive process to a proactive, data-driven process (Kazeem et al., 2023).

Tackling the challenge of supply chain disruptions arising from global events, natural disasters or geopolitical issues is critical in ensuring the timely availability of resources. A strategic solution involves a proactive approach that includes diversifying suppliers, maintaining buffer stocks and devising contingency plans for alternative resource sources. By establishing strategic alliances with suppliers and subcontractors and cultivating robust relationships, construction companies can navigate the complexities of resource shortages and supply chain disruptions. They can also foster a dynamic ecosystem where suppliers and subcontractors align with the project's unique demands (Sundquist et al., 2018).

Huemann et al. (2007) argued that project-oriented companies do have specific requirements regarding HRM because of specific features such as the temporary nature of projects, dynamism and 'managing by projects' as their strategy. Additionally, employees in project-oriented companies can have multiple roles in different projects at the same time, which may create human resource challenges such as multi-resource allocation and role conflict (Huemann et al., 2007).

Another major challenge construction companies face in HRM is skill shortages. The construction industry worldwide faces a persistent challenge of skilled labour shortage, exerting a substantial impact on project timelines and overall productivity. Implementing targeted training programs, fostering apprenticeships and establishing partnerships with educational institutions has become instrumental in nurturing and developing a skilled workforce. By providing ongoing education, skill enhancement programs and mentorship, construction companies can bolster the capabilities of their teams and cultivate a versatile and adaptable workforce ready to navigate the intricacies of diverse projects (Raiden et al., 2008). Additionally, offering competitive salaries and other employment benefits can help companies attract and retain talented employees.

Similarly, improving diversity, equity and inclusion (DEI) in construction companies is both a challenge and an opportunity for HRM in construction companies. While DEI efforts are often met with resistance due to the male-dominated work culture at construction workplaces, successful DEI initiatives can help companies overcome skill shortage issues and improve their performance, decision-making and competitiveness due to a diverse workforce, as discussed in the chapter on DEI in this book. HRM in construction companies also needs to focus on occupational health and safety, work–life balance and the well-being of construction workers and professionals, as construction workplaces are known to be risky and involve long work hours, including work during the weekends. Moreover, construction companies must provide a psychologically safe and respectful work culture to retain workers.

7.10 Conclusion

Construction companies rely heavily on various resources to complete projects successfully. These resources can be broadly categorised as human, material, equipment and financial resources. Effective resource management is fundamental to successfully managing a construction company. However, resource management in the construction industry is a multifaceted challenge that necessitates meticulous planning, collaboration and adaptability. Establishing a comprehensive resource management strategy, strategic partnerships with suppliers, collaboration across the supply chain, investments in workforce development and adoption of modern inventory management practices and technologies can help construction companies maintain a competitive advantage in a business environment characterised by resource shortages and supply chain disruptions.

Robust industry networks and good relationships with supply chain partners, stakeholders and regulatory bodies are crucial for effective resource management in construction companies. Construction companies must establish reliable supply chains with strong connections with suppliers and vendors to help them successfully deliver construction projects and grow their business. Moreover, collaborative networks with other organisations bring together a broader pool of resources and knowledge. Being in a resource-intensive industry, construction companies must also pay attention to both the use of sustainable resources and the sustainable use of resources to minimise their adverse effects on the environment.

References

Addai, P., Kyeremeh, E., Abdulai, W., & Sarfo, J. O. (2018). Organizational justice and job satisfaction as predictors of turnover intentions among teachers in the Offinso south district of Ghana. *European Journal of Contemporary Education, 7*(2), 235–243.

Australian Government. (2024). *National employment standards.* https://www.fairwork.gov.au/employment-conditions/national-employment-standards

Barney, J. (1991). Firm resources and sustained competitive advantage. *Journal of Management, 17*(1), 99–120.

Basheer, M., Elghaish, F., Brooks, T., Rahimian, F. P., & Park, C. (2024). Blockchain-based decentralised material management system for construction projects. *Journal of Building Engineering, 82*, 108263.

Becker, B., & Gerhart, B. (1996). The impact of human resource management on organizational performance: Progress and prospects. *Academy of Management Journal, 39*(4), 779–801.

Caldas, C. H., Menches, C. L., Reyes, P. M., Navarro, L., & Vargas, D. M. (2015). Materials management practices in the construction industry. *Practice Periodical on Structural Design and Construction, 20*(3), 04014039.

Čech, M., Yao, W., Samolejova, A., Li, J., & Wicher, P. (2016). Human resource management in Chinese manufacturing companies. *Perspectives in Science, 7*, 6–9.

Chen, P. C., Liu, K. H., & Ma, H. W. (2017). Resource and waste-stream modeling and visualization as decision support tools for sustainable materials management. *Journal of Cleaner Production, 150*, 16–25.

Dallasega, P., Rauch, E., & Linder, C. (2018). Industry 4.0 as an enabler of proximity for construction supply chains: A systematic literature review. *Computers in Industry, 99*, 205–225.

Elghaish, F., Hosseini, M. R., Kocaturk, T., Arashpour, M., & Ledarim, M. B. (2023). Digitalised circular construction supply chain: An integrated BIM–blockchain solution. *Automation in Construction, 148*, 104746.

Gao, L., Luo, X., Wang, Y., Zhang, N., & Deng, X. (2024). Retention in challenging international construction assignments: Role of expatriate resilience. *Journal of Construction Engineering and Management, 150*(2), 04023158.

Guo, F., Jahren, C. T., & Turkan, Y. (2021). Electronic document management systems for the transportation construction industry. *International Journal of Construction Education and Research, 17*(1), 52–67.

Hasan, A., Baroudi, B., Elmualim, A., & Rameezdeen, R. (2018). Factors affecting construction productivity: A 30-year systematic review. *Engineering, Construction and Architectural Management, 25*(7), 916–937.

Hasan, A., & Kamardeen, I. (2022). Occupational health and safety barriers for gender diversity in the Australian construction industry. *Journal of Construction Engineering and Management, 148*(9), 04022100.

Hazeen Fathima, M., & Umarani, C. (2023). Fairness in human resource management practices and engineers' intention to stay in Indian construction firms. *Employee Relations: The International Journal, 45*(1), 156–171.

He, D., Li, Z., Wu, C., & Ning, X. (2018). An e-commerce platform for industrialized construction procurement based on BIM and linked data. *Sustainability, 10*(8), 2613.

Honic, M., Kovacic, I., Sibenik, G., & Rechberger, H. (2019). Data and stakeholder management framework for the implementation of BIM-based material passports. *Journal of Building Engineering, 23*, 341–350.

Howarth, T., & Watson, P. (2012). *Construction quality management: Principles and practice.* Routledge.

Huemann, M., Keegan, A., & Turner, J. R. (2007). Human resource management in the project-oriented company: A review. *International Journal of Project Management, 25*(3), 315–323.

Jaśkowski, P., Sobotka, A., & Czarnigowska, A. (2018). Decision model for planning material supply channels in construction. *Automation in Construction, 90*, 235–242.

Karaa, F. A., & Nasr, A. Y. (1986). Resource management in construction. *Journal of Construction Engineering and Management, 112*(3), 346–357.

Kazeem, K. O., Olawumi, T. O., & Osunsanmi, T. (2023). Roles of artificial intelligence and machine learning in enhancing construction processes and sustainable communities. *Buildings, 13*(8), 2061.

Loosemore, M., Dainty, A., & Lingard, H. (2003). *Human resource management in construction projects: Strategic and operational approaches.* Taylor & Francis.

Ma, Z., Lu, N., & Wu, S. (2011). Identification and representation of information resources for construction firms. *Advanced Engineering Informatics, 25*(4), 612–624.

Omotayo, T., Tan, S. W., & Ekundayo, D. (2023). Sustainable construction and the versatility of the quantity surveying profession in Singapore. *Smart and Sustainable Built Environment, 12*(2), 435–457.

Payán-Sánchez, B., Labella-Fernández, A., & Serrano-Arcos, M. M. (2021). Modern age of sustainability: Supply chain resource management. In C. M. Hussain & J. F. Velasco-Muñoz (Eds.), *Sustainable resource management* (pp. 75–98). Elsevier.

Perreira, T. A., Berta, W., & Herbert, M. (2018). The employee retention triad in health care: Exploring relationships amongst organisational justice, affective commitment and turnover intention. *Journal of Clinical Nursing, 27*(7–8), e1451–e1461.

Pfeffer, J. (1994). *Competitive advantage through people: Unleashing the power of the work force.* Harvard Business School Press.

Puranam, P. (2018). *The microstructure of organizations.* Oxford University Press.

Raiden, A. B., Dainty, A. R. J., & Neale, R. H. (2008). Understanding employee resourcing in construction organizations. *Construction Management and Economics, 26*(11), 1133–1143.

Sinesilassie, E. G., Tripathi, K. K., Tabish, S. Z. S. & Jha, K. N. (2019). Modeling success factors for public construction projects with the SEM approach: Engineer's perspective. *Engineering, Construction and Architectural Management, 26*(10), 2410–2431.

Srivastava, S. K. (2007). Green supply-chain management: A state-of-the-art literature review. *International Journal of Management Reviews, 9*(1), 53–80.

Sundquist, V., Hulthén, K., & Gadde, L. E. (2018). From project partnering towards strategic supplier partnering. *Engineering, Construction and Architectural Management, 25*(3), 358–373.

Tabassi, A. A., & Abu Bakar, A. H. (2009). Training, motivation, and performance: The case of human resource management in construction projects in Mashhad, Iran. *International Journal of Project Management, 27*(5), 471–480.

Tetteh, S. D., Osafo, J., Ansah-Nyarko, M., & Amponsah-Tawiah, K. (2019). Interpersonal fairness, willingness-to-stay and organisation-based self-esteem: The mediating role of affective commitment. *Frontiers in Psychology, 10*, 1315.

Úbeda-García, M., Marco-Lajara, B., Sabater-Sempere, V., & Garcia-Lillo, F. (2013). Training policy and organisational performance in the Spanish hotel industry. *The International Journal of Human Resource Management, 24*(15), 2851–2875.

Umar, T. (2022). The impact of COVID-19 on the GCC construction industry. *International Journal of Service Science, Management, Engineering, and Technology, 13*(2), 1–17.

Wilkinson, A. (2022). *Human resource management: A very short introduction.* Oxford University Press.

8 Financial Management

Asheem Shrestha

8.1 Introduction

Effective financial management is essential to operating a successful construction company. Generally, financial management in construction is fraught with challenges. The construction industry is risky – there are low barriers to entry and high competition, and the failure of a single project can impact the entire company. Project uncertainties are often exacerbated by external factors such as the economic environment, market volatility, regulatory changes and competition. Numerous building firms collapse every year due to financial difficulties. For instance, according to the Australian Securities and Investments Commission (ASIC, 2023) data, 2,911 construction businesses entered voluntary administration during the 2021–2022 financial year, the highest number in any industry. Although construction companies comprise the largest number of businesses, their rate of insolvency is greater than that of other industries.

Construction companies are project oriented, largely reliant on subcontractors and cope with complex payment structures (Peterson, 2013). Moreover, multiple projects running concurrently, each at different stages with distinct capital requirements, impose a substantial financial burden on the company. External variables, such as market and regulatory conditions, and internal factors, such as the business's capacity for proper planning and management, all play important roles in shaping the pattern of a high number of insolvencies in the construction industry. While various factors may cause company failures, most can be attributed to financial issues (Khosrowshahi & Kaka, 2007). The leading causes of insolvencies in Australia are (a) inadequate cash flow or high cash use, (b) trading losses, (c) poor strategic management, (d) poor management of accounts receivables and (e) under-capitalisation (ASIC, 2023). These challenges, which, in reality, are linked to company finances, underline the need for construction companies to have a strong understanding of financial management.

Avoiding and predicting financial issues require efficient cash flow management, comprehensive planning and timely financial reporting and analysis. However, many companies lack the resources and expertise to maintain sound financial systems. As a result, they face the risk of failure. Construction companies need good financial management practices not only to increase profits but also to evaluate their performance and put in place the necessary safeguards against potential risks so that they can make sound decisions that can help them survive and be successful in a highly competitive industry.

This chapter focuses on financial management, specifically in the context of construction companies. While different types of construction companies have different corporate structures and serve different functions in the construction industry, this chapter focuses mainly on medium to large primary contractors unless specifically mentioned. The chapter is organised as follows: it starts by explaining the role of financial management in the construction industry and

DOI: 10.1201/9781003223092-8

how it applies to the short- and long-term planning of a company. The following sections then look at funding sources and the accounting function for construction companies, followed by a discussion on financial analysis, focusing specifically on cash flow and financial performance measurement. Finally, a case study is presented to illustrate some of the concepts discussed in this chapter.

8.2 Financial Management Challenges in the Construction Industry

To achieve success, businesses must find a delicate balance between short-term and long-term objectives. In the short term, companies often prioritise immediate financial stability, which includes profitability, delivering returns to stockholders and maintaining a sound financial position. On the other hand, long-term goals have a more strategic focus, such as attaining a strong market share, maintaining high growth and ultimately establishing industry leadership. Financial planning and management have a direct association with both the short- and long-term goals of the company and relate to how a construction company manages its day-to-day operations, as well as how it makes critical decisions about investments and resource allocation for its future success. Moreover, financial analysis helps construction companies measure their financial performance, compare it with their rivals, benchmark it against industry standards and formulate plans based on credible financial data.

In the short term, a company's main objective is often to maximise its profit to ensure that revenues produced outweigh expenses, resulting in a positive return on investment. This objective is supported by careful cost control and other measures to improve efficiency. By carefully setting and monitoring their budgets, companies can detect and correct any inconsistencies or inefficiencies in their financial activity. However, the construction business is unusual. One of the distinct properties of the industry is related to the complicated payment systems.

Unlike many sectors, where transactions are relatively straightforward, construction involves a host of parties and complex work structures with varying payment cycles. A delay or default by one party may have a domino effect, affecting timelines, delaying subsequent project phases and making cash flow management more challenging. Moreover, companies are often pulled into insolvency by the failure of another company they were doing business with (Lowe, 1997). For instance, if the client becomes bankrupt, the main contractor, who is owed large sums of money, may also be pulled into insolvency, which might have a domino effect on subcontractors and suppliers down the chain. Given these complex interdependencies, effective coordination and comprehensive financial planning are necessary to ensure the smooth progression of projects and minimise any financial risks.

The relationship between financial planning and project schedules is also complex. As a project proceeds, its expenditures and income should conform to predetermined timelines. Typically, clients pay for finished work in instalments over the duration of a project. This staged payment mechanism ties cash inflow to specified milestones or stages of completion, which are structured to ensure that the client pays only for demonstrable progress. Both the client and the contractor gain from this arrangement since it provides progressive payment for ongoing construction expenses. It also helps contractors maintain sufficient liquidity or, in other words, the capacity to pay off their current obligations, which generally consist of debts and expenses due in the short term.

However, in reality, there are substantial time lags between when costs are incurred and when income is received. In most cases, contract clauses also allow the client to retain a certain percentage of the payments until the work is delivered to the satisfaction of all parties involved. So, when expected payments are not received on time, it can put a strain on the contractor's budget and cash flow. Furthermore, it impacts suppliers and subcontractors, who will be compensated only when the main contractor has been paid for its progress claim by the client. Construction companies often negotiate trade credits with their suppliers and subcontractors, which allows them to delay their payments. However, when payments are not made within agreed time frames, there may be negative impacts resulting in project delays, increased costs and damaged relationships.

In Australia, for example, a report by CreditWatch (2022) identified the construction industry as having the worst late payment records of any industry, with 1 in 10 builders owing their suppliers more than 60 days in arrears on their payments. The prevalence of such payment practices has serious implications for the industry. So, construction companies must carefully undertake their financial planning to ensure that they have sufficient finances to meet their operating expenses. This entails regularly monitoring receivables, pursuing overdue payments and sometimes arranging for short-term funding to manage cash flow shortfalls.

Contractors are often responsible for financing the first phases of a project, though they receive some mobilisation costs from the client. The period prior to the first payment from the client may be capital intensive, requiring either the use of capital reserves or the acquisition of finance from external parties (i.e., borrowing). This may put a strain on a company's liquidity since money flows out before it flows back in. In some contracts, such as design-and-build contracts, the contractor is responsible for both project design and construction, meaning they have exposure to financial risks for a longer duration, from the original design stages to the completion of construction.

Construction companies are susceptible to both external and project risks. External risks are those beyond the direct influence of the risk bearer, and project risks are those specific to the project. When these risks occur during the project, the budget is often stretched, either as a direct impact of the risk or as a result of putting measures in place to mitigate them. During times of inflation, for example, prices may increase rapidly, affecting the company's cash flow. Similarly, when cost overruns occur and the company lacks enough contingency reserves, it may find itself in a difficult financial position. Sometimes, project payments are delayed or reduced. In this case, companies may experience additional financial stress. In severe circumstances, a cumulative effect of such situations could lead to business failure.

Smaller and medium-sized companies were reported to have a higher insolvency rate than larger firms (ASIC, 2023). It is generally common for large construction companies to have dedicated finance and accounting teams. They may also have the resources and tools to perform strong financial analyses and risk assessments. However, this may not be the case for smaller companies, which is why poor financial management is often considered a primary reason why smaller and medium-sized companies fail.

Selecting the right project is often a strategic choice for a construction company and requires careful evaluation of the costs and expected financial returns from available projects. Companies also need to ensure that every project they undertake has the potential to offer a favourable return on investment. Financial assessment can be tricky, particularly for projects with longer durations, because forecasting future costs and revenue while considering future risks is challenging. It requires specialised financial knowledge and tools to help companies perform the analysis and make important business decisions.

Financial assessments are also required when a firm makes investment decisions, whether in technology, equipment or skills. The assessments need to ensure that investments align with the company's short- and long-term goals. Frequently, difficulty may arise when these goals are in tension, requiring a careful strategic balance in decision-making. For example, companies are often confronted with 'make or buy' decisions where they must consider both the short- and long-term implications of their choices. Outsourcing, for instance, may provide operational flexibility and immediate financial relief through reduced expenditures. However, in the long term, it may result in a lack of trained personnel, affecting the company's competitiveness. So, companies should carefully plan their operations by scrutinising the costs and benefits in financial and other terms.

8.3 Main Sources of Finance

The success of construction companies is contingent on their access to appropriate funding under favourable conditions for the execution of planned projects. In the short term, companies must have enough liquidity to guarantee they have enough funds to meet their financial commitments. So, companies resort to bank overdrafts and leveraging cash from creditors, trade credits and their own finances to manage immediate financial needs and maintain liquidity.

Bank overdrafts are fairly common for construction companies, allowing them to draw money through their bank accounts up to a maximum amount when their account reaches a low threshold, which then has to be paid back within a short period of time (Shash & Qarra, 2018). Companies in the construction industry also rely heavily on trade credit, where the provider of goods and services agrees to delay payment. According to Ive and Murray (2013), trade credit is by far the most important, continuous and widespread source of finance for construction operations. Construction companies may also seek short-term loans from banks to meet their short-term obligations, although these loans may have a higher interest rate than medium- and long-term loans.

Construction companies may seek medium-term loans to meet their medium-term obligations. The typical repayment schedule for medium-term loans is 1 to 5 years. These loans are often used for funding specific projects, purchasing equipment or expanding operations. For long-term financial needs, a diverse array of sources are considered. These may include issuing shares to attract equity investors, securing mortgage loans on assets, relying on retained profits from previous projects, seeking venture capital for growth potential, issuing bonds to tap into the debt market and utilising project finance. Each financing option offers unique advantages and disadvantages and is selected based on the strategic goals and the nature of the projects the company is involved in.

Companies often use debt and equity financing to secure funds for larger and more capital-intensive projects, to invest in resources or to grow their business. Equity finance typically involves raising capital by selling company shares or ownership interests, which can include stakes in specific projects. Equity may also come from a company's retained earnings when profits from previous projects are re-invested in future ventures. External investors, such as private investors, venture capitalists or even other construction companies, may also provide capital seeking partnerships in projects. These investors contribute capital in return for a share in the project or business.

Debt finance entails borrowing funds from financial organisations such as banks, credit unions or even specialist construction finance companies. The borrowed money is subject to principal and interest repayment over a certain period of time. Debt instruments may range from short-term loans for urgent cash needs to long-term loans or bonds for substantial projects. With debt financing, the company maintains complete ownership and control over the project. However, the company must guarantee that it is capable of meeting its debt commitments.

The important question for construction companies is which option to choose. Deciding between debt and equity financing requires looking at various factors. For debt financing, the cost of borrowing needs to be considered. Interest and principal repayments are mandatory, regardless of business performance. But, more important, considerations need to be made regarding the potential implications this has on cash flow stability, credit ratings and financial risk. On the other hand, while repayments are not necessary in equity financing, equity providers will want a return on their investment. Moreover, it is vital that the company be aware of the trade-offs associated with sacrificing a fraction of its ownership, which would give equity investors the power to influence the operations and direction of the firm.

Debt financing is generally a viable option for established businesses with stable and predictable cash flows. However, it is crucial for companies to keep monitoring their debt levels. They must also ensure that they have a robust debt repayment strategy that takes their cash inflows and outflows into account. Monitoring interest rates and even refinancing when more favourable rates become available are also essential components of debt management. Maintaining a good debt-to-equity ratio is vital since excessive debt may place a firm at risk of insolvency, particularly during times of downturn in the industry or the economy.

Financial institutions and creditors, however, exercise discretion in lending money, and not all businesses may be deemed eligible. A company's track record will significantly influence the terms of any external funding. The performance of the company in winning new contracts, its ability to manage costs and its overall financial success will be scrutinised by lending entities, which typically conduct thorough risk assessments before approving loan applications or credit arrangements. They are more willing to finance a company if they have reason to believe that it will continue to succeed in the future and if they stand to gain from its success. Thus, the financial statements of a company are key components used to evaluate its business potential (Murray & Kulakov, 2022).

In some instances, the government may offer assistance to construction companies in the form of grants, direct loans or land contributions. This kind of support is often provided to programmes that bring social benefits. The benefits of public financing include financial assistance on more favourable terms than banks. However, publicly sponsored projects may be subject to more stringent regulatory requirements, more scrutiny and longer approval processes.

Project finance is the method used for funding long-term projects utilising a non-recourse or limited-recourse financial structure. In this approach, the repayment of the debt and equity comes from the cash flow generated by the project itself. Construction companies obtain project-specific financing, which is structured around the specific needs of the project and its cash flow projections. Typically, the interest on such loans is considered a project cost and must be accounted for in the project's budget and pricing strategy.

8.4 Types of Construction Costs

Cost management is a fundamental responsibility of the managers of construction companies. To manage costs well, they must comprehend the relevance of different types of costs. There are two basic categories of costs: variable costs and fixed costs. Variable costs are those that vary based on the level of activity and fluctuate with construction volume. These costs can vary based on several factors, including the scope of projects, the materials and equipment required and the number of workers involved. Theoretically, if there is no construction activity, there should be no variable costs.

In contrast, fixed costs are those overhead expenses that remain constant throughout the short term. Here, it is important to stress the term *short term* because all costs can become variable in the long term. Fixed costs include expenses like office rent, salaried administrative staff wages and annual software licensing fees. Fixed and variable costs consistently impact the financial standing of construction companies.

Construction companies may also incur substantial costs associated with the purchase of construction materials. Material prices are often affected by global markets and can fluctuate considerably in response to events such as geopolitical tensions, trade disputes and natural disasters. Moreover, certain environmentally friendly or sustainable materials may come at a higher price. To maintain a project's financial sustainability, construction companies must continually track the pricing of materials, predict trends in price changes and strategically procure materials. According to Murray and Kulakov (2022), construction companies, particularly larger ones, often purchase large volumes of materials and supplies to achieve greater economies of scale and avoid future price hikes. However, high material inventories reduce available cash; therefore, the company may suffer from insufficient liquidity.

Wages and compensation are also among the most significant costs. They involve not only on-site labour but also administrative, management, design and engineering personnel. Due to the specialised nature of many roles in the industry, these salaries may often be high. Similarly, subcontractor claims may incur high costs. A lot of the work in construction is performed by subcontractors, comprising specialist teams or professionals with specialised expertise hired for specific tasks – these range from electrical and plumbing work to excavation and landscaping. Payments to subcontractors can vary based on the scope, complexity and duration of the subcontracted work, and any delays or modifications in the project can lead to additional claims, impacting the overall budget and profitability of the project.

The cost of construction plant and machinery (P&M) may also be significant for construction companies. These costs are related to the purchase or lease of equipment. Large equipment such as cranes, excavators and backhoes are often leased rather than purchased, especially if their usage is not frequent. Leasing may sometimes be more cost-effective than the ongoing ownership, maintenance, and depreciation costs.

Companies must also consider the fees associated with borrowing funds. If a construction company relies on borrowed capital for operations or specific projects, the interest payments can be substantial. The interest rates may also fluctuate, and for construction companies with variable interest rate loans, these fluctuations can lead to variable costs. The fluctuations can be seen as a risk, especially for long-term projects where consistent financial planning is crucial. Other costs include taxes, licences and rent that need to be paid for office space and the storage of materials and equipment.

8.5 Financial Accounting

Accounting may be described as the act of gathering, classifying and processing a company's financial data to produce financial statements. These statements are presented in financial reports, which reflect the financial position of the company. There are two main financial statements that are used to evaluate a business: the profit-and-loss statement and the balance sheet.

8.5.1 Profit-and-Loss Statements

The objective of the profit-and-loss statement (also known as an income statement) is to present a snapshot of a company's construction activity within a certain accounting period. It sets out the money collected less the expenses incurred to produce a profit or loss and encompasses three main account categories: revenues, expenses and profit. A construction company's revenue structure is designed to balance its expenses, ensure sustainable operations and secure profit.

Typically, construction revenue comes from client payments for completed projects. These may be arranged in a variety of ways, including fixed-price contracts, cost-plus contracts and anything in between. The objective is to ensure that the revenue not only covers variable and fixed costs but also leaves a profit margin. The profit-and-loss statement is used to look at a construction company's financial performance over a given period using this information. It offers stakeholders a clear insight into how effectively the company manages its costs and revenue. A successful construction company will exhibit an upward trend in net profit over time, indicating effective financial management.

In most cases, revenue is generated from progress claims invoiced to the client for work completed during a specific time period, and the expenses associated with that revenue are incurred during the same period. Matching revenue with expenses is known as *accrual accounting*. Accrual accounting documents financial transactions based on when they were earned or incurred, as opposed to when cash is actually traded. *Cash* is a generic term for transactions that are increasingly taking place online. So, profit generally represents the difference between revenues and costs for an accounting period, regardless of whether the money has actually been received or paid out.

Accrual accounting introduces specific accounts such as accounts receivables and accounts payable. Accounts receivables represent money owed to the company for goods or services provided but not yet collected. Conversely, accounts payable denotes expenses the company has recognised but has not settled. By understanding these liabilities and expected inflows, construction companies can perform their financial forecasting and make informed budgetary adjustments.

8.5.2 Balance Sheets

The balance sheet provides a snapshot of a company's financial standing at a specific point in time. It includes three account categories: assets, liabilities and owner's equity. The accounting equation for the balance is structured as

$$\text{Assets} = \text{Liabilities} + \text{Owner's Equity}$$

Assets include both current and non-current assets, while liabilities include both current and long-term liabilities. Current assets include cash, inventory and accounts receivables. Work in progress is also considered a current asset for construction firms. Non-current assets are those that cannot be easily converted into cash in the short term (generally within a year) but provide economic benefit to the company over time. They include P&M, land, buildings, etc.

Normally, depreciation is subtracted from assets such as P&M to obtain the value for their inclusion in the balance sheet. Depreciation recognises that a purchased asset has a fixed life span and may lose value over time. It also recognises that the ownership of an asset contributes to income generation in each accounting period during the asset's useful life. For instance, if the company did not own the asset, it would need to rent equipment or outsource the task. While cumulative depreciation is a liability account, it is often shown as a negative asset on the balance sheet alongside the depreciating asset account.

While intangible or non-physical assets, such as intellectual property, goodwill, reputation, trademarks, patents, etc., are also assets, they may not be included in the balance sheet as their value is difficult to determine in dollar terms. However, if these include something that has been acquired for a fee or has an identifiable dollar value, then it may be included in the balance sheet.

Liabilities are amounts that are owed to others. Among other things, they include mortgages, bank overdrafts, hire–purchase debts, supplier and subcontractor accounts, wages owing to employees and future long-service leave obligations. Current liabilities are those payments that must be made in the short term (generally within a year) and include accounts payable, wages, interest and taxes. The cost of leasing or renting equipment and plant is included as a liability. Owner's equity (also referred to as net assets) is the balance remaining after deducting all liabilities from assets. It represents the owners' or shareholders' financial claim on the company's resources. When shareholders put more money into the business or when the company earns more profit, the owner's equity also grows. The balance sheet not only shows the financial health of a construction company but also serves as a lens through which management, investors and other stakeholders assess its capacity to generate profits, manage debts and ensure sustainable growth.

8.6 Cash Flow Management

Cash flow refers to the flow of cash in and out of a company, as well as the amount of money that is available at any given moment. Cash flowing in is termed positive cash flow, and cash flowing out is termed negative cash flow. The difference between the positive and negative cash flows is the net cash flow. Here, cash is a generic term for payments, and many of the cash transactions are carried out online. The company's ability to manage cash flow is one of the most important indicators of its financial health (Shash & Qarra, 2018), and the lack of cash has been recognised as the most prevalent reason for construction company failures (Odeyinka et al., 2008).

The cash flow statement is a financial statement that shows how changes in the balance sheet and income accounts affect cash and cash equivalents. Generally, the movement in cash flows can be classified into three segments: (a) operating activities, (b) investing activities and (c) financing activities. Operating activities represent the cash flow from the primary activities of a business, investing activities represent the cash flow from the purchase and sale of assets other than inventories and financial activities represent the cash flow generated or spent on raising and repaying share capital and debt together with the payments of interest and dividends. Cash flow statements provide the information required to optimise cash flow management.

The cash flow of a construction firm is illustrated in Figure 8.1. The figure shows that construction firms manage cash flow at both the business and project levels. However, as they are project centric, the success of a construction company ultimately depends on the success of the projects within its portfolio of works (Murray & Kulakov, 2022). So, the financial health of a construction company is generally determined by the cumulative cash flow from its individual projects. If one project becomes too expensive, it may impact the entire business. Thus, construction companies must establish accurate budgets, monitor costs and perform financial reviews at each project stage (Navon, 1996). Proper cash flow management requires monitoring, analysing and adjusting a company's short-term and long-term cash flows to efficiently use available cash resources. It ensures that sufficient cash is always available to meet financial obligations and any financial surpluses are used optimally.

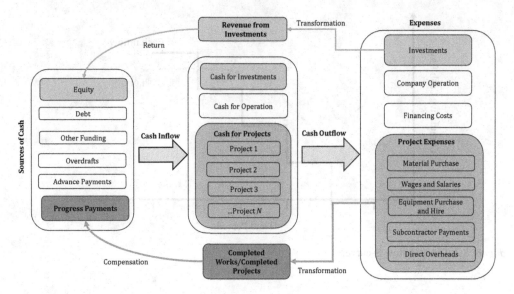

Figure 8.1 Cash flow of a construction firm.

Source: Adapted from Shash and Qarra (2018)

8.6.1 *Cash Flow Forecasting in Construction Projects*

Forecasting cash flow throughout the duration of a project is essential for efficiently controlling project finances. For construction companies, forecasting future cash flows from their projects is an important part of their strategic planning. To accurately determine the cash required for a project, companies must carefully evaluate the timing of cash outflows in relation to expected cash inflows throughout the project's duration. So, the timeline of the client's progress payments is a major component in establishing the cash requirements for a project (Peterson, 2013). However, payments are only received for the work completed; additionally, clients often retain part of the progress payments as a form of security to ensure that the project is completed to the agreed standards. This means the cash inflow is reduced, and the company may require additional working capital to manage a project (Shash & Qarra, 2018). Moreover, construction companies often face surges in capital requirements for their projects, leading to uneven cash flows.

Construction companies generally rely on an 'S-curve' to monitor and plan their cash flow in projects. The S-curve is a graphical representation of the cumulative project expenses as the project progresses. The idea of the S-curve is that when the project begins, the cash requirement increases gradually, but in the construction stage, it rises significantly before slowing down in the final stages. As a result, the cumulative curves assume the form of the letter 'S'. This curve is sometimes also the basis for progress payments from the client to the contractor (Cui et al., 2010).

Figure 8.2 shows an example of the S-curve. As shown in the figure, the progress payments lag behind the project expenses. While payments are received relatively late, suppliers are typically expected to be paid much sooner, and workers need to be paid even earlier. This payment structure often results in negative cash flow, and the discrepancies need to be covered by the construction company, which is often done through overdrafts as well as various short-term financing methods.

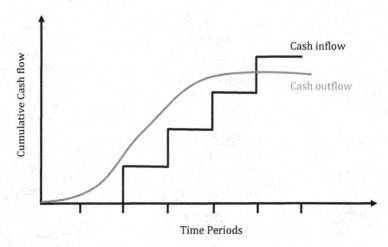

Figure 8.2 The cash flow S-curve.

8.6.2 *Strategies to Manage Short-Term Cash Flow*

Construction companies utilise various strategies to improve their cash flow. One of these is front loading, where they receive advanced payments or a higher portion of the payments at the early project stages. According to Cui et al. (2010), typical front-loading techniques include mobilisation costs, unbalanced pricing and overbilling. Mobilisation costs are associated with receiving advanced payments for site setup and transferring P&M and workers to the site. Unbalanced pricing involves inflating the costs of early-stage project activities while reducing the costs of later activities. Overbilling refers to contractors billing for materials that have not been installed. As long as these do not present additional costs or any significant risk to the client, they may be allowed by the client.

However, when front-loading strategies are used, the company may find itself in a position where it does not have adequate cash at the end of the project. So, they have to rely on securing a new contract, whose advance payments are used to complete the project. The cycle then gets repeated over and over and can turn into a serious problem if the company does not secure a new project on time (Mutti & Hughes, 2002). Moreover, Ive and Murray (2013) highlighted that the construction industry allows under-capitalised companies to enter the market, and these companies depend on the working capital generated from their operations, leading to a culture where contractors manipulate the payment system through various means.

One common strategy construction companies use is managing cash flow through trade credits. It is a prevalent practice in the construction industry where suppliers offer materials to contractors with a deferred payment option. This allows construction companies to delay their payments to their suppliers, which enables contractors to use these funds to manage their cash flow more effectively. Often, they can hold off on paying suppliers until they have received payments from clients to ensure that they have sufficient liquidity to cover operational costs. It also offers a source of short-term funding for companies, which may be especially useful for contractors working on large projects with significant upfront costs. Suppliers may extend trade credit for various reasons, such as to build relationships with contractors, to be more competitive than their rivals or even to increase revenue by charging higher prices. Ive and Murray (2013) found that the larger the company, the more trade credit they receive. In other words, suppliers allow larger companies more time to make their payments.

Similarly, construction companies may also delay payment to their subcontractors while utilising the labour, materials and P&M of the subcontractors to construct the project. In larger projects, where there is a heavy reliance on subcontracting work, the payment amounts could be substantial, which the firm can use as short-term funds for their projects. A 'pay-when-paid' provision allows them to withhold cash payments until they receive payment from the client (Cui et al., 2010). Moreover, they may retain a portion of the subcontractor's payment until the project is fully completed. However, according to Murray and Kulakov (2022), there is a trade-off, as subcontractors generally raise their prices when they have to fund the additional costs associated with extending trade credit.

The choice of financing option is generally influenced by a company's capacity to forecast cash flow throughout the project and the level of risk associated with the financing approach. The objective of the company will be to find the most cost-effective way to fund a project, so a key approach would be to limit borrowing cash with interest (Halpin & Senior, 2009). Shash and Qarra (2018) found that retained profit, equity and trade credit were the common strategies contractors in Saudi Arabia used to limit their short-term borrowing. Similarly, Odeyinka et al. (2003) identified and ranked the strategies in order of their use by contractors to manage cash deficits in the United Kingdom. They found that the use of the company's cash reserves ranked first, followed by unbalanced pricing, delayed payment to subcontractors, deferred payment to suppliers, the use of company assets and, finally, the use of borrowed funds.

8.6.3 Forecasting Long-Term Cash Flow and Planning

Securing long-term financing is important for a company's growth. Especially with recent developments in the construction industry, companies require capital to invest in sustainable practices, advanced technologies and workforce upskilling. Capital investments are also essential for established companies seeking to diversify into new areas and enter new markets. Typically, developers and construction companies that compete for large-scale projects also have higher capital needs, as larger infrastructure and commercial projects are inherently more capital intensive and require extensive funding that cannot be covered by short-term financing options. Given that these projects can take years to complete, long-term financing is essential to guarantee that the company has adequate funds during the entire project duration to manage its cash and contingencies. It also gives the company the stability it needs to establish future business strategies and plans.

Forecasting over a longer period may be more challenging due to the unpredictability of future transactions and the need to take into consideration the changing value of money over time. The accuracy of long-term forecasts is critical, as they underpin the strategic decisions companies make regarding their financial future. Essentially, companies employ capital budgeting to project their cash flows, which is integral to their long-term planning and investment decisions.

Capital budgeting is the process of selecting a company's medium- and long-term capital investments and expenditures. It is a critical function that enables construction companies to evaluate potential projects or investments and determine their potential to create value for the firm and its shareholders (Lam & Oshodi, 2015) and involves the identification of the potential returns on investments relative to their costs, considering the time value of money. By effectively applying capital budgeting, a construction company can allocate its limited resources to projects that promise the highest returns over time. According to Batra and Verma (2014), these capital budgeting decisions are crucial for securing a company's long-term viability in today's highly competitive environment.

In capital budgeting, the net present value (NPV) is a fundamental concept that converts the future cash inflows and outflows linked to a particular project or investment into today's dollar value. As the value of money changes with time, NPV ensures that comparisons are conducted by taking this into account. The process involves forecasting the cash flows from the investment and then discounting them back to their present value using a predetermined discount rate. It helps in determining the attractiveness of an investment by assessing whether the NPV of expected cash flows is greater than the initial capital outlay. Each project or investment is assessed based on its potential to deliver value in today's terms and ensure that the returns are appropriately weighed against the cost of capital over the investment period. It provides the framework for strategic financial planning and decision-making for companies. The rule of thumb is that the NPV must be greater than 1 before an investment is approved. Similarly, when evaluating potential projects, the one with the higher NPV should be selected, considering all other factors are equal. The formula for calculating NPV is as follows:

$$NPV = \sum_{t=0}^{n} \frac{R_t}{(1+i)^t}$$

where R_t is net cash flow during the period t, i is the discount rate or the return that could be earned via different investment, and t is the number of time periods.

Other common capital budgeting methods used in the construction industry include the internal rate of return (IRR), the average accounting rate of return and the payback period. IRR evaluates the rate of growth an investment is expected to generate. It is calculated using the same concept as NPV, except it sets the NPV equal to 0, and the interest rate that produces that is the IRR. It is ideal for analysing projects to understand and compare potential rates of annual return over time. The payback period is the length of time required to recover the initial investment in a project, calculated by dividing the initial investment by the annual cash inflows from that project. While these methods all look at specific metrics, the decisions made using these methods are all founded on forecasts of cash flows. While construction companies often use capital budgeting, questions have been raised about its reliability, particularly regarding the assumed certainty of future cash flow (Ye & Tiong, 2000). So, there are various mechanisms within capital budgeting that are devised to address these issues. Many models that extend the traditional methods to incorporate uncertainty factors within the traditional capital budgeting methods have been developed. However, these models are more complex, and only larger construction companies that possess the necessary resources and expertise to perform complex analyses use them (Lam & Oshodi, 2015).

8.7 Performance Measurement and Financial Ratios

The financial performance of a company may be assessed via two basic forms of analysis: vertical analysis, which examines key balance sheet and income statement indicators, and horizontal analysis, which examines trends across time. Both types of analysis have value, and both should be used when examining financial accounts (Halpin & Senior, 2009).

Accounting and financial management are closely related yet involve distinct tasks within a company. Accounting examines the financial history of a company, recording and reporting prior events and transactions. On the other hand, financial management is mostly concerned with planning for the future. It employs accounting insights collected from the past record as the basis for future financial plans and decisions.

Financial analysis is the process of selecting, evaluating and interpreting financial data, with other related information, to formulate an assessment of a company's present and future financial condition and performance. While financial reports like the balance sheet and income statement do provide information regarding a company's financial health, performing a comprehensive analysis of its performance requires an assessment of the company's earnings stability, debt management and competitive performance.

Financial disclosures along with market and economic data are the main sources of information used in the financial analysis (see Figure 8.3). Financial disclosures from the company's internal reports serve as a main source of information, providing insights into a company's financial health, performance and prospects. In addition, market data, which include factors such as the market price of stocks, trading volume and bond values, provide essential real-time information that assists investors and analysts in making informed decisions. Finally, economic statistics, such as GDP numbers, inflation and consumer spending patterns, serve as broad indications of the general economic environment, enabling macro-level evaluations that may be vital for financial analysis and forecasting. The integration of these varied information sources is essential for undertaking thorough financial analysis, allowing businesses to navigate complex financial markets and corporate performance with greater accuracy.

8.7.1 *Ratio Analysis*

Information from the balance sheet and income statement of a company is used to produce financial ratios. Financial ratios summarise a vast quantity of financial data to allow more informed

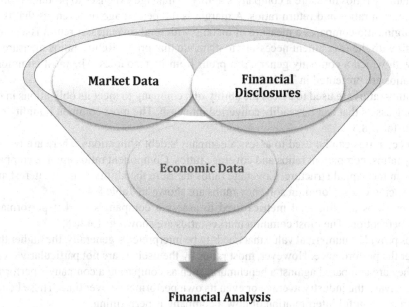

Figure 8.3 Information sources for financial analysis.

decision-making. They provide insight into possible issues or areas requiring more investigation by highlighting patterns that may not be obvious when examining individual elements in isolation (Halpin & Senior, 2009). These indicators assist in the identification of undesirable situations and provide a foundation for comparison, facilitating the evaluation of a company's financial health.

The assessment of a company's performance relies on a range of accounting ratios tailored to the specific needs of the firm. There are many different types of ratios, each used to look at specific information. Some are more important than others, and some are better suited for analysing construction companies (Öcal et al., 2007). Different user groups in the construction industry will have distinct information needs, so each group may use different ratios to analyse different aspects of company performance. For example, managers may evaluate a company's performance in comparison to its competitors or its historical performance, while investors may be more concerned with facts related to the company's ability to bring in future profit (Higham et al., 2016).

While ratios can be used to analyse a range of performance measures, Öcal et al. (2007) identified five factors that are the most important for construction companies: liquidity, capital structure and profitability, activity efficiency, profit margin and growth and the asset structure of companies. In general, ratios can be categorised into five fundamental domains: (a) activity ratios, (b) profitability ratios, (c) liquidity ratios, (d) solvency ratios and (e) market ratios. Liquidity, activity and solvency ratios assess a company's risk; profitability ratios focus on return; and market ratios reflect both risk and return (Gitman et al., 2015).

Activity ratios, also referred to as turnover ratios, are financial metrics that measure how efficiently the company utilises its assets. A simple way to look at the activity ratio is to consider the investment that is being put to work as the denominator and the result of that effort as the numerator. There are various types of activity ratios, each of which measures a distinct attribute. Common examples of activity ratios are presented in Table 8.1.

Profitability ratios measure a company's ability to manage expenses to produce profits. They include margin ratios and return ratios. A margin is the percentage of revenues that is a profit. Each margin ratio compares a measure of income with total revenues. A return is a comparison of a profit with the investment necessary to generate that profit. Return ratios measure the efficiency with which a company generates a profit from its resources. The most common profitability ratios are presented in Table 8.2.

Liquidity ratios are used to assess the ability of a company to meet its obligations in the short term using assets that can be readily converted into cash. The most common liquidity ratios are shown in Table 8.3.

Solvency ratios can be used to assess a company's debt obligations. There are two types of solvency ratios: component ratios and coverage ratios. Component ratios involve comparing the elements in the capital structure. Coverage ratios measure the ability to meet interest and other fixed financing costs. Common solvency ratios are shown in Table 8.4.

Market ratios are financial metrics used to assess a company's market performance and investor perception. The most common market ratios are shown in Table 8.5.

Ratios provide a numerical value that needs to be interpreted – generally, the higher the value, the better the performance. However, most ratios by themselves are not particularly meaningful unless they are compared against a benchmark, such as comparing a company's performance to that of its rivals, the industry average or even its own performance over time. These benchmarks provide a meaningful interpretation of how a company is performing.

Table 8.1 Examples of activity ratios

Type of ratio	Formulae	Remarks
Inventory turnover ratio	$\dfrac{\text{Cost of goods sold}}{\text{Average inventory}}$	How many times inventory is created and sold during the period
Receivables turnover ratio	$\dfrac{\text{Total revenue}}{\text{Average receivables}}$	How many times accounts receivables is created and collected during the period
Total asset turnover ratio	$\dfrac{\text{Total revenue}}{\text{Average total assets}}$	Extent to which total assets create revenues during the period
Working capital turnover ratio	$\dfrac{\text{Total revenue}}{\text{Average total assets}}$	Efficiency of putting working capital to work

Table 8.2 Examples of profitability ratios

Type of ratio		Formulae	Remarks
Margin ratios	Gross profit margin	$\left(\dfrac{\text{Gross profit}}{\text{Revenue}}\right) \times 100\%$	Production efficiency and pricing strategy
	Operating profit margin	$\left(\dfrac{\text{Operating income}}{\text{Revenue}}\right) \times 100\%$	Operational efficiency and the ability to control costs
	Net profit margin	$\left(\dfrac{\text{Net income}}{\text{Revenue}}\right) \times 100\%$	Overall financial health, efficiency and profitability
Return ratios	Return on investment	$\left(\dfrac{\text{Net return on investment}}{\text{Cost of investment}}\right) \times 100\%$	How efficient is the investment's profitability
	Return on assets	$\left(\dfrac{\text{Net income}}{\text{Average total assets}}\right) \times 100\%$	How effectively assets are utilised to produce profits
	Return on equity	$\left(\dfrac{\text{Net income}}{\text{Average shareholders' equity}}\right) \times 100\%$	How effectively is equity financing utilised to fund operations

Table 8.3 Examples of liquidity ratios

Type of ratio	Formulae	Remarks
Current ratio	$\dfrac{\text{Current assets}}{\text{Current liabilities}}$	Ability to satisfy current liabilities using current assets
Quick ratio	$\dfrac{\text{Cash + Short-term investments + Receivables}}{\text{Current liabilities}}$	Ability to satisfy current liabilities using the most liquid of current assets
Cash ratio	$\dfrac{\text{Cash + Short-term investments}}{\text{Current liabilities}}$	Ability to satisfy current liabilities using only cash and cash equivalents

Table 8.4 Examples of solvency ratios

Type of ratio		Formulae	Remarks
Component ratios	Debt to asset ratio	$\dfrac{\text{Total debt}}{\text{Total assets}}$	Proportion of assets financed with debt
	Debt to equity ratio	$\dfrac{\text{Total debt}}{\text{Total shareholders' equity}}$	Debt financing relative to equity financing
	Financial leverage	$\dfrac{\text{Total assets}}{\text{Total shareholders' equity}}$	Reliance on debt financing
Coverage ratios	Interest coverage ratio	$\dfrac{\text{Earnings before interest and taxes (EBIT)}}{\text{Interest payments}}$	Ability to satisfy interest obligations
	Fixed charge coverage ratio	$\dfrac{\text{EBIT} + \text{Lease payments}}{\text{Interest payments} + \text{Lease payments}}$	Ability to satisfy interest and lease obligations
	Cash flow coverage ratio	$\dfrac{\text{Cash flow from operations} + \text{Interest payments} + \text{Tax payments}}{\text{Interest payments}}$	Ability to satisfy interest obligations with cash flows

Table 8.5 Examples of market ratios

Type of ratio	Formulae	Remarks
Price to earnings ratio	$\dfrac{\text{Market price per share}}{\text{Earnings per share}}$	How much investors are willing to pay per dollar of earnings
Price to book ratio	$\dfrac{\text{Market price per share}}{\text{Book value per share}}$	How much investors are willing to pay per dollar of earnings
Dividend yield	$\left(\dfrac{\text{Annual dividends per shares}}{\text{Market price per share}}\right) \times 100\%$	The return investors get for each dollar invested in the company's stock

8.7.2 Altman's Z Score

Several complex models have also been developed to examine company performance in the construction field. For example, Koksal and Arditi (2004) examined several factors to determine the financial health of construction companies in the United States, Chan et al. (2005) used selected ratios to evaluate the financial performance of construction firms in Hong Kong and Abidali and Harris (1995) developed a model to predict construction company failure using several ratios in the UK.

However, perhaps one of the most well-known models is Altman's Z-score (Altman, 1968), which, according to Murray and Kulakov (2022), has become a template for new models to forecast the viability of a company. This model, developed by Edward Altman in 1968, is a linear combination of five common ratios weighted by coefficients. It uses data from a company's balance sheet and income statement to produce a single number, indicating how the company is performing. It comprises the following formula:

$$Z \text{ score} = 1.2T1 + 1.4T2 + 3.3T3 + 0.6T4 + 0.999T5$$

where

$T1$ = Working capital/Total assets
$T2$ = Retained earning/Total assets
$T3$ = EBIT/Total assets
$T4$ = Market value of equity/Book value of liability
$T5$ = Sales/Total assets

According to the model, if the computed Z-score value is more than 2.99, the company is performing well and is in a safe zone; if the value is between 1.8 and 2.99, the company is in a grey zone, which means that the company may be in some financial stress but not quite close to bankruptcy; and if the value is below 1.8, then the company is in a distress zone, which means that it will most likely become insolvent. The model has been tested on several large-scale data sets of failed companies globally and has proved reliable (Altman et al., 2017). Several studies, including Ng et al. (2011), have also used the model to examine insolvencies in the construction industry.

However, it is important to emphasise here that the reliability of any model or ratio analysis depends on data accuracy. Financial data used in such analysis could be inaccurate for various reasons: there may be incorrect calculations, the data may be outdated or accounting standards might vary across businesses and internationally, potentially affecting the comparability of financial ratios. Additionally, when analysing information included in a company's annual report, for instance, it is important to exercise caution. Company accounts are generally a year old and often need to be revised, and the information provided from one specific point in time may not accurately reflect its overall financial standing. It can lead to decisions based on outdated data, impacting project planning and budgeting.

Financial accounts, while crucial for understanding the financial health of construction companies, are also susceptible to manipulation. Construction companies, facing intense competition and the pressure to appear financially robust, may manipulate accounts to present a more favourable picture to investors, lenders and stakeholders (Mutti & Hughes, 2002). The manipulation can include practices like recognising revenue too prematurely or deferring expenses. Moreover, companies might use creative accounting techniques to obscure debts or overstate assets, making it challenging for external parties to ascertain the true financial position of the company. So, to rely solely on financial account and ratios drawn from them without considering the context under which they have been produced could lead to a misinformed assessment of a company's financial health, which underscores the importance of a careful review of the inputs used in the analysis.

8.8 Case Study: Carillion

Carillion, a major construction and services company in the United Kingdom, collapsed in January 2018. It cost UK taxpayers an estimated £148 million, and it has officially been labelled as one of the largest trading liquidations in the United Kingdom (Hajikazemi et al., 2020). With contracts for the construction and delivery of projects ranging from hospitals and schools to transportation projects, Carillion was the second biggest construction company in the

United Kingdom and a key strategic supplier to the government. Carillion's initial success was founded on its identity as a provider of integrated solutions, including construction, facility management and support services. It engaged in public–private partnerships extensively, establishing long-term revenue streams and government contracts. When it collapsed, Carillion was working on 420 government contracts.

According to a briefing paper by Conway and Mor (2018), the collapse of Carillion was a result of several financial missteps and poor management decisions over the years. Its demise has been attributed to the massive amounts of debt the firm had accumulated. From December 2009 to January 2018, Carillion's total debt rose from £242 million to approximately £1.3 billion.

Additionally, there were problems, such as falling revenues and a lack of cash flow. Carillion's revenue dropped by 2% between 2009 and 2016, and forecasts for 2017 indicated a further drop. This decline in revenue, coupled with increasing debt, created a very risky financial situation for the company. Moreover, Carillion had some serious problems with project delays, cost overruns and conflicts with clients, which all contributed to a fall in the company's declining profitability. However, they kept increasing their debt levels and increasing their risk-taking in an effort to grow. The briefing paper by Conway and Mor (2018) highlighted that despite the significant increase in borrowing, this debt was not used efficiently. The value of its long-term assets grew by only 14% between 2009 and 2017, compared to a 297% increase in debts. When Carillion announced in 2017 that their profits would be significantly impacted, this led to a subsequent 70% drop in share value.

Carillion was criticised for aggressive bidding on contracts and overly optimistic accounting practices. Their revenue and profits were mostly based on forecasts before the actual realisation of these funds. This approach led to significant issues when the anticipated profits did not materialise, as expected profits from several large projects turned into substantial losses. Even with all the issues it was facing, Carillion paid out £554 million in dividends in the years leading up to its collapse, which was substantially more than the cash it generated. This practice of paying dividends based on expected profits rather than actual cash flow essentially meant the company was borrowing to pay its shareholders, a strategy that was unsustainable in the long term (Hajikazemi et al., 2020). These factors combined to create a situation where Carillion was over-leveraged, lacked liquid assets and could not sustain its operations.

The company had developed a bad reputation for delaying payments to subcontractors and creditors to manage their short-term liabilities (Hajikazemi et al., 2020). The supply chain finance scheme launched by the UK government in 2012 allowed companies to submit invoices to any participating banks in return for early payment for a fee; Carillion used this scheme and would owe the money to the banks. However, no provision for this was recorded in the consolidated financial statements. Consequently, the number of trade creditors increased considerably.

Furthermore, Carillion created the image of being a successful company despite its unstable financial performance. Their annual report painted a picture to the audience in the investment market that they were a dynamic, growing organisation. A report by S&P Global (2018) that came out after Carillion's collapse highlighted the inaccuracy of their financial statements. The finding from the report suggested that accounting discretion, particularly with respect to the presentation of current liabilities, was distorted. According to Murray and Kulakov (2022), if all current liabilities were to be included, Carillion's current ratio could have been below 1, indicating a high risk of failure.

The collapse of Carillion, which was caused by financial mismanagement and over-optimism, serves as an example of the risks of poor financial management. Their aggressive expansion, supported by unrealistic accounting and debt accumulation, led to an unsustainable business model. Despite projecting success, the reality of its financial instability and reliance on future profits, not actual cash flow, was a major contributor to the company's failure. The case study highlights the importance of good financial management practices for construction companies involved in long-term projects.

8.9 Conclusion

Good financial management is fundamental for construction companies, as it enables them to navigate the industry's unique challenges and risky environment. Construction companies operate within a complex landscape where traditional financial methods need to be employed in greater depth due to the industry's distinct payment structures and project-based nature. It requires a tailored approach recognising the importance of strategic planning, efficient resource utilisation and contingency planning. Effective financial management for construction companies is thus more than maintaining profitability; it also ensures the sustainability of the business in the face of changing market conditions and project uncertainties.

This chapter reviews some of the fundamental concepts of financial management that are relevant in the context of construction companies. It includes discussions on funding sources, financial accounting, cash flow management, forecasting and long-term financial planning and ratio analysis. Understanding and effectively implementing these concepts is fundamental to managing successful construction companies. The case study of Carillion illustrates the consequences of poor financial management and the perils of neglecting fundamental financial principles.

As the construction sector undergoes ongoing transformations, the financial strategies implemented by companies must also adapt accordingly. Future success is contingent upon embracing innovation, adopting best practices and learning from past errors. In essence, effective financial management serves as the critical element that binds the diverse components of a construction company, guiding it towards enduring profitability and long-term sustainability.

References

Abidali, A. F., & Harris, F. (1995). A methodology for predicting company failure in the construction industry. *Construction Management and Economics, 13*(3), 189–196.

Altman, E. I. (1968). Financial ratios, discriminant analysis and the prediction of corporate bankruptcy. *The Journal of Finance, 23*(4), 589–609.

Altman, E. I., Iwanicz-Drozdowska, M., Laitinen, E. K., & Suvas, A. (2017). Financial distress prediction in an international context: A review and empirical analysis of Altman's Z-score model. *Journal of International Financial Management & Accounting, 28*(2), 131–171.

Australian Securities and Investments Commission. (2023). *Insolvency statistics.* https://asic.gov.au/regulatory-resources/find-a-document/statistics/insolvency-statistics/insolvency-statistics-current/

Batra, R., & Verma, S. (2014). An empirical insight into different stages of capital budgeting. *Global Business Review, 15*(2), 339–362.

Chan, J. K., Tam, C., & Cheung, R. K. (2005). Construction firms at the crossroads in Hong Kong: Going insolvency or seeking opportunity. *Engineering, Construction and Architectural Management, 12*(2), 111–124.

Conway, L., & Mor, F. (2018). *The collapse of Carillion* (House of Commons Briefing Paper 08206).

CreditWatch. (2022). *Cracks in the foundation.* https://creditorwatch.com.au/webinars/topic/cracks-in-the-foundation-construction-white-paper-2022/

Cui, Q., Hastak, M., & Halpin, D. (2010). Systems analysis of project cash flow management strategies. *Construction Management and Economics, 28*(4), 361–376.

Gitman, L. J., Juchau, R., & Flanagan, J. (2015). *Principles of managerial finance.* Pearson Higher Education AU.

Hajikazemi, S., Aaltonen, K., Ahola, T., Aarseth, W., & Andersen, B. (2020). Normalising deviance in construction project organizations: A case study on the collapse of Carillion. *Construction Management and Economics, 38*(12), 1122–1138.

Halpin, D. W., & Senior, B. A. (2009). *Financial management and accounting fundamentals for construction.* John Wiley & Sons.

Higham, A., Bridge, C., & Farrell, P. (2016). *Project finance for construction.* Taylor & Francis.

Ive, G., & Murray, A. (2013). *Trade credit in the UK construction industry: An empirical analysis of construction contractor financial positioning and performance.* Department for Business, Innovation and Skills. https://assets.publishing.service.gov.uk/media/5a7b87c240f0b645ba3c4e9f/bis-13-956-trade_credit-in-uk-construction-industry-analysis.pdf

Khosrowshahi, F., & Kaka, A. (2007). A decision support model for construction cash flow management. *Computer-Aided Civil and Infrastructure Engineering, 22*(7), 527–539.

Koksal, A., & Arditi, D. (2004). Predicting construction company decline. *Journal of Construction Engineering and Management, 130*(6), 799–807.

Lam, K. C., & Oshodi, O. S. (2015). The capital budgeting evaluation practices (2014) of contractors in the Hong Kong construction industry. *Construction Management and Economics, 33*(7), 587–600.

Lowe, J. (1997). Insolvency in the UK construction industry. *Journal of Financial Management of Property and Construction, 2*(1), 83–107.

Murray, A., & Kulakov, A. (2022). Using financial concepts to understand failing construction contractors. In R. Best & J. Meikle (Eds.), *Describing construction* (pp. 249–275). Routledge.

Mutti, C. d. N., & Hughes, W. (2002, September 2–4). *Cash flow management in construction firms* [Paper presentation]. 18th Annual ARCOM Conference, New Castle, United Kingdom.

Navon, R. (1996). Cash flow forecasting and updating for building projects. *Project Management Journal, 27*, 14–23.

Ng, S. T., Wong, J. M., & Zhang, J. (2011). Applying Z-score model to distinguish insolvent construction companies in China. *Habitat International, 35*(4), 599–607.

Öcal, M. E., Oral, E. L., Erdis, E., & Vural, G. (2007). Industry financial ratios – Application of factor analysis in Turkish construction industry. *Building and Environment, 42*(1), 385–392.

Odeyinka, H. A., Kaka, A., & Morledge, R. (2003, September 3–5). *An evaluation of construction cash flow management approaches in contracting organisations* [Paper presentation]. 19th Annual ARCOM Conference, Brighton, United Kingdom.

Odeyinka, H. A., Lowe, J., & Kaka, A. (2008). An evaluation of risk factors impacting construction cash flow forecast. *Journal of Financial Management of Property and Construction, 13*(1), 5–17.

Peterson, S. J. (2013). *Construction accounting and financial management* (Vol. 2). Pearson.

Shash, A. A., & Qarra, A. A. (2018). Cash flow management of construction projects in Saudi Arabia. *Project Management Journal, 49*(5), 48–63.

S&P Global. (2018). *Carillion's demise: What's at stake?* https://www.infrapppworld.com/report/carillion-s-demise-what-s-at-stake

Ye, S., & Tiong, R. L. (2000). NPV-at-risk method in infrastructure project investment evaluation. *Journal of Construction Engineering and Management, 126*(3), 227–233.

9   Risk Management

Kumar Neeraj Jha

9.1 Introduction

Effective risk management is critical for any business to sustain its competitiveness and resilience. The dynamic nature of the construction industry, marked by technical advancements, supply chain disruptions, regulatory environment changes and evolving market needs, presents complex issues that require a robust risk management framework and practices.

Al-Bahar and Crandall (1990) defined *risk* as a function of the uncertainty of an event and the potential gain or loss resulting from it. Ward and Chapman (2003) argued that risk management should analyse all prospects of possible risks, not only threats. The Project Management Institute (PMI, 2019), incorporating both dimensions – that is, threat and opportunity – defined risk as an uncertain event or condition that has a positive or negative effect on one or more organisational objectives if it occurs.

An organisation's long-term objectives, such as becoming a market leader, achieving sizeable/target market share, etc., are supported by short-term and medium-term objectives, such as annual financial growth targets, year-on-year customer growth benchmarks, etc. These objectives help management evaluate the organisation's productivity levels, business strategy efficacy and performance. Positive risks are opportunities that, upon occurrence, support the organisational objectives, and negative risks, on the other hand, are threats that, upon occurrence, will act against the organisational objectives.

Risks, especially in construction companies, commonly impact their performance and threaten to limit their growth potential. The organisation's performance, in different contexts, is defined based on two significant aspects; that is, economic performance, such as profits, sales, return on investments for shareholders, etc., and operational performance, such as customer satisfaction, timely completion, within budget, fit-for-purpose, customer satisfaction, zero accident and planned deliverable output. Risk, if it occurs, can impact the organisation's performance directly. In other words, every unplanned or unprecedented event will have repercussions that may improve or impair the organisational performance parameters.

Risk has a cause and, if it occurs, has a consequence (PMI, 2019). Causes are the circumstances or events, either existing or anticipated, that might give rise to risk. On the other hand, a consequence, effect, or impact comprises the conditional future events or conditions that affect the organisation's objective(s) directly if the associated risk occurs. In an organisation, a risk can arise from all possible sources, including individual portfolios, project domains and other internal and external factors. Organisational risks include uncertain but distinct events that can be clearly defined and less specific or non-distinct conditions that may cause uncertainty.

The chapter focuses on both enterprise-level and project risk management as they go hand in hand in project-based construction companies. It begins by explaining what risks mean for

DOI: 10.1201/9781003223092-9

construction companies and the importance of risk management. Next, it discusses various processes involved in managing risks, followed by a review of the statutory and regulatory norms governing risk management. Finally, the chapter presents a case study of one of the largest construction conglomerates, highlighting its risk assessment processes.

9.2 Typical Risks for Construction Companies

Risk in the context of a construction company can find its genesis in various sources, such as political influence, financial uncertainties, environmental constraints, legal liabilities, technological issues, strategic management errors, accidents and natural disasters. It could arise from factors internal or external to the construction company. Internal risks are those that find their inception within the construction company; for example, quality issues, delays in decision-making and insufficient capital. On the other hand, external risks are those whose genesis is beyond the company's boundaries, such as technological disruptions, law or policy reforms and delays in government approvals. The root cause lies in inherited variability, such as skill and specialisation requirements, engagement with multiple stakeholders, supply chain constraints, outdoor work environment and a wide range of activities and stakeholders requiring coordinated effort.

Construction companies increasingly rely on enterprise risk management (ERM) to manage risks effectively at the organisational level. ERM offers a structured and holistic approach by taking into account risks that emerge in projects as well as those that are related to the company's strategic goals and operations (McGeorge & Zou, 2012). Construction enterprises may be confronted with a wide variety of risks. ERM-focused risks can be divided into strategic risks, market risks, operational risks, financial risks and compliance risks (Liu et al., 2011).

Strategic risks are those that can impact a company's long-term goals. Construction companies may face strategic risks when their long-term plans are threatened by any changes in the industry or external environment. These risks may also be related to other factors that hinder companies from implementing their strategies. An example of a strategic risk for a construction company might be the potential loss of market share to competitors who are more successful in adopting sustainable building practices, potentially making the company's traditional construction method-focused strategy less appealing to environmentally conscious clients.

Market risks in construction involve the uncertainty of the external market environment, including changes in demand and price fluctuations. Market risks are directly related to economic conditions, client preferences and technological advancements in the industry. Risks associated with supply chains may also fall under market risks when fluctuations in the supply chain directly impact the market dynamics of the industry. For instance, a shortage of key building materials like steel or timber due to global supply disruptions can increase material costs, which then affects the pricing strategies of construction companies, potentially making their services less competitive in the market.

Operational risks are those that may impact the company's day-to-day operations. Operational risks in construction are generally related to internal processes, systems and human resources. They may also relate to the cumulative risks from ongoing projects that might affect the smooth functioning of the company. A simple example of operational risk is a natural disaster that damages the warehouse where a company's inventory is stored. Another example might be the risk posed by integrating new technologies into existing processes. While new technology may provide efficiency, its implementation can be disruptive, requiring significant training and adjustments in operational processes.

Financial risks involve a wide range of challenges related to the management of finances, both at the project level and across the company's entire portfolio. Key aspects include securing adequate project financing, managing cash flow effectively and meeting debt obligations. For example, a construction company may face financial risk if it overextends itself by simultaneously taking on multiple large projects without securing sufficient capital. This situation can lead to cash flow issues. Similarly, if the company has taken on significant debt to finance these projects, fluctuations in interest rates could put a strain on the company's finances.

Compliance risks for construction companies involve the risk of failing to adhere to laws, regulations and standards. An example is the evolving landscape of environmental regulations where governments impose stricter sustainability standards. Construction companies must comply with these new regulations, which might involve using new materials, standards and practices. Non-compliance may not only lead to legal repercussions but can also impact the company's reputation and chances of securing new contracts.

9.3 Risk Management

The roots of the risk and its management lie deep in history. However, the formal study of risks and their management began after World War II. Risk management has a long association with protecting individuals and companies from various losses associated with accidents through market insurance (Faber, 1979). It has comprehensive coverage across multiple industries and has been popular among researchers globally. Construction risk management was relatively untapped until the mid-1980s. It gained popularity during the mid-1990s. Risk assessment and modelling gained momentum in the 1990s, and, since then, risk analysis has been one of the major research areas in construction (Edwards & Bowen, 1998).

In the context of project management, Al-Bahar and Crandall (1990, p. 534) defined *risk management* as a 'formal orderly process for systematically identifying, analysing, and responding to risk events throughout the life of the project to obtain the optimum or acceptable degree of risk elimination or control'. In the context of construction management, Banaitiene and Banaitis (2012) explained risk management as a comprehensive and systematic way of identifying and analysing risk, improving the construction management process and effectively using resources. Risk management involves strategies, tools and techniques that help construction professionals and companies maximise the probability and consequence of positive events and minimise the likelihood and impact of adverse events.

PMI (2019) defined risk management as a systematic planning process for identifying, analysing, responding to and monitoring project risks. It aims to develop means to achieve corporate objectives, realise strategic vision and value creation and predict risks and their impact. Risk management is an integrated framework that covers all organisational levels. Its planning and implementation can be at any level; that is, enterprise, portfolio, program or project level; however, objectives and techniques may vary.

Construction companies are tasked with the dual responsibility of managing risks at both the project and enterprise levels. Given the inherently risky nature of construction projects, these companies must confront and manage the cumulative risks stemming from all their ongoing projects. This complex challenge demands the implementation of sophisticated risk management frameworks capable of addressing the multi-faceted risks cohesively and effectively.

The aim of risk management should not be to eliminate all the threats, as it is not always feasible or possible in practice; instead, it must identify and analyse the risks for informed decision-making and reduce the severity of various risks. Many business and operational concerns can be addressed through proper risk management, involving proactive planning and mitigating or avoiding consequences (Schieg, 2006).

9.4 Importance of Risk Management

Among many others, the weakness in implementing prevailing risk management practices was one of the significant contributors to the 2008 global recession. Better risk management practices could have shortened the duration and reduced its severity (Harner, 2010). Similarly, a global survey on infrastructure projects during COVID-19 indicated that risk management processes, irrespective of the country's development status, were not adequately adapted or were ineffective in meeting project objectives (McKinsey, 2020).

Many construction companies fail to implement a comprehensive and systematic approach to risk management. Previous studies indicate that only specific projects that are high value or innovative include risk management during project appraisals (Burchett et al., 1999; Gupta, 2011; Lundy & Morin, 2013; Pai & Varma, 2020; Pike & Ho, 1991). In 2018, a KPMG report concluded that lack of comprehensive upfront planning and risk management, lack of maturity of project management processes, scarcity of skilled workforce and non-collaboration across stakeholders were among the main factors contributing to time and cost overruns in the construction industry (PMI-KPMG, 2019). Furthermore, the report indicated that the unavailability of trained staff, inadequate technology usage and ad hoc planning contributed to nearly 41% of cost overruns. In comparison, 29% of cost overruns were due to weak risk management and delayed decision-making. McKinsey's report on infrastructure projects during COVID-19 found similar results (McKinsey, 2020).

PMI (2019) specified risk management as one of the core requirements of project management in construction, irrespective of the nature or value of the project. It recommended having risk management planning at all levels of organisation operations; that is, enterprise, portfolio, program and project. Developing an appropriate project risk management plan and its successful implementation could improve decision-making. Further, it could support enhanced profitability and earnings, better risk reporting and communications, better resource allocations, reduced earning volatilities and improved project performance parameters (PMI, 2019).

Project risk management and ERM are two interrelated domains of risk management that are crucial for construction companies. Project risk management focuses on identifying and mitigating risks specific to individual construction projects. ERM functions at the organisational level, providing a holistic approach to managing various risks, including strategic, market, operational and financial risks, impacting the company's short- and long-term objectives. Integrating project risk management into ERM is vital for effectively managing a construction company's overall risk portfolio. Various insights into project-specific risks enhance ERM processes, facilitating the development of a comprehensive risk strategy and better resource allocation.

On the other hand, ERM frameworks can improve project risk management by offering efficient risk information sharing across projects, departments and portfolios. This relationship ensures that risk management is not reactive but a strategic tool aligned with the company's overall goals, risk attitude and risk management capabilities. The general consensus is that companies can enhance performance by implementing the ERM framework (Liu et al., 2011).

In short, implementing a proper risk management process allows companies to improve decision-making and performance, minimise management by crisis, anticipate and manage change, proactively implement preventive actions, increase chances to realise opportunities and support organisational agility and resilience (Caltrans, 2022; PMI, 2019).

9.5 Risk Management Process

An organisation is exposed to various risks and uncertainties in its operations and requires planning to manage them to maximise opportunities and minimise threats (PMI, 2019). Risk management requires meticulous planning and effort towards risk identification, assessment and response planning. Organisations must be able to change their strategies to address the challenges that can deter their performance and objectives. In many countries, organisations are mandated to define a risk management policy and establish processes for practical implementation, irrespective of the nature of the business. Moreover, the policy must be reviewed periodically rather than as a one-time exercise (Committee of Sponsoring Organisations [COSO], 2009).

PMI (2019) recommended developing a risk management plan for each project in line with the enterprise policy and its implementation in every project (irrespective of its nature or value) to identify potential uncertainties and probable impacts on the project performance. It also defined the typical stages of risk management as risk management planning, risk identification, risk analysis, risk response planning, risk monitoring and control and risk closure. These management stages are consistent across both ERM and project-level risk management; however, while the focus at the project level is on specific risks inherent to individual projects, ERM adopts a macro, holistic approach, concentrating on risks at the organisational level.

9.5.1 Risk Management Planning

Risk management planning involves the development of a comprehensive risk management strategy that enables the risk management process implementation while integrating the same with the enterprise or project objectives. The risk management plan development is a systematic and iterative process subjected to periodic reviews and revisions to keep the plan aligned with the organisational goals and objectives and suit the project-specific requirements.

The risk management plan must cover the fundamental aspects of the business; that is, strategy, operations, marketing and finances. It defines an exhaustive list of anticipated risks, tools, techniques and methods for risk analysis and reporting, appropriate mitigation measures and the people responsible for addressing them. It should also define the risk tolerance and risk appetite of the organisation.

Risk tolerance can vary significantly from organisation to organisation; for example, a large construction company may have a higher tolerance for operational risks than a smaller company. PMI (2019) explained risk appetite as the degree of uncertainty an organisation or an individual is willing to accept in anticipation of a reward. A construction company with a high risk appetite will be prepared to accept a high level of risk in pursuit of its objectives before taking any actions to reduce the risk. A clear understanding of the company's risk tolerance and risk appetite will help decide the appropriate risk response strategy and actions.

Ward and Chapman (2003) argued that both threats and the opportunities should be considered in risk planning. Nasirzadeh et al. (2008) concluded that risks have a cascading impact; hence, analysing risks and their interactions will help quantify the implications. Risk management planning should be done at the beginning of an activity or a decision so that the team is ready with all alternatives to deal with the risk as and when it occurs. Looking for the option after the risk (opportunity or threat) has arisen could result in a significant loss. Timely action could prevent the anticipated loss or provide unexpected benefits. For example, developing a detailed risk management plan during the project appraisal stage is crucial for its success. Further, the plan must be followed and monitored for the entire project life cycle; that is, pre-construction/appraisal, procurement, execution, closure and operation and maintenance phase.

Therefore, construction companies must develop and follow a robust risk management plan to manage various risks appropriately as and when they occur. The risk management plan should be a comprehensive document defining all the possible risks, mitigation strategies, action plans in case of their occurrence and the roles and responsibilities of relevant persons or teams for appropriate redressal. It is essential to review the risk management plan frequently to incorporate and improvise upon the current situation.

9.5.2 Risk Identification

Risk identification is a systematic process to identify all possible risks that may arise and could prevent the organisation from achieving its objectives. Risk identification should incorporate operational attributes, stakeholder attitudes, strategic goals and eventualities that could impact business operations. The process recommends identifying risks to cover the broadest range possible. It would help analyse risk impact to a greater extent and to assign suitable and adequate measures. Risk identification should be a continuous exercise and must involve a variety of expertise to arrive at an exhaustive list.

Some of the risk identification techniques are described below (Siraj & Fayek, 2019).

1. Brainstorming: Brainstorming is an interactive and team-based approach where a facilitator moderates discussion among the members with the domain expertise to arrive at an exhaustive list of risks. The activity aims to cover all the potential uncertainties without judging their importance at this stage. The successful outcome depends upon the expertise and skills of the members involved in the activity and the facilitator.
2. Checklist analysis: The checklist-based risk identification process is helpful to identify standard and routine risks. In this case, the facilitator already has a definitive list of risks. The user will have to mark the threats applicable to the specific context. In the case of a non-standard situation, this process may constrain creative thought and may not be effective.
3. Diagramming technique: Techniques such as cause–effect diagrams, process flow charts and influence diagrams can help identify risks not readily identified in verbal discussions. These techniques are also helpful in the identification of the risk genesis.
4. Delphi technique: This method systematically collects responses or judgements from anonymous respondents/experts on the subject through a carefully designed questionnaire survey. The results are collected, summarised, analysed and revisited in multiple rounds until the achievement of consensus. It is an effective exercise for obtaining the results without bias, as the respondents remain anonymous throughout the process.
5. Surveys and interviews: Surveys and interviews can help obtain information on risks from different people. These can supplement other techniques to improve the effectiveness of the risk identification process if effectively designed and implemented.
6. Documentation review: The process involves the study of relevant documents, such as government policies, technical papers and document records, and identifying potential areas that could generate risks.

The risk identification process does not always need to include only the individuals within an organisation; participants can be from outside the organisation. Sometimes organisations engage special agencies or consultants to facilitate the process. Techniques such as brainstorming, the Delphi technique and deliberation involve a high level of experience and subject expertise through practising professionals; hence, they can be useful tools in the risk identification process, especially in construction companies. However, these sessions should involve a cross-functional group of individuals of varying experience levels because mixing up the functions and the experience level could offer the widest array of ideas (Becker, 2004).

9.5.3 Risk Analysis

After identifying various risks, it is essential to have a detailed analysis of the risks and their bearing on the organisation's objectives and performance. Risk analysis mainly focuses on two aspects of a risk, namely, the probability or likelihood of its occurrence and the likely impact, to determine the exposure of a risk. Likelihood refers to the chances of a risk happening with the current controls in place. In contrast, risk impact or severity or consequence refers to the expected harm or adverse effect that may occur due to exposure to the risk. The severity of certain risks varies across projects and companies. For example, an insignificant risk in a house construction project could be catastrophic in a refinery construction project.

The risk analysis categorises and prioritises risks by determining their importance, impact on the objectives and likelihood of occurrence. The resulting risk matrix reveals the priority of all the identified risks by assessing the risk exposure and assigning the risk rating. The risk rating is the rank given to a risk by evaluating its impact and determining the likelihood of its occurrence. The risks with a high probability of occurrence and high impact value will have a high risk exposure and, thus, a higher risk rating. Table 9.1 provides a typical categorisation of risks and risk ratings.

The enterprise or project risk matrix provides a guiding principle for risk prioritisation. It is an inherent component of the organisation's ERM policy document. It provides a clear and visible distinction among risks with different priorities and ratings by providing a colour code. For instance, red (or dark grey) can be assigned to high-priority risks, while green (or light grey) refers to low-priority risks. Figure 9.1 provides a typical risk matrix. Using the matrix, the level of risk can be calculated by finding the intersection between the likelihood and the consequences.

A proper risk analysis provides a solid basis for developing appropriate organisational risk responses. More attention and resources are devoted to managing the risks that have higher risk ratings. Depending on the nature of the business, organisations adopt different analysis techniques. Quantitative risk analysis techniques can provide numerical estimates for the combined effect of identified risks on the organisational or project objectives.

Table 9.1 Typical risk severity categories and ratings

Category	Rating	Description
Catastrophic	5	**Catastrophic** concern and impact on strategy or operation activity
Major	4	**Major** concern and impact on strategy or operation activity
Moderate	3	**Moderate** concern and impact on strategy or operation activity
Minor	2	**Minor** concern and impact on strategy or operation activity
Insignificant	1	**Insignificant** concern and impact on strategy or operation activity

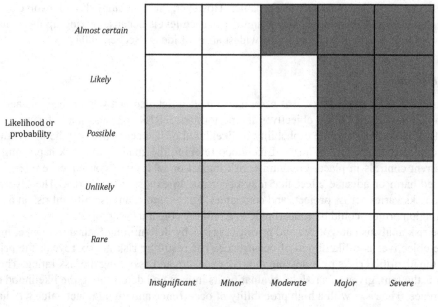

Impact or severity or consequence

Figure 9.1 A typical risk matrix.

It may be noted that finding probabilities for risk occurrence and their respective consequences on organisational objectives and performance can be subjective if they only rely on the competency of the people involved in the process. Using past event records and quantitative data could eliminate some of the subjectivity involved in the expertise or experience-based risk assessment and provide a more objective approach towards risk management. For smaller projects, qualitative assessments are generally accepted, but for larger and more complex projects, more rigorous assessments using quantitative methods are preferred.

Companies should maintain a risk register that provides comprehensive coverage of the risk management plan. It incorporates all the identified risks, their detailed description, likelihood, impact or consequence, risk rating, risk indicators, risk mitigation measures and the person, team or stakeholder responsible for its mitigation, monitoring and reporting. The risk register includes all the possible risks to the construction company and its project. Periodic review of the risk register is necessary to incorporate new risks and keep it updated. Maintaining separate risk registers for each project is recommended to incorporate project-specific risks and associated information.

9.5.4 Risk Response

The risk matrix, determined as a part of the risk analysis process, provides a list of all the risks with their associated risk criticality or rating and forms the basis for developing appropriate risk responses. Risk response planning intends to explore options and determine actions to enhance opportunities and reduce threats to the organisation's objectives. It includes identifying and assigning roles and responsibilities to different individuals, teams, groups and stakeholders for different risks to ensure the appropriate redressal of the risks, assign adequate resources and fix accountability. The response strategy broadly entails choosing the appropriate risk mitigation for each risk – that is, avoidance, transfer, reduction, and retention or acceptance – either individually or in combination.

1. Risk avoidance or elimination: Preventable risks are generally internal risks arising from within the company. Examples include the risks from employees' illegal or unethical actions (e.g., bribing) and the risks associated with routine operational processes. Companies can eliminate such risks through active prevention, such as enforcing proper policies and procedures to guide employees' professional conduct and monitoring operational processes. Companies receive no strategic benefits from taking these risks as, over time, these will diminish the company's value (Kaplan & Mikes, 2012). Although it is not always possible to completely eliminate all risks, the possibility should not be overlooked.
2. Risk reduction: Risk reduction involves implementing a pre-defined appropriate action plan to minimise the likelihood of the risk or its impact or both; for example, identifying alternate technology, subcontractor screening, competitive vendor selection process, etc.
3. Risk transfer: In some cases, risk can be transferred to other parties, usually by contract. It may involve re-assigning the accountability and responsibility to another party willing to accept it through means such as insurance and contractual clause negotiations, etc.
4. Risk acceptance: Sometimes, based on the likelihood and severity of a risk, retaining it completely or a portion of it may be cost-effective even though other methods of handling the risk are available. Acknowledging a specific risk and its deliberate acceptance without making any efforts for its control can be an effective response, especially when the probability of occurrence and the associated impact or the resulting risk rating is quite low. However, it is essential to monitor such risks as they may require a dynamic or more elaborate response to mitigate them if the circumstances surrounding them change, increasing their risk rating.

It may be noted that the above approaches are not mutually exclusive. Depending on the risk's nature and attributes, the risk response options may be applied individually or in combination. Furthermore, it is important to remember that the response strategy for risk could vary in different scenarios; accordingly, construction companies must perform independent risk management planning and impact evaluation for every project or business decision.

9.5.5 *Risk Monitoring and Control*

Risks require regular monitoring, and appropriate control measures should be developed and implemented for effective mitigation, as discussed in the previous section. Responding to the threat as soon as it emerges is essential to minimise its impact and suppress its consequences. Thus, deploying an adequate reporting and review mechanism and documentation system is crucial to control the risks effectively.

A risk treatment or response may introduce additional risks known as secondary risks (PMI, 2019). In other words, secondary risks are generated once the risk response plan is implemented. Similarly, residual risks are those risks that remain after the planned response of risk has been implemented, as well as those that have been deliberately accepted by the company (PMI, 2019). Companies accept residual risks if they fall within their risk tolerance level or if they do not have a reasonable response. Therefore, companies must continuously review and monitor their risk response practices to identify the efficacy of risk responses as well as secondary and residual risks.

Risk monitoring and control involve keeping track of the identified and emerging risks, monitoring residual risks, ensuring the execution of the risk response plans and evaluating their effectiveness in maintaining the organisation's objectives. It requires the development of the necessary risk governance structure, allocating resources and implementing proper formats, checklists, protocols and reporting.

9.5.6 Risk Closure

Risk closure involves formally completing or closing the risk management process in a project, program or business activity. It includes recording the information and recommendations for improving the overall risk management process (i.e., lessons learned). The recorded data should be made available to other employees and teams to provide a basis for risk management decision-making in the future. The data can help develop data-driven models for a more precise approach to risk management.

9.6 Regulatory Framework and Governance for Enterprise Risk Management

Globally, risk management is recognised as an integral part of corporate governance. Companies operating in a country, irrespective of the nature of business, come under the purview of corporate laws stipulated by the statute through the respective governing acts and rules for operation. These acts and regulations define the framework for risk management in an enterprise. The core of the regulations is to provide a mechanism to protect the organisation from the risks that can derail its objectives, functions and performance.

The regulatory requirements are mentioned as part of the company law (Austria, Germany, Turkey and Japan), listing rules and regulations (India, the United Kingdom and the United States) and stock exchange laws (Mexico), usually in connection with the audit or internal control function (Organisation for Economic Co-operation and Development [OECD], 2014). Different countries adopt different approaches to governance towards ensuring targets. The regulatory approach could be centralised, decentralised, dual model and decentralised with coordinating agencies. For example, countries like China, Finland, France, Sweden, etc., adopt a centralised approach through a ministry or holding company; in contrast, countries like Argentina, Colombia and Mexico adopt the decentralised governance approach through multiple entities (OECD, 2018). In India, the Department of Public Enterprise acts as a coordinating agency and governs the operation of all public sector enterprises.

The governing agency or entity, in line with the requirements stipulated under the statute, sets principles and regulations specific to the functioning of the enterprises. Moreover, these governing entities issue and modify these regulations to ensure that the functioning of enterprises is current with the market requirements and aligned with the government's vision. COSO (2009) recommends that the risk management plan be defined to address the enterprise's primary objectives; hence, it is the responsibility of the company's top management or board. COSO advocates that the involvement of board members and senior management to further reinforce the implementation. COSO also recommends the appointment of a chief staff or leader and defines its responsibilities in the ERM framework. Most countries mandating the practice of ERM follow COSO's recommendations on the chief leader's appointment (OECD, 2018).

In any case, the regulatory framework should consider the entire organisation and its functions as a unified entity focusing on the robustness of enterprise-level risk management and recognise the need to expand the horizon from financial to other aspects of business such as governance, operational, corporate behaviour, cybersecurity, etc. The regulatory framework could enable and demand a company to have a well-defined risk management policy and framework, including organisational objectives, processes, roles and responsibilities, formal risk management committee formation, implementation mechanisms and review and control for effective risk identification, analysis and mitigation. For instance, Figure 9.2 shows the governance framework used for enterprise-level risk management in India (Securities and Exchange Board of India, 2015).

The general framework recommends following a top-down approach, allowing clear communication, quick decisions and faster implementation. However, the effectiveness of the approach and its implementation solely depends upon the company's leadership, and it assumes that the company's leadership is aligned, on board, and open to the processes.

The risk management committee constitutes most of its members from the board of directors and reports to the board's audit committee. The committee is responsible for developing the risk management policy at the enterprise level, considering all organisational objectives and devising an organisation-specific framework for its implementation. The committee is empowered to seek information and feedback from employees and obtain legal or professional advice from an outside organisation. It may hire consultants with relevant expertise as deemed necessary. Furthermore, the committee reviews the effectiveness of the risk management implementation and improvises if necessary.

The chief risk officer (CRO) is a designated officer responsible for executing the risk management plan. It is the responsibility of the CRO to develop the strategy and processes in line with the risk management policy for its effective implementation. The CRO is a dedicated resource and exclusively conducts the organisation's risk management functions. The CRO administers the entire policy and establishes the system necessary for its performance, including risk response planning, developing risk strategies, monitoring and reporting and ensuring corporate

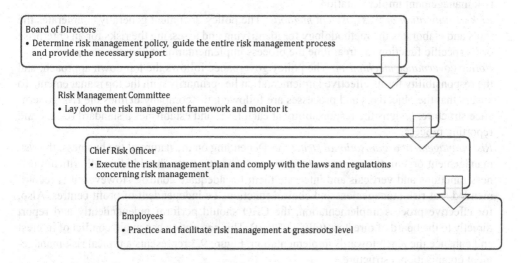

Figure 9.2 General risk governance framework.

Source: Securities and Exchange Board of India (2015)

governance–related statutory compliance. Concerning risk management, a CRO acts as a link between the operational-level employees and the top management of the organisation.

9.6.1 Risk Management Policy

Risk management policy development is primarily an enterprise function of ERM and is often informed by the statutory or regulatory guidelines for corporate governance. Usually, the policy is defined and documented at the enterprise level, while the functions, program and project-specific risk management guidelines align with it. Typically, an organisation's risk management policy depicts its risk management objectives, structure, processes and control mechanism. The approach, however, may vary with the nature of the business. The policy is deemed comprehensive and should provide clarity for its implementation. Also, the policy must include a review and reporting structure for a smooth flow of information to enable swift response.

Implementing an enterprise-wide risk management policy is crucial for managing construction companies successfully. Generally, the risk management policy, irrespective of its applicability at the enterprise level or the project level or the nature and volume of business, consists of the following:

Objective(s): The objectives of a risk management policy include establishing enterprise-wide processes and implementation, systematic assessment, compliance, communication, etc. They are stipulated to keep sustainable business growth and stability in the company.

Key areas of implementation: Companies perform various functions, operations, programs and projects to achieve their business goals. However, some operations or functions can be critical for sustaining the business's functionality and contributing directly to the business goals. Uncertainties and disruptions in these areas could be devastating for achieving business goals and therefore require systematic and robust risk management. Depending on the nature of the business, companies should identify such vital areas for more focused risk management implementation to have more effective outcomes. For example, the finance and accounting functions or pre-bid project risk assessments can be considered critical areas for risk management implementation.

Risk identification and assessment methods: The policy document generally enumerates the tools and elaborates the methodology for identifying and assessing the risks for an enterprise or its specific function to streamline the process implementation within the company.

Policy governance mechanism: The policy governance follows the top-down approach, and the responsibility for its effective implementation lies primarily with the top management. To ensure that the objectives and processes are followed, the company defines the risk governance structure, assigns the responsibility at each level and establishes a standard review and reporting process.

Risk management organisational structure: Depending on the nature of the business, the risk management organisational structure may vary. The structure should align all critical business functions and verticals and integrate them for adequate control. However, it is recommended that risk management and control functions be independent of profit centres. Also, for effective process implementation, the CRO should perform independently and report directly to the board of directors (OECD, 2010). This separation can avoid conflict of interest and enhance the reach towards implementation. Figure 9.3 represents a typical risk management organisational structure.

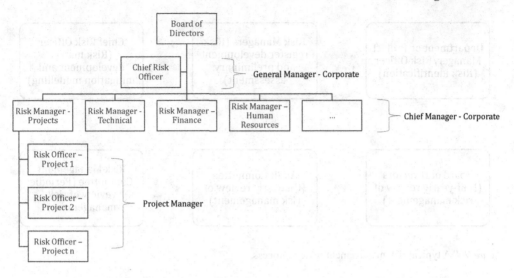

Figure 9.3 A typical risk management organisational structure.

The role and responsibilities of each member in the risk management organisational structure must be explicitly mentioned and communicated for better governance. The board of directors is responsible for developing the policy and its processes, establishing the necessary systems, ensuring enterprise-wide implementation, providing necessary resources and maintaining alignment with the organisation's objectives. Depending on the nature of the business, the board of directors performs the business functions through numerous sub-committees such as audit committees, remuneration committees, project committees, etc. For instance, the audit committee is responsible for managing financial oversight and reporting, safeguarding internal controls, minimising organisational risks, protecting shareholders' interests, etc. Sometimes, the audit committee operates through a dedicated sub-committee, the risk management committee, that administers the policy and its implementation. The project risk officers are responsible for identifying, assessing, monitoring and reporting risk in their respective project sites. The project manager can be assigned the responsibilities of a risk officer. The risk manager aggregates the risks from each project site related to their respective domains, performs the risk assessment and provides risk rating, monitoring and reporting. The risk cell, led by the CRO, performs a key role in risk planning and implementation and maintains the repository of risk-related data for the entire organisation.

Process review and reporting mechanism: The management information system reports the details of risk events and their consequences to relevant stakeholders in a time-bound manner. The system should enable quick identification of risks and deployment of appropriate risk response plans. It requires developing reporting formats, checklists, communication protocols (internal and external) and reporting frequency. Depending on the nature of the business and project value, the board or CRO may decide the review frequency. Figure 9.4 shows a typical risk management review process flow.

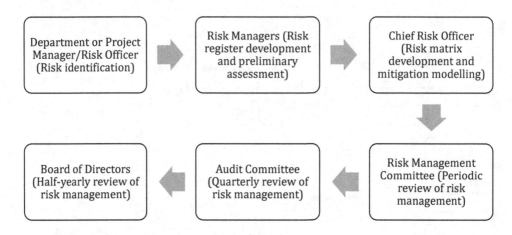

Figure 9.4 A typical risk management review process.

Policy review frequency and control: Developing a risk management policy and its implementation is continuous and requires periodic review. The policy document may also need regular updates to sustain business in the changing market conditions. The policy should define the policy review frequency. Furthermore, the policy document is central to the risk management process, so modifications and improvisations in the policy document are also controlled and must follow the document change control process. The policy document should define the custodian of the policy document. Usually, the risk management committee or designated CRO, as applicable, is the custodian of the policy document and is responsible for maintaining revision history and suggestive changes.

9.7 Risk Management in a Large Construction Company: A Case Study

Construction companies adopt various risk management strategies and techniques depending on the nature of their business, the nature of the projects and other attributes. Construction giants, which simultaneously execute several projects across various businesses and geographies, perform massive risk management and monitoring tasks for all projects. The following case study describes the risk management practices of a major construction conglomerate in India.

The organisation has an annual turnover of more than $20 billion. It executes over 1,000 projects concurrently across various businesses, such as infrastructure, material handling, hydrocarbons, defence, manufacturing, IT, etc., in different geographies ranging from low-income to advanced economies. As a result, it faces the immense task of assessing and monitoring risks for all the projects throughout their life cycles. The organisation has devised a standardised methodology for identifying and classifying risks to facilitate risk assessment across its businesses. The process is based on over 800 reviews conducted over the last several years.

The company considers risk management a central function and objective for all business verticals to achieve. Each business vertical of the organisation appoints a risk officer responsible for identifying and assessing possible threats to its construction projects, workforce and the organisation, including financial, legal, environmental and reputational risks. The officer works closely with project managers, environment, health and safety teams, human resources and legal teams and reports to the business unit head. While doing so, the risk officer develops the policies to protect assets and minimise mistakes, budget loss or public liability and thus plays a crucial role in bridging the risk management process gaps across various functions and roles. The ERM system was developed after integrating the inputs from all business verticals to highlight the top risks faced by various business units, independent companies, locations and the entire company. Based on large data sets of risks accumulated over time, predictive models have been developed for forecasting the success or failure of a project based on the likelihood of occurrence and impact of risks. Accordingly, mitigation plans are developed.

The company also surveys its senior employees to develop a list of generic risks. These risks are perceived as general business risks and are often included in the risk management framework as enterprise-level risks. In contrast, the project-specific risks are determined during the bidding stage and evaluated along with the generic risks. The critical analysis of survey responses is performed to reach a consensus on risk perception among the different business verticals, roles and domains. The company concluded one such online survey in December 2018. It intended to showcase the risk register to the stakeholders, such as the project and business managers, top management and investors, for their views on major risks and the adoption of standard taxonomy of risk identification in their ERM system.

A total of 458 construction managers responded to the survey. These respondents belonged to businesses such as building, defence, heavy civil, transportation, water treatment, power, heavy engineering, mineral handling, etc. The respondents were part of project management, finance and risk management, marketing, design and engineering teams of the abovementioned businesses.

The company identified a large number of specific risks and categorised them into 14 categories: (a) client-related; (b) contractual; (c) estimation; (d) approvals or clearances; (e) financial or commercial; (f) construction and operation; (g) engineering; (h) project resources; (i) supply chain; (j) strategic; (k) organisational issues; (l) political and regulatory; (m) health, safety, environment and quality (HSEQ); and (n) partners. The top 10 specific risks are shown in Table 9.2.

The statistical analysis of the risk responses provided various inferences about the risk perception by the respondents. Plotting averages of ranks of all risk categories against the standard deviation explored the degree of convergence in risk perception among the respondents. Furthermore, for analysis, the plot was represented in four quadrants representing the importance of risk. A high degree of convergence and high rank described high significance for the particular risk, and so on. Figure 9.5 presents rank averages for all risk categories on the Y-axis while the degree of convergence is on the X-axis.

For some of the risks, such as HSEQ risk, the degree of convergence is very high, which means that the perception of all the participants converged on the assessment or ranking of that risk. For risks such as political and regulatory, the participants' views varied widely and, as a result, the degree of convergence is low. This indicates that the impact of political and regulatory risks can be very project specific and therefore is not ranked similarly across the company. The bottom left quadrant, representing high convergence and high ranking, contains the most critical risks, such as financial or commercial, contractual and client-related risks.

The results also showed a difference in the perception among different functions. For example, marketing team members considered strategic risk to have a high impact. In contrast, the project management team perceived approvals or clearances as a high-impact risk compared to the marketing team. However, there was a convergence in perceptions across the various roles in the company for some risks related to HSEQ, partners, financial and contractual.

Table 9.2 Top 10 specific risks with their risk categories – Case study company

Rank	Specific risk	Risk category
1	Delays in getting site or land acquisition from the client	Client-related
2	Changes in design due to client requirements	Client-related
3	Estimation errors	Estimation
4	Incorrect site investigation	Estimation
5	Approval from local government or statutory bodies	Approvals or clearance risks
6	Delays in payment	Financial or commercial risks
7	Non-availability of workers	Project resources
8	Aggressive pricing	Strategic
9	Constraints due to internal processes	Organisational issues
10	Non-compliance with safety requirements	HSEQ risks

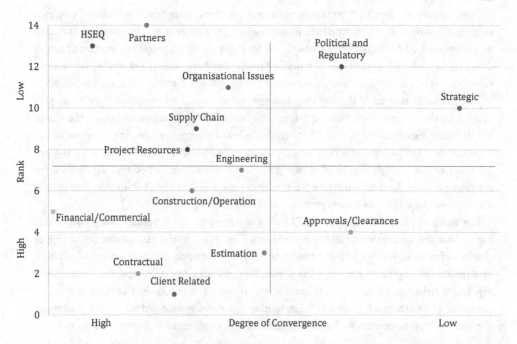

Figure 9.5 Risk ranking perception and degree of convergence among risk categories – Case study company.

9.8 Conclusion

Construction companies are subjected to a multitude of risks and uncertainties, as discussed earlier in the chapter. The aggressive competition across companies further amplifies business risks. Bid parameters and project timelines are increasingly becoming stringent, and projects are affected by economic volatility, supply chain disruptions, global events and inherent project-specific risks. Moreover, organisational risks have grown more complex than ever with the rapid pace of globalisation and the development of digital technology. Risks – for example, climate change, pandemics, environmental regulations, cyberthreats, etc. – pose various challenges in maintaining organisational performance and growth. Therefore, construction companies must implement robust risk mitigation strategies. Proper risk management is critical for success in the construction industry, and ineffective risk management may prevent construction companies from achieving their organisational objectives and optimum performance.

The chapter also discussed regulatory framework and governance. Risk management should be embedded into the organisation's culture. It should follow a comprehensive approach where the top management and board of directors, in line with the organisation's vision, goals and objectives, define the risk management policy for adoption by the company's programs, portfolios and projects. The management must ensure its enterprise-wide implementation through a proper policy and governance structure, allocation of sufficient resources, periodic reviews and changes when necessary. At the same time, employees are responsible for exercising the risk management processes in their work domains and providing feedback for improvements.

References

Al-Bahar, J. F., & Crandall, K. C. (1990). Systematic risk management approach for construction projects. *Journal of Construction Engineering and Management, 116*(3), 533–546.

Banaitiene, N., & Banaitis, A. (2012). Risk management in construction projects. In N. Banaitiene (Ed.), *Risk management – Current issues and challenges* (pp. 429–448). InTech.

Becker, G. M. (2004, October 23–26). *A practical risk management approach* [Paper presentation]. PMI® Global Congress 2004, Anaheim, CA, United States.

Burchett, J. F., Rao Tummala, V. M., & Leung, H. M. (1999). A world-wide survey of current practices in the management of risk within electrical supply projects. *Construction Management & Economics, 17*(1), 77–90.

Caltrans. (2022). *Project risk management. Project delivery directives.*

Committee of Sponsoring Organisations. (2009). *Strengthening enterprise risk management for strategic advantage.* North Carolina State University.

Edwards, P. J., & Bowen., P. A. (1998). Risk and risk management in construction: A review and future directions for research. *Engineering, Construction and Architectural Management, 5*(4), 339–349.

Faber, W. (1979). *Protecting giant projects: A study of problems and solutions in the area of risk and insurance.*

Gupta, P. K. (2011). Risk management in Indian companies: EWRM concerns and issues. *Journal of Risk Finance, 12*(2), 121–139.

Harner, M. (2010). Barriers to effective risk management. *Seton Hall Law Review, 40*(4), 1323–1365.

Kaplan, R. S., & Mikes, A. (2012, June). Managing risks: A new framework. *Harvard Business Review.* https://hbr.org/2012/06/managing-risks-a-new-framework

Liu, J. Y., Low, S. P., & He, X. (2011). Current practices and challenges of implementing enterprise risk management (ERM) in Chinese construction enterprises. *International Journal of Construction Management, 11*(4), 49–63.

Lundy, V., & Morin, P. P. (2013). Project leadership influences resistance to change: The case of the Canadian public service. *Project Management Journal, 44*(4), 45–64.

McGeorge, D., & Zou, P. X. W. (2012). *Construction management: New directions.* John Wiley & Sons.

McKinsey. (2020). *The next normal in construction.*

Nasirzadeh, F., Afshar, A., & Khanzadi, M. (2008). Dynamic risk analysis in construction projects. *Canadian Journal of Civil Engineering, 35*(8), 820–831.

Organisation for Economic Co-operation and Development. (2010). *Corporate governance and the financial crisis: Conclusions and emerging good practices to enhance implementation of the principles.* OECD Publishing.

Organisation for Economic Co-operation and Development. (2014). *Risk management and corporate governance.* OECD Publishing.

Organisation for Economic Co-operation and Development. (2018). *Ownership and governance of state-owned enterprises: A compendium of national practices.* OECD Publishing.

Pai, S., & Varma, R. (2020). Risk management in Indian infrastructure projects. *China Business Law Journal.* https://law.asia/risk-management-indian-projects/

Pike, R. H., & Ho., S. (1991). Risk analysis in capital budgeting: Barriers and benefits. *Omega, 19*(4), 235–245.

Project Management Institute. (2019). *The standard for risk in portfolios, programs, and projects.*

PMI-KPMG. (2019). *Revamping project management: Assessment of infrastructure projects and corrective recommendations for performance improvement.*

Schieg, M. (2006). Risk management in construction project management. *Journal of Business Economics and Management, 7*(2), 77–83.

Securities and Exchange Board of India. (2015). *Listing obligations and disclosure requirements.*

Siraj, N. B., & Fayek, A. R. (2019). Risk identification and common risks in construction: Literature review and content analysis. *Journal of Construction Engineering and Management, 145*(9), 03119004.

Ward, S., & Chapman, C. (2003). Transforming project risk management into project uncertainty management. *International Journal of Project Management, 21*(2), 97–105.

10 Digital Transformation

Abid Hasan

10.1 Introduction

Effective communication, information management and collaboration are the cornerstones of successful construction businesses and are often considered the primary steps in achieving efficiency and productivity. However, communication, information management and collaboration in the construction industry are relatively complex and much more challenging than in some other industries due to the temporary nature of work, dispersed locations and how the job, workers, teams and companies are organised.

At the industry level, a wide range of organisations must collaborate, communicate effectively and access and share accurate information on time despite working in a heterogeneous, highly fragmented and project-oriented business environment. At the organisational level, communication and information management within an organisation are challenging due to geographically dispersed teams and projects. At the project level, the mobile nature of construction professionals and multiple data generation points make communication and timely access to information difficult for individuals and teams and affect their ability to deliver construction projects successfully. Project and corporate head offices are often located in different regions, states or even countries and thus deal with spatial and temporal differences in communication and collaboration.

Consequently, to make timely decisions and achieve efficiency, the decision-making process in construction companies is often decentralised. Project teams are entrusted with the execution and management of construction projects with little control from the head office. They are also responsible for dealing with a multitude of stakeholders with varied needs and expectations. However, outdated practices and deficiencies in communication and information management in construction projects result in bad decisions with profound implications for individuals, teams and organisations (Hasan et al., 2018). Poor communication, low levels of collaboration and inefficient information management also lead to a lack of transparency and mistrust among the project participants, negatively affecting project performance and working relationships.

Industry data show that inaccuracies and inefficiencies in information and communication management in the construction industry can be very costly for companies. For instance, delays in accessing up-to-date data cause rework of up to 30% of the total work and waste more than one-third of construction professionals' time looking for project information, documents and other non-value-adding activities (Autodesk, 2021). The wasted time, effort and resources could be used for productive work. Therefore, effective communication, timely access to accurate information and a collaborative work environment are essential to ensure that decisions are informed by correct data and consider inputs from relevant stakeholders.

DOI: 10.1201/9781003223092-10

One of the solutions to improve communication and information management in construction companies and enhance collaboration among industry stakeholders is using digital technologies. Digital transformation of construction companies can remove information and communication management inefficiencies, resulting in considerable productivity gains and cost reductions (Ellis, 2020). As projects become more complex, construction companies operate in an information-intensive business environment. As a result, adopting digital technologies and advanced means of communication, collaboration and information management has become necessary to improve efficiency and productivity in various construction activities and business processes. The increased accessibility and affordability of various available technological solutions and the internet offer construction companies an excellent opportunity to transform their business practices digitally to overcome the challenges posed by fragmented supply chains, decentralised structures, transient teams and information silos within and across teams and organisations.

The chapter will first discuss the communication and information management challenges and opportunities individuals, teams and companies face in the construction industry to underpin the discussion on the need for digital transformation. Next, it will offer insights into key elements of digital transformation from the perspectives of technology, people, processes and the broader context. In addition, it will reflect on digital maturity and some of the tools used to measure the digital maturity of construction companies. Finally, the chapter will discuss data privacy challenges, cybersecurity concerns and legal considerations.

10.2 Communication and Information Management Challenges and Opportunities in Construction: Need for Digital Transformation

Information is a general term meaning "knowledge" or processed data that typically influences recipients by increasing knowledge, skill and awareness, while *communication* describes the transmission of information (Cheng et al., 2001; Emmitt & Gorse, 2007). Information in the construction industry can take different forms, such as collecting data about the schedule or work progress, labour and equipment productivity, material or plant logistics, project cost, quality, waste generation, energy consumption, occupational health and safety incidents, etc. Furthermore, the collected data or information must be shared or communicated to relevant teams and shareholders in different written and verbal forms using emails, reports, phone calls and various information and communication technologies (ICT).

Inter-organisational communication sometimes involves communication concerning bidding, contracts, business transactions and stakeholder matters and may contain sensitive personal and organisational data. Considering the increased vulnerability of some technologies and transmission media to cyberattacks and information breaches, the sensitivity of the confidential data and the implications of any disclosure to unintended parties must be considered while determining the most suitable ICT solution. As a result, technologies and media utilised for communication within a team or company could differ from those used with external groups or organisations. The contractual provisions also dictate the format and timing of communication with different stakeholders. Preferred communication and information exchange methods must consider these complexities attached to the project and organisational characteristics, data, stakeholder requirements and contractual provisions. Essentially, the choice of the preferred ICT in the construction industry should be based on criteria such as (a) the amount and type of data; (b) sensitivity of data; (c) accuracy, accessibility and transmission speed of the technology; (d) cybersecurity measures; (e) contractual conditions; (f) stakeholder requirements; and (g) project and organisational norms. Thus, construction companies must employ multiple channels and digital platforms to meet these criteria and effectively communicate and share information with various stakeholders in different contexts.

However, construction companies have been long criticised for their slow adoption of technology and innovation (Klinc & Turk, 2019). The uptake and implementation of digital technology and processes in the construction industry have often proved challenging and, in some cases, slow, despite the perceived and demonstrated benefits of digitalisation in facilitating effective communication and improving business processes and productivity. Construction companies are among some of the least digitised businesses (Koeleman et al., 2019). McKinsey (2017) found that information technology (IT) spending in the construction industry is less than 1% of construction revenue. The fragmented supply chain, presence of many small and medium-sized enterprises (SMEs) and information asymmetries imply substantial differences in digital infrastructure and capabilities. While large companies could invest in sophisticated software and platforms, hardware and devices and research and development activities and employ dedicated IT professionals and support staff, SMEs often lack these resources. For instance, while the adoption and use of building information modelling (BIM) have increased in the construction industry in the last few years, SMEs and developing countries lag in BIM implementation. Similarly, many companies do not use BIM to its full potential. The NBS *Digital Construction Report* (NBS, 2021) noted that more than two-thirds of respondents had adopted BIM, a substantial increase over the last 10 years from just over 10% in 2011. However, small companies (with 15 staff or less) were less likely to have adopted BIM (55%). The report also revealed that for almost a third of respondents, BIM adoption means working with 3D models. Moreover, many companies use BIM only during the design phase. They do not fully utilise the developed model in pre-construction, site construction, operations and maintenance, with minimal usage in facilities management (Ellis, 2019). Therefore, many construction companies adopt technology solutions in an ad hoc way or use them in a limited way.

It can be argued that, in the past, commercially available ICT solutions often failed to meet the diverse expectations of stakeholders and requirements of construction projects, including harsh outdoor environments of worksites, data sharing and security challenges, contractual conditions and multi-actor decision-making processes, which led to slow adoption of ICT in the construction industry. Additionally, network-related factors such as weak signal strength and slow data transmission speeds reduced the efficiency and capabilities of digital technologies on construction sites. For instance, it was challenging to communicate and share information with geographically dispersed teams using landline phones, wired computers or information stored on local servers. In addition, wired desktop computers had limited applications on construction sites as they confined the user to a specific location. The asynchronous communication through emails and messages on wired devices caused delays in accessing critical information. Nonetheless, many companies still experienced dramatic productivity gains from networked computers that supported communication and collaboration among teams, business units and enterprises (Jurison, 2003).

In contrast, recent advancements in digital capabilities, the proliferation of various digital technologies, and the speed and reach of the internet offer several opportunities for construction companies to transform their manual and outdated information management and communication processes and practices. Developments in machine learning and artificial intelligence (AI) and innovations in the form of wearable devices, virtual reality (VR), augmented reality (AR), mixed reality (MR), cloud computing (CC), drones, the Internet of Things (IoT), digital twins and big data have the potential to disrupt traditional ways of working, communicating and managing data and information in the construction industry.

These newer technologies and digital tools enable on-demand and real-time access to information, facilitate synchronous communication and address the spatial and temporal limitations of previous generations of ICT, allowing construction professionals to be mobile. Accessing up-to-date information could help construction companies prevent re-work and disputes from individuals or teams working on redundant data or superseded drawings or documents. Moreover, the latest and emerging digital technologies integrate technology, work processes, project teams and actors, businesses and supply chains and blend digital and physical assets (Casini, 2021). Many industry reports show that digital transformation leads to improved operational efficiency, productivity, customer satisfaction, innovation and growth across the organisation (Accenture, 2023; Koeleman et al., 2019). For instance, the following example from the website of a large construction company, John Holland (2024), highlights some of the advantages and benefits (see text in *italics*) construction companies can derive from digitalisation.

Digital Construction Leaders

One of our *key differentiators* is our industry leadership in digital construction, and our pioneering advances in digital engineering, innovation and emerging technologies.

We focus on *'building digitally first'*, and our *integrated approach* to the way we create, capture and share information across the project lifecycle is revolutionising how we deliver projects.

We are one of the very few companies globally *certified and accredited against ISO19650–1* – organisation and digitisation of information about buildings and civil engineering works.

Important benefits of this include *greater certainty* for our clients and stakeholders, *greater collaboration* through digital solutions, *enhanced project outcomes* and driving *greater success* on our projects.

Leveraging a digital approach allows us to *better understand spatial complexities, reduces the risks* on projects, and provides *greater control and certainty* during the design, construction and operation. Critically, our advanced approach enables us to create *safer working conditions* for our teams, enhances our *focus towards environmental and sustainable outcomes*, and improves our *customer outcomes and experiences*.

Our digital delivery has a principle focus on the integration of our BIM (Building Information Management) & VDC (Virtual Design Construction), GIS & Geospatial, Surveying and Emerging Technology functions. We are one of a few companies in Australia that has developed an *integrated management solution*. The result is a more streamlined and integrated approach that *leverages digital data and intelligence* more effectively, with a multitude of benefits to our clients and their customers.

(Source: John Holland, 2024)

Therefore, digital transformation presents immense opportunities for construction companies to improve their information and communication management practices, make informed decisions and enhance productivity. At the same time, it could increase client satisfaction, improve project outcomes and offer companies a competitive advantage over rivals. Companies that are digital leaders and drive digital transformations before their competitors do or quickly follow the leads of those that do stand to reap the most gains in the current market conditions compared to slower-acting companies (Koeleman et al., 2019). However, successful digital transformation is not limited to technology adoption only. It can be much more complex, demanding integrated processes and efforts from individuals, teams, companies and stakeholders, as discussed in the subsequent sections of this chapter.

10.3 Digital Transformation – Technology, People and Process Imperatives and Broader Context

Technology has the potential to change our lives drastically. However, in an organisational context, complex disruption stemming from a convergence of technological and non-technological trends will be needed to disrupt construction business practices more than any technology on its own (Evans-Greenwood et al., 2019). Digital transformation differs from digitisation and digitalisation, as shown in Figure 10.1. *Digitisation* is the phase where organisations 'use digital format to store information', whereas digitalisation corresponds to 'the use of the technologies to alter existing business processes' (Verhoef et al., 2021, p. 891). In contrast, *digital transformation* is defined as the 'company-wide change leading to the development of new business models' (Verhoef et al., 2021, p. 891). Digital transformation includes digitalisation efforts; that is, using digital technologies to change business processes and projects. However, it goes beyond the project level and affects the entire organisation (Accenture, 2023).

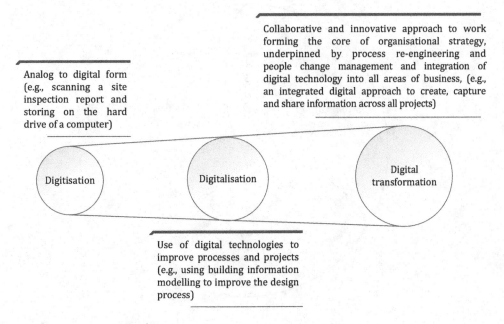

Figure 10.1 Digitisation, digitalisation and digital transformation.

Saldanha (2019) defined digital transformation as the migration of organisations from the third industrial revolution (use of PCs and the internet) to the fourth (emergence of massive computing capacity), utilising digital technology to create new products and services, ways of operation and business models. Digital transformation in construction companies involves coordinated changes in technology, people and processes to integrate digital technologies into all business areas and bring fundamental changes to how they operate and deliver value to different stakeholders. In other words, a successful digital transformation would require integrated and cohesive systems based on the broader context of technology use, users and conditions of usage, not only digital technologies. Moreover, integrating technology, people and processes will be guided by the constraints imposed by the broader contexts of project, organisation, stakeholders and regulatory environments, as depicted in Figure 10.2.

A holistic approach to digital transformation is essential because the long-term usage and associated productivity gains largely depend on how digital technologies are utilised and whether they become a routine part of organisational activities (Bhattacherjee, 2001). Tabrizi et al. (2019) argued that possibilities for efficiency gains will not be realised if digital transformation efforts lack the right mindset and are accompanied by flawed organisational practices. Saldanha (2019) also cautioned that digital transformations could fail in established enterprises because it is hard to drive and manage organisational and cultural change. Failure of digital transformation efforts or improper execution could result in significant economic losses and lost business opportunities (Tabrizi et al., 2019).

Construction companies therefore must re-engineer their business processes, establish a culture of innovation, train existing employees and find people with the right digital skills and knowledge to drive digital transformation efforts. At the same time, they must respond to ever-evolving digital transformation opportunities and challenges posed by internal and external

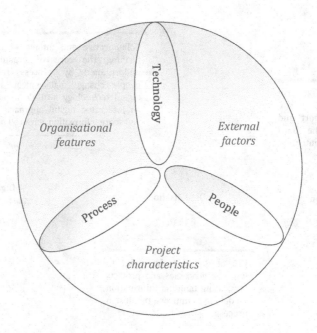

Figure 10.2 Elements of digital transformation.

environments. Digital leadership underpinned by the use and adoption of digital technologies and the ability to react to these opportunities could provide a competitive advantage to construction companies if changes in their business strategy, models, culture and people accompany it (Morgan & Papadonikolaki, 2022).

We will now discuss the technology, people and process imperatives of digital transformation in the construction industry and how the broader context or conditions imposed by the internal and external environments inform the digital transformation and digital practices of construction companies.

10.3.1 Digital Transformation – Technology Imperative

The successful adoption of Industry 4.0 technologies such as sensors, IoT, blockchain, virtual reality, robotics, machine learning and AI, 3D printing and intelligent machines in sectors such as manufacturing and health care inspired the construction industry to adopt them. Consequently, the term Construction 4.0 was coined to represent the digitalisation of the construction industry (Demirkesen & Tezel, 2021). Sawhney et al. (2020) classified Construction 4.0 technologies into three groups: (a) industrial production (prefabrication, 3D printing, etc.), (b) cyber–physical systems (robots, drones, etc.) and (c) digital technologies (BIM, IoT, AI, CC, etc.). They also identified three transformative processes, viz. product transformation, digital transformation and transformation in project delivery processes and related business processes under the Construction 4.0 framework.

Construction 4.0 represents a vast landscape of technologies and platforms, novel manufacturing or production methods and transformative techniques and strategies that could help construction companies rapidly move towards comprehensive connectivity and collaboration, bringing higher efficiency and transparency to all business processes. Additionally, several web-based technologies such as electronic procurement (e-procurement) and enterprise resource planning (ERP) have been used by construction companies for quite some time for performing procurement and supply chain management functions such as material and supplier selection and sourcing, bid management, logistics management and other project management tasks. These web-based technologies offer efficiency, coordination, resource optimisation and process integrity and help integrate sustainability into procurement and supply chain management processes (Yevu et al., 2021). We will briefly discuss some promising digital technologies and cyber–physical systems expected to play an essential role in the digital transformation of construction companies.

Mobile Technology

Delays in access to information and other resources concern various construction project actors. For instance, construction professionals are highly mobile. They often roam extensively on construction sites and between different project locations, leading to poor communication and delays in the transfer of information. However, mobile technology allows them to be easily contacted and consulted on important matters, exchange project information and communicate regardless of location, resulting in faster decision-making. Moreover, specific mobile applications (e.g., Aconex Field, CoConstruct, PlanGrid and Procore) can improve data and document management, safety management, defect management and other site management processes (Hasan et al., 2019). It is not uncommon nowadays to see construction supervisors and site personnel using tablets for various functions, including accessing digital drawings, online documents and forms. In addition, many construction companies frequently use social media platforms for recruitment, corporate news dissemination, project updates and stakeholder communication and engagement.

Cloud Computing

Asynchronous processes of dealing with information and storing, managing, sharing and retrieving a large volume of data and paper documents constantly produced in construction projects delay decision-making, cause miscommunication and confusion and affect productivity. Cloud storage and computing could facilitate real-time access to project data and documents and allow collaboration among dispersed individuals and teams. The National Institute of Standards and Technology defines *cloud computing* as 'a model for enabling ubiquitous, convenient, on-demand network access to a shared pool of configurable computing resources (e.g., networks, servers, storage, applications, and services) that can be rapidly provisioned and released with minimal management effort or service provider interaction' (Mell & Grance, 2011, p. 2). The CC model comprises three service models (Software as a Service or SaaS, Platform as a Service or PaaS and Infrastructure as a Service or IaaS) and four deployment models (private, community, public and hybrid cloud). In other words, CC offers a virtual desktop infrastructure platform that securely delivers virtual desktops and remote apps and allows users to store and retrieve data from remote servers.

CC enables hosting and providing services over the internet. It reduces reliance on information stored locally on computers or hard drives and servers and the need to be physically present at a particular location to access the stored data. For instance, readily available applications such as Google Docs, OneDrive and Dropbox could enable online collaboration and document management. Storing and sharing documents in digital formats on the cloud also allows users to readily search and access relevant information without going through thousands of pages of hard copies of documents. CC also enables users to work remotely and in different zones, reducing the wait time for information. Consequently, geographically dispersed individuals and teams could collaborate and access or update information in real time and stay updated on various developments (Hasan et al., 2019). Construction companies, especially SMEs, could also reduce the costs of acquiring, delivering and adopting future technologies by investing in CC services rather than setting up a complex or expensive IT infrastructure. Many emerging software and digital platforms require a collaborative online work environment and high computing power, which are not locally available on construction sites or may require considerable investments. In such cases, CC can help companies reduce the cost of IT resources, IT support and hardware and server maintenance.

Building Information Modelling

Building information modelling is a 'set of technologies, processes, and policies enabling multiple stakeholders to collaboratively design, construct and operate a facility in virtual space' (BIMe Initiative, 2022). ISO 19650 Part 1 defines BIM as the 'use of a shared digital representation of a built Asset to facilitate design, construction, and operation processes to form a reliable basis for decisions' (International Organisation for Standardisation [ISO], 2018). A BIM model differs from a 3D geometric model created in CAD as it allows vertical and horizontal integration of data and a collaborative digital environment. Moreover, BIM ensures access to stored physical and functional data beyond the project handover to enable facility management and reuse of building elements for a circular economy. Therefore, BIM is a collaborative process and could be used by different project stakeholders to store and manage data, communicate and collaborate during the design and construction stages, manage facilities after the project handover and reuse and recycle materials and components beyond the building or structure's life cycle.

Along with a cloud-based common data environment, it provides modelling and simulation features and stores and shares project life cycle data via a single shared platform. A cloud-based federated BIM model could allow all project stakeholders to operate and manage data within a common data environment. Thus, the federated BIM model is a trustworthy digital model representing the latest or updated information. BIM could also be used for creating and managing information concerning construction schedules, costs, facility management, sustainability and health and safety. Additionally, BIM can be integrated with other digital platforms and technologies such as VR, AR, AI and IoT to extend its capabilities and functions. Standards (e.g., ISO 19650 and 12006 series of standards) and frameworks (e.g., the RIBA Plan of Work) offer guidance on the BIM process and associated data structures.

Virtual Reality, Augmented Reality and Mixed Reality

Virtual reality offers users an immersive and interactive environment and can help construction stakeholders experience and understand construction designs and site environments. Construction companies could use VR technologies for preliminary design review, building geometry, site layout and workflow planning optimisation, design issue identification, occupational health and safety and workforce training. For instance, the VR-based walkthrough during the design stage could allow users to visualise the design and space and collaborate on optimising design, layout and functionality. VR can also be used in training construction workers and professionals by observing their behaviour, decisions and actions in different situations. Creating virtual construction site environments and construction hazard scenarios with the help of VR could help train construction workers concerning hazard and risk identification and risk management without exposing them to any danger or harm through the simulation of virtual scenarios.

On the other hand, AR, unlike VR, does not provide an immersive or interactive experience or exclusive virtual environment as the user continues to view themselves in the real world and, therefore, is not detached from the real world. Instead, it allows the user to overlay information, such as text, images and interactive graphics, onto real-world objects. As a result, the user can visualise the digital object in the physical world context to help them inspect object alignments, check work progress, assist in the assembly of components and retrofit work and make other decisions concerning the site and work planning. In addition, mobile devices such as smartphones and tablets, apps and head-mounted devices allow users to experience AR. Therefore, VR solutions can be most relevant in the planning and design phases when the asset or structure does not exist. In contrast, AR technology could be more useful during the construction and operation phases where a building – or parts of it – already exists (Hussien et al., 2020).

Finally, MR brings the virtual and real worlds together, where the virtual and real elements can interact, creating new ways for teams on complex projects to visualise, share ideas and manage change rather than just working on 2D drawings and designs. Therefore, MR merges the best aspects of VR and AR and allows the co-existence of physical and digital objects and their interaction in real time.

Internet of Things

The IoT connects physical items embedded with software and sensors, smart devices and network connectivity to allow the collection and exchange of data (Oke & Arowoiya, 2021). IoT can effectively monitor and analyse the performance of materials (e.g., development of strength or cracks in concrete) and structures (e.g., structural health of a bridge, building energy performance) to facilitate preventive and predictive maintenance. Additionally, IoT has been used in the construction industry for remote machine operation, material and equipment tracking, equipment servicing and repair, fleet management and site monitoring purposes (Oke & Arowoiya, 2021).

IoT enables real-time monitoring and access to data, facilitating timely decision-making and emergency response and reducing human error or omissions. IoT has also been utilised to improve lean construction, smart assembly, energy management, waste management and prefabrication practices (Mohammed et al., 2022). Real-time information regarding the structural health of buildings, bridges and other structures and their components using IoT sensors is now used by many companies (Ghosh et al., 2021). It is important to remember that IoT cannot work as a stand-alone solution but rather serves as an enabling platform for digital twins, smart buildings, smart cities, AI, big data analytics, etc.

Digital Twins

A digital twins model is a dynamic and real-time digital representation or virtual mirror of the physical asset. It can provide valuable insights into what is happening with physical assets. For example, it can be used during the construction and operation phase of the building to simulate and predict various aspects such as work progress, resource management and maintenance requirements. The model is updated with the real-time information collected from the physical asset using multiple sensors (IoT) and documents everything related to the asset. For on-demand prediction services, digital twins must continuously acquire, integrate, analyse, simulate and synchronise data (Kor et al., 2023).

Combining and processing real-time and historical data collected from various sources and technologies, as well as AI features, allows planning and predictive services. Digital twins could help in planning and maintenance activities and enable users to do scenario planning. The digital model could be trained using AI and machine learning to predict future events and asset performance during facilities management. Therefore, construction companies can use digital twins to collect data from connected sensors to track past, current and future performance throughout the asset's life cycle. It can help them make informed decisions and explore feasible solutions to design, construction and operational problems before they occur.

Blockchain and Smart Contracts

The spread of data and information across several organisations and digital platforms could result in a lack of transparency and inaccuracies, leading to disputes, corruption and mistrust among project participants, teams and construction stakeholders. Blockchain technology, a digitised distributed ledger, could address these inefficiencies by recording and sharing information or transactions that cannot be tampered with. In other words, it serves as a ledger of transactions resistant to duplicitous dealings.

Blockchain could be used to create smart contracts (based on 'if A, then' conditions). Smart contracts are immutable, meaning that they cannot be altered or deleted because the information on the blockchain cannot be changed once published. As a result, just one single source of information or single source of truth is shared with all network parties. Once a centralised tracking system is created, the smart contract automatically enforces the pre-defined rules, regulations, deadlines or milestones, payments and penalties. For instance, the contractor, after reviewing the work completed by the subcontractor or materials supplied by a vendor, if satisfied, can approve the transaction or their payment request, which will transfer payment to the receiver block. Smart contracts therefore remove the middleman from payouts between two parties once specific criteria have been met, eliminating reliance on others to follow through on their commitments. Specifically, incorporating BIM into smart contracts could automate progress payments when the construction work progresses according to the digital plans. Such an integration could address the chronic issue of late payments in the construction industry. Additionally, the availability of relevant information to all stakeholders would create accountability, trust and transparency.

Blockchain can also create unique digital identities and asset registries to trace material movement, identify custody and verify vendors and suppliers (Ellis, 2020). The blockchain technology timeline developed by ARUP using several blockchain case studies highlights potential applications of blockchain in technologies for digitising procurement of supply chains, title records and lease agreements, smart cities, circular economy, material passports, etc. Based on their maturity level, many of these technologies are expected to transition through the product development cycle from idea, concept, demonstration and commercialisation to early adoption over the next decade or so (ARUP, 2022).

Unmanned Aerial Vehicles

Construction companies can use unmanned aerial vehicles (UAVs) or commercial drones for various purposes, such as site surveys or aerial mapping of the construction site and project route, feasibility studies, safety inspection and hazard identification, quality inspection, damage assessment and project progress data gathering (Irizarry & Costa, 2016). UAVs can easily access remote and hard-to-reach places to capture important data, helping companies make informed decisions. The most used onboard sensor in UAVs is the camera (Gheisari et al., 2020). The laser scanner can also be used for point cloud generation, but weight and cost considerations make them less preferable. Drones can fly autonomously and operate with the help of a tablet or a preloaded flight chart to take high-resolution images and videos, which can be directly transmitted to the user or uploaded to cloud platforms to enable further data processing (Bogue, 2018).

e-Procurement Platforms

The use of e-procurement could help construction companies overcome the inefficiencies and irregularities associated with conventional manual or paper-based bidding processes. e-procurement platforms could offer seamless management of tendering/bidding, bid evaluation, vendor selection and contract award. Still, in many countries, the adoption of e-procurement is lagging. In developed countries, e-procurement implementation has faced barriers such as unready business partners, lack of flexibility of the technology and lack of trust and confidentiality. In developing countries, resistance to change attitudes, unreliable internet, weak IT infrastructure and capability, lack of regulations and insufficient management support have affected the uptake and use of e-procurement services (Yevu et al., 2023). Nevertheless, the adoption of e-procurement is increasing across the industry. Many clients, especially in the public sector, have replaced paper-based procurement processes with e-procurement to award construction work packages and projects.

An e-procurement software or platform allows users to advertise a job and share tender documents, send out bidding invitations to contractors, compare multiple bids from contractors, communicate with contractors and award the tender in a timely and transparent manner, replacing the time-consuming and effort-intensive manual tendering processes. Similarly, e-platforms for construction material and equipment procurement offer a one-stop solution by connecting manufacturers, distributors and other partners, enabling the search and purchase, simplifying the quote, purchase and qualification process for vendors and improving the visibility, efficiency and transparency in the flow of information. In addition, digital platforms speed up the procurement process and reduce the lead time and the amount of paperwork. Some platforms also provide a start-to-end solution from purchase to receipt, including tendering and purchase tasks, real-time procurement tracking, visibility across projects, centralised work libraries and live support service.

Enterprise Resource Planning

Enterprise resource planning is one of the most popular and widely used enterprise software solutions companies use to manage business activities such as finance and accounting, resource management, vendor management, procurement and purchase, logistics, payroll and project management. In ERP, users can access the aggregate comprehensive, real-time data dashboard from different departments and projects, which helps companies track expenses, progress and project resources across multiple projects. In addition, the central database of business data in ERP can be seamlessly accessed and integrated across departments, improving efficiency, transparency, productivity and competitiveness. However, ERP implementation is a serious undertaking as it requires considerable investment and resources and drastic changes in the workflows and business processes. Consequently, it is mainly implemented by large construction companies with several large projects and a high volume of work.

10.3.2 Digital Transformation – People Imperative

People involved in the development, adoption and implementation of technology (e.g., technology developers, service providers, leaders, users, etc.) form the core of the digital transformation journey of construction companies. If the people who design, define, regulate and amend the capabilities and use of technology and those affected by the technology are not an integral part of the process, the digital transformation efforts will not be successful. Therefore, it is important to remember that people are and will likely remain central to the digital transformation journey of construction companies. Lucas (1975) argued that the failure of technology implementation might be attributed to an excessive focus on technology and the neglect of its impact on the people it affects. Excluding or ignoring users and their requirements or input in technology implementation decisions could result in a gap between what users need or want and what technology can offer, eventually leading to less than the intended technology usage or complete abandonment (Hasan et al., 2019).

Therefore, engaging with stakeholders and incorporating their feedback at different stages of the technology implementation process is crucial. Ensuring that potential users are on board with the technology solution to be adopted by the company is a primary step towards successful technology implementation. Regular consultation with the users could help construction companies identify and better understand the root issues concerning communication and information management and how they could be overcome with the help of technological solutions. In turn, companies can make better decisions concerning the choice of technology and the implementation process.

Research shows that resistance to change is a critical barrier to the implementation of Construction 4.0 technology at both individual and organisational levels (Demirkesen & Tezel,

2021). How users perceive the usefulness of technology and the satisfaction derived from their use of technology determine their intention to use it (Bhattacherjee, 2001). Technologies that users deem useful are likely to be used more frequently and for longer durations. Similarly, the perceived ease of use predicts user satisfaction and technology usage behaviour (Zhou, 2011). Users avoid technology that does not meet their expectations, is ill-suited to tasks or work environments or when they do not understand how to use them effectively to perform various tasks (Hasan et al., 2019). While optimism and innovation drive the use of new technologies, inhibitors such as insecurity and discomfort affect the intention to continue using them (Son & Han, 2011). As digital transformation may initially require additional work and effort from already-overloaded construction workers and professionals due to training or upskilling and change in the existing processes and their working style, they may resist it.

Construction companies must also carefully consider how they will meet the demand for digital skills to drive digital transformation. While the industry is moving towards higher adoption of digital technologies, professionals and workers with skills and expertise in emerging technologies are in short supply in many countries. The successful implementation of new technologies would require new skills. Companies will need people with expertise in specific technologies to drive their strategic implementation and integration into organisational processes. Moreover, emerging technologies are creating new opportunities and roles. For instance, BIM implementation has created specialised BIM roles, such as BIM coordinators and managers, which were not conventional career opportunities for construction management graduates a few years ago.

If construction companies cannot access such skills and expertise, their digital transformation journey will be slow. Skills and knowledge gaps could force the death of many companies in the construction industry (Farmer, 2016). Therefore, the upskilling and reskilling of experienced professionals and workers over the next few years to prepare the industry for the full adoption of Construction 4.0 technologies is a challenging but essential task for construction stakeholders. While large companies are good at creating in-house training opportunities to upskill their employees and have access to a larger pool of skilled professionals, many SMEs rely on external providers for upskilling and training opportunities. The lack of professionals confident in utilising Construction 4.0 and other emerging technologies is a significant concern for construction companies aiming to implement these technologies.

Research shows that training and change management interventions are critical as they allow organisations to benefit from previous learnings and adjust to ongoing changes in the work system (Jasperson et al., 2005). The level of training influences ease of use and the understanding of usefulness and system characteristics. The presence of a technology champion in teams and projects to encourage, motivate and lead others in implementing technology could play a vital role in a successful digital transformation journey. Additionally, the support and commitment of the senior management teams are critical to ensuring that implementation-related tasks stay high on everyone's agenda and that adequate resources are made available to support the digital transformation processes. Failure to allocate sufficient resources to support the users or inadequate training could result in the unsuccessful implementation of technology or its suboptimal usage in the post-adoption phase. In an ever-evolving digital space, the capacity of employees to learn, change, adapt and evolve is crucial for successful technology implementation in construction companies.

There is also an ongoing debate about whether digital technologies and automated work processes using innovations in AI and robotics will replace manual workers in construction. While automated processes are likely to reduce the number of manual workers, there will always be tasks that will not be fully automated and require workers to monitor and control automated processes (García de Soto et al., 2022). In addition, there will be increased demand for digital skills and tech-savvy workers and professionals. A lack of soft skills such as communication, creativity and problem-solving and inadequate digital leadership could also impact the digital transformation of construction companies. To fill the digital skill gap, the construction industry must work with other stakeholders, such as the higher education sector and training institutes, to integrate digital skills into built environment curricula and training. Governments could offer incentives and education subsidies to help individuals and companies reskill to improve their digital skill competencies. A collaborative approach involving various stakeholders could help identify the current and future digital skill gaps in knowledge and practice and train the current and next generation of workers and professionals to address these issues.

Finally, the role of top management in providing information about new technologies and cultural and technical support is crucial for modelling the use of technologies and encouraging peer collaboration (Barrette, 2015). Digitally mature leaders are knowledgeable, exhibit entrepreneurial behaviours and know how to promote and support digital transformation (Mugge et al., 2020). Morgan and Papadonikolaki (2022) recommended that digital leaders in the construction industry clearly articulate the long-term benefits of technology; understand the magnitude of the change and resource requirements; provide vision, processes and procedures; and create and support highly collaborative teams and 'digital champions' within the company to support the digitalisation journey of their organisation. In addition, they need to support supply chain partners in their digital transformation to create an ecosystem of innovation, collaboration and knowledge sharing.

10.3.3 Digital Transformation – Process Imperative

Digital transformation affects all aspects of the business. Implementing digital technologies offers new ways of collaboration and often changes the communication and information management processes and how work is organised or performed in organisations. However, technology does not predetermine how it will be used in the organisational context. The users, organisational policies and processes and external constraints may decide when, where and how frequently various functions of technology will be used. Unless technology is embedded in organisational processes and integrated with the existing technologies and workflow, its adoption will have a limited impact on productivity and efficiency. Users must routinely accept and engage with the technology for continued use. In addition, they should be willing to adapt it to their work and vice versa.

Successful digital transformation requires restructuring the organisation, business strategy and processes to incorporate digital technology successfully (Kitsios et al., 2023). If construction companies adopt technology but do not intend to disrupt the institutional status quo regarding how work and project management activities are performed, the half-baked approach to digital transformation may result in the limited use of technology and negatively affect the existing processes. They could fail to leverage the full system capabilities when their workflow and business needs are not fully aligned with the system. In fact, in many cases, a few features of the technology are used more often than others due to a lack of strategic and integrated approach. For instance, research shows that practitioners may use some BIM functions, such as 3D visualisation and clash detection, more often than other functions, such as facility management and energy analysis (Gholizadeh et al., 2018).

Similarly, when companies force-fit technologies into construction work, work processes and work environments without a clear strategic plan, they are likely to face user resistance. Many construction companies deploy technology before understanding whether and how it can improve their operations (Koeleman et al., 2019). Using technology without clear instructions or a road map on how users can integrate it into their work can lead to productivity loss and user frustration (Hasan et al., 2021). Real performance benefits are most likely when individuals recognise a fit between the task and technology and use the technology to improve the execution of the task. Technology implementation results in productivity gain only when used in an integrated way in organisations (van der Vlist et al., 2014). Without a clear, comprehensive and pragmatic plan for digital transformation, companies will not fulfil their implementation expectations or realise the intended benefits of technology. Therefore, the existing business and work processes must be upgraded or re-engineered to take full advantage of digital tools and technology.

Enabling operational changes using technology by adopting a process-centred approach is critical to meeting real business needs (Koeleman et al., 2019). Technology should be considered an enabler of business process redesign or re-engineering with an aim at increasing organisational efficiency and productivity. However, process re-engineering may involve significant changes in organisational structures and workplace policies and procedures, demanding considerable organisational resources, management commitment and employee support. For instance, process re-engineering tied to BIM implementation occurs at several levels to realise the full benefits of BIM implementation. Some pre-operative formats, such as spreadsheets and CAD-generated drawings, will be rendered obsolete. At the same time, new processes must be established to allow the use of BIM in a multi-stakeholder collaborative environment.

The digital ways of working could also demand a shift from a traditional, linear design and construction process to a more agile and collaborative approach (Koeleman et al., 2019). The process re-engineering would also require technical and managerial support in the form of new job positions and roles (e.g., BIM coordinator, BIM manager) and changes in business strategy, governance and contractual terms and conditions. Reconfiguring the tasks and sub-tasks that make up the process to align with the technology can be tedious but an essential step towards successful digital transformation. Employees and stakeholders should be convinced about the benefits of process re-engineering and the need to eliminate obsolete processes so they do not resist the change and derail the whole thing. Consequently, timely communication of various changes minimises confusion and disruption.

10.3.4 Digital Transformation – Contextual Factors

The internal and external factors embedded in the contexts in which digital transformation occurs also influence it in several ways. For technologies used in organisational contexts, institutional policies and structure define the conditions of use of the technologies (Hasan et al., 2021). Orlikowski (1992), in her seminal work on the structuration model of technology, argued that users and organisational dimensions such as structural arrangements, business strategies, culture, and standard operating procedures, along with external factors such as government regulations and professional norms, mediate the context-dependent use of technology. García de Soto et al. (2022) suggested that construction companies should consider restructuring from fragmented organisational structures to platform-based structures to support digitalisation.

Similarly, the regulatory environment can have significant implications for the digital transformation of construction companies. When the government mandates the use of a particular technology in construction projects, companies implement and use that technology, as seen in the case of BIM implementation in some public projects in countries such as the United Kingdom. Additionally, the regulations and guidelines concerning technology usage may vary across countries and influence how it will be used in different contexts. For instance, the regulations concerning UAV operations may dictate pilot certification and licencing requirements, safety requirements, aircraft requirements, the timing of operations, flying zone and drone speed, etc., influencing their application in construction projects (Gheisari et al., 2020).

Other stakeholders could also profoundly influence digital transformation efforts in the construction industry due to contractual relationships and collaborative work environments. A lack of strong client demand, inadequate support from supply chain partners and the absence of contractual requirements concerning digital technologies may discourage contractors from investing in digital transformation processes, technology and employees. In contrast, the adoption of digital technology is likely to increase if clients stipulate using a particular technology to achieve better value (Hasan et al., 2019).

Additionally, bringing other stakeholders on board can be arduous, particularly if they lack the capacity to embark on a digital transformation journey. Subcontractors are typically SMEs with limited digital capabilities and may not be willing to adapt to new ways of performing work. Similarly, other stakeholders, such as architects, consultants, suppliers, etc., could influence the speed and extent of a company's digital transformation efforts in a collaborative work environment.

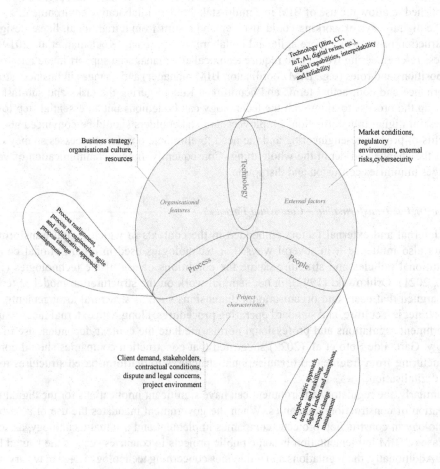

Figure 10.3 A holistic approach to digital transformation.

While the evidence on the influence of individuals' and firms' age on resisting changes and adopting innovative practices is inconclusive (Hemström et al., 2017), some studies have shown that younger teams and companies are more likely to adopt new technologies and change their practices to grow (Czarnitzki & Delanote, 2013; Demirkesen & Tezel, 2021). For digital transformation to work, the organisational culture must also be aligned to embrace and support the accompanying changes. There is always a risk that implementing new technology may disrupt existing work processes if not done correctly. However, a supportive culture could help individuals and teams navigate new challenges.

Figure 10.3 reiterates the need for a holistic approach to digital transformation based on our discussion of technology (Section 10.3.1), people (Section 10.3.2), process (Section 10.3.3) and contextual factors (Section 10.3.4).

10.4 Digital Maturity

Digital maturity measures an organisation's ability to take advantage of developments in technology to maintain a competitive advantage over rivals. It represents the current status quo of the digital capabilities of an organisation against a set of benchmark variables and indicators for different maturity stages. More digitally mature companies are found to outperform their less digitally mature peers. They are also more resilient during crises and create better value for their stakeholders (Boston Consulting Group, 2021).

Several digital maturity models are available to determine the level of digital maturity based on different criteria. These models can help companies in their transformation journey by enabling them to assess where they are in their transformation journey (e.g., lower or higher maturity). The assessment then allows them to develop short-term and long-term strategic goals and plans to improve their level of maturity. Essentially, digital maturity models consist of dimensions and criteria against which maturity stages are defined following an evolution path from lower to higher maturity, encouraging companies to take the next step towards their digital transformation journey.

However, it may be difficult to define common benchmarks for measuring digital maturity, given the numerous types and mix of digital technologies used in construction companies. The rate of implementation of different digital technologies in companies can vary. Moreover, a company's different departments or projects can be at different levels or stages of digital transformation. Similarly, a company can have different maturity levels against different dimensions of digital maturity, such as strategy, technology, organisational culture and operations. Therefore, digital maturity could vary considerably across and within organisations. To address this, companies can use a maturity framework that focuses on a specific technology or process to establish a common benchmark. For instance, Succar (2009) defined five maturity levels for BIM as ad hoc, defined, managed, integrated and optimised to capture gradual and continual improvement in BIM capability.

Another BIM maturity assessment, known as organisational BIM assessment, developed by Pennsylvania State University, measures various items under the categories of strategy (five items), BIM uses (two items), process (two items), information (three items), infrastructure (three items) and personnel (five items) on a scale of 0 to 5 (0 = *non-existent*, 1 = *initial*, 2 = *managed*, 3 = *defined*, 4 = *quantitatively managed*, and 5 = *optimising*) and the total score represents the total maturity score for the organisation (The Pennsylvania State University, 2023). A high score does not mean a successful implementation, as areas with low scores can hinder the effective use of BIM. Therefore, companies can identify specific areas where their score is low to set the target score and ultimately reach the optimising (Level 5) maturity level across all items. Williams et al. (2019), based on a detailed systematic literature review of 25 digital maturity models for SMEs, identified six essential maturity model dimensions for SMEs: (a) strategy, (b) products/services, (c) technology, (d) people and culture, (e) management and (f) processes. Some benchmarking and diagnostic tools, such as the Digital Acceleration Index, also allow comparisons between an organisation's digital maturity level, competitors and industry averages (Boston Consulting Group, 2021).

Saldanha (2019) proposed a digital transformation roadmap or Digital Transformation 5.0 model consisting of the following five stages.

1. *Foundation* (Stage 1) involves automating or digitizing internal processes using digital platforms such as SAP and Oracle and lays the foundation for future transformation.
2. *Siloed* (Stage 2) involves using disruptive technologies for particular functions or by individual departments or business units to create new business models but without an overall company strategy (e.g., the use of IoT for asset management and logistics, blockchain in contracting).
3. *Partially Synchronized* (Stage 3) transformation involves recognising the disruptive power of digital technologies and initiating the organisation-wide transformation by creating a vision and strategy for a digital future.
4. *Fully Synchronized* (Stage 4) transformation refers to a state where an organisation-wide digital transformation has taken place.
5. *Living DNA* (Stage 5) is where the organisation becomes a disciplined innovator and digital leader, setting industry trends.

Both anecdotal and empirical evidence suggest that the implementation of digital technologies in many construction companies falls into either Stage 1 or 2 of the above roadmap as they either adopt a technology to digitise some of their processes or transform a particular function or department. Following a digital maturity model or digital road map could help construction companies think about and plan the next step in their digital transformation journey.

However, in an evolving landscape of maturity models for different technologies, projects, companies and industries, it is important to note that calculating a score for digital maturity should not be the sole purpose of using maturity models. Instead, the focus must be on creating awareness regarding the current status and continuous improvement (Wernicke et al., 2023). Understanding their digital maturity could help teams and companies take the necessary steps to achieve the next level in their digital transformation journey and continuously improve their processes and digital capabilities to stay ahead of their competitors. Therefore, in most digital maturity models, the highest level corresponds to optimisation and continuous improvement.

10.5 Data Privacy and Cybersecurity Concerns

As more and more digital technologies are adopted and used by construction companies, the associated risks of cybersecurity breaches are likely to increase. Data privacy, unauthorised access to confidential data and cyberattacks must be significant concerns for construction companies undergoing digital transformation. Digital platforms such as e-tendering or digital procurement portals contain highly confidential or proprietary information such as project specifications, financial data, employee information and banking records, which could be exposed in the event of a cybersecurity breach (Pärn & de Soto, 2020). Similarly, companies using digital supply chain and procurement technologies should be concerned about the security of online data, documents with confidential information and transactions (Yevu et al., 2021).

Cybersecurity is one of the most significant risks associated with BIM and digital twins, which may contain useful information and data critical for the integrity and security of the structure or system (Halmetoja, 2022). Similarly, the large volume of information and data points collected by the IoT raises security and privacy concerns, given the recent rise in data breaches and cybersecurity threats (Ghosh et al., 2021). Therefore, in most cases, the data collected by construction companies using Construction 4.0 technologies and digital platforms could contain highly confidential, financial or proprietary information that could make companies vulnerable to cyberattacks. In addition, the stored data may include the personal data of employees or workers.

However, awareness and investment in high-level cybersecurity measures and efforts in the construction industry are still very low, which could present a significant challenge in the successful digital transformation of construction companies (Pärn & de Soto, 2020). Many construction companies overlook the threat of cybercrime within the built environment. As a result, they do not have definite plans or standards and the necessary infrastructure, resources and capabilities to protect their digital asset and data in the event of serious data security breaches (Mantha & de Soto, 2019).

Construction companies must proactively implement adequate measures to ensure data safety. They must devise and implement cybersecurity strategies, policies and safe digital work practices, as well as educate and train employees to protect individual and organisational data against cyberthreats. Moreover, they should invest in cybersecurity infrastructure and resources capable of preventing and handling cybersecurity risks. The cybersecurity strategy should include management and security aspects of physical and digital assets and information, responsibility and accountability, and processes, procedures and governance (Tezel et al., 2022). Ghosh et al. (2021) stressed the need for robust data management due to the massive amount of data generated and collected through digital technologies.

ISO 19650:2020 (Part 5) – Security-Minded Approach to Information Management specifies the principles and requirements for security-minded information management of sensitive information that is obtained, created, processed and stored as part of, or in relation to, any other initiative, project, asset, product or service throughout the life cycle. The document helps cultivate an appropriate and proportionate security mindset and culture across organizations with access to sensitive information, including the need to monitor and audit compliance (ISO, 2020). Moreover, alongside investment in cybersecurity and developing robust processes to maintain data integrity and confidentiality, companies must ensure that external teams and supply chain partners involved in digital data exchange and collaborative platforms have also put adequate measures. Often, security measures are only as strong as the weakest link in the system. In any case, sensitive and confidential data must be protected by all stakeholders at any cost against cybersecurity breaches that can put individuals and companies at risk.

10.6 Disputes and Legal Concerns

In the absence of relevant legislation, common standards and practices, the implementation of digital technologies could also present legal challenges to construction companies. For instance, the use of blockchain and smart contracts in the construction industry needs common standards and frameworks for the users and stakeholders. Some companies using blockchain for construction might be concerned about how much information should be shared with others and whether they are allowed to withhold or remove certain data from the transaction record. At the same time, others could demand full disclosure (Thomson Reuters, 2018). How smart contracts and blockchain technology for various purposes comply with privacy, data use and legal requirements will be a crucial factor determining their wide adoption in the highly regulated construction industry in many countries.

Similarly, standard or conventional contract documents must be amended with provisions concerning digital data, collaborative tools and model ownership, copyright and intellectual property.

As several stakeholders and supply chain partners will share data and information to create a collaborative digital environment, a lack of clarity or confusion on data ownership could lead to copyright violations and legal disputes. It is also essential to clarify how information and data shared by different project partners will be used, stored or destroyed after project completion. Therefore, construction contracts must clearly define the roles and responsibilities of different parties regarding digital data management and how disputes will be handled, along with potential implications for contractual breaches. Moreover, laws and regulations need to be amended appropriately to cover various aspects concerning the enforceability or interpretation of the smart contract.

It must be acknowledged that in an ever-changing landscape of digital technologies and their capabilities and applications in construction, it would be challenging to remove or address all sources of vagueness or open interpretation clauses in the contractual documents. It will also be problematic when the clauses are revisited in the future to resolve a dispute due to the moving nature of digital tools and data understanding. Nonetheless, comprehensive guidelines, standards and protocols and detailed records could aid the understanding of different stakeholders, clarify roles and responsibilities and minimise the chance of misunderstandings, data misuse and contractual disputes (Winfield, 2022).

It is also recommended that construction companies consider insurance with adequate cover to protect them from the potential risk or different liabilities emerging from the increased use of digital technologies. While using some digital tools and platforms presents data security and ownership issues, others can lead to injuries, property damage and privacy issues. For instance, drones and robots could injure workers and the public and damage properties if there are programming or hardware issues and the operator is at fault. Similarly, data- and image-capturing devices could accidentally record data of the public and neighbouring properties without their consent, resulting in ethical and legal implications for construction companies. The issues of accountability, responsibility and liability in such situations can be contentious and therefore are essential considerations for companies in their digital transformation journey.

10.7 Key Questions

It is crucial that construction companies carefully consider various internal and external factors before embarking on a digital transformation journey. Table 10.1 shows some of the questions construction companies may consider before and during digital technology implementation. Please note that the list of questions is not exhaustive and only summarises some of the critical issues raised in the chapter.

Table 10.1 Considerations for a successful digital transformation journey

S. No.	Domain	Key questions
1	Strategy	Is there a need to implement digital technology, or what is the motive or rationale for digital transformation? Is it to solve a problem or process, or is it in response to a regulatory requirement or the client's demand, or is it for efficiency, productivity gains and competitive advantage? How can digital processes help, and how will the company benefit from digital transformation? Is there a strong business case for technology adoption? Is it a part of the core business strategy?

(Continued)

Table 10.1 (Continued)

S. No.	Domain	Key questions
2	Digital technology	Have the existing workflows and processes been reviewed to identify which technologies would be best suited? What are the initial and ongoing costs? Is the technology mature or in the early phases of development? Are there any interoperability issues with the existing digital tools? Is the technology reliable and safe? Does the technology offer flexibility and agility to respond to changes quickly?
3	Organisational factors (people and processes)	Are there standards, policies and procedures to support technology implementation? Does the company have the right skills and expertise to implement the technology? Does the company have the right digital leadership? Were users consulted while making decisions concerning technology adoption? Does the company have enough resources to oversee and monitor the implementation of technology to enable successful transformation and business growth? How should roles and responsibilities evolve?
4	Market and stakeholder factors	Is it the right time for technology implementation – is there a demand in the market? Do other supply chain partners use it, or are they willing to use it? Have ownership and intellectual property rights been discussed with shared users? What rights does a party have over the data it provides, receives or manages? Do the contractual conditions cover various aspects concerning the use of digital technology? What support would be needed to help supply chain partners in their onboarding?
5	External environment	What are the regulatory requirements and conditions of technology use? What are the legal implications? What are the moral and ethical duties concerning digitalisation? What are the cybersecurity concerns? Is the company capable of protecting data privacy and confidentiality from cybersecurity breaches?

10.8 Conclusion

Innovation and process improvements using digital platforms and technologies in the construction industry are paramount to remain competitive in the digital economy. Many construction companies are excited at the prospect of what Construction 4.0 technologies and digital transformation can do for them, and rightly so, given the immense potential of these technologies to transform how the industry has functioned for several years and address communication, collaboration and productivity concerns. The way companies collect, use and manage data and information and deliver projects, products and services is rapidly changing. But like any disruptive tool or process, construction companies on the path of digital transformation need to consider their readiness from a people perspective, the appropriateness and usefulness of the digital tools from a technology perspective and integration into their business activities from a process perspective. At the same time, companies must recognise the opportunities and constraints posed by the external environment. It is essential to follow the regulatory requirements that apply to the uptake and use of technologies in the project and organisational contexts.

Digital transformation can be a complex process if it is not accompanied by necessary changes in people and processes. Moreover, the ethical, legal, financial and cybersecurity challenges associated with emerging technologies require serious consideration to protect individuals, teams and organisations. An elaborate risk management strategy concerning digital technologies must be developed for both project and organisational risks. Finally, not all the digital tools and technologies available in the market would be useful for a construction company and its goals. Asking fundamental questions concerning what clients, customers and employees need most and how digital processes could help overcome collaboration, communication and information management challenges that we raised earlier in this chapter is crucial to success.

References

Accenture. (2023). *Digital transformation: Understand digital transformation and how our insights can help drive business value.* https://www.accenture.com/us-en/insights/digital-transformation-index

ARUP. (2022). *Blockchain technology timeline.* https://www.arup.com/perspectives/publications/promotional-materials/section/blockchain-technology-timeline

Autodesk. (2021). *Connected construction: A better way to build, together.*

Barrette, C. M. (2015). Usefulness of technology adoption research in introducing an online workbook. *System, 49,* 133–144.

Bhattacherjee, A. (2001). Understanding information systems continuance: An expectation-confirmation model. *MIS Quarterly, 25*(3), 351–370.

BIMe Initiative. (2022). Building information modelling (BIM). In *BIM dictionary.* Retrieved March 6, 2023, from http://bimdictionary.com/building-information-modelling/

Bogue, R. (2018). What are the prospects for robots in the construction industry? *Industrial Robot: An International Journal, 45*(1), 1–6.

Boston Consulting Group. (2021). *The leaders' path to digital value.* https://www.bcg.com/publications/2021/digital-acceleration-index

Casini, M. (2021). *Construction 4.0: Advanced technology, tools and materials for the digital transformation of the construction industry.* Woodhead.

Cheng, E. W. L., Li, H., Love, P. E. D., & Irani, Z. (2001). Network communication in the construction industry. *Corporate Communications: An International Journal, 6*(2), 61–70.

Czarnitzki, D., & Delanote, J. (2013). Young innovative companies: The new high-growth firms? *Industrial and Corporate Change, 22*(5), 1315–1340.

Demirkesen, S., & Tezel, A. (2021). Investigating major challenges for industry 4.0 adoption among construction companies. *Engineering, Construction and Architectural Management, 29*(3), 1470–1503.

Ellis, G. (2019). *Blockchain in construction: A look inside the future of BIM* [infographic]. Autodesk Construction Cloud. https://constructionblog.autodesk.com/future-of-bim-infographic/

Ellis, G. (2020). *Blockchain in construction: 4 Ways it could revolutionize the industry.* Autodesk Construction Cloud. https://constructionblog.autodesk.com/blockchain-in-construction/

Emmitt, S., & Gorse, C. (2007). *Communication in construction teams.* Routledge.

Evans-Greenwood, P., Hillard, R., & Williams, P. (2019). *Digitalizing the construction industry: A case study in complex disruption.* Deloitte Insights. https://www2.deloitte.com/us/en/insights/topics/digital-transformation/digitizing-the-construction-industry.html/#endnote-2

Farmer, M. (2016). *The Farmer review of the UK construction labour model.* Construction Leadership Council.

García de Soto, B., Agustí-Juan, I., Joss, S., & Hunhevicz, J. (2022). Implications of Construction 4.0 to the workforce and organizational structures. *International Journal of Construction Management, 22*(2), 205–217.

Gheisari, M., Bastos Costa, D., & Irizarry, J. (2020). Unmanned aerial system applications in construction. In A. Sawhney, M. Riley, & J. Irizarry (Eds.), *Construction 4.0: An innovation platform for the built environment* (pp. 264–288). CRC Press.

Gholizadeh, P., Esmaeili, B., & Goodrum, P. (2018). Diffusion of building information modelling functions in the construction industry. *Journal of Management in Engineering, 34*(2), 04017060.

Ghosh, A., Edwards, D. J., & Hosseini, M. R. (2021). Patterns and trends in Internet of Things (IoT) research: Future applications in the construction industry. *Engineering, Construction and Architectural Management, 28*(2), 457–481.

Halmetoja, E. (2022). The role of digital twins and their application for the built environment. In M. Bolpagni, R. Gavina, & D. Ribeiro (Eds.), *Industry 4.0 for the built environment methodologies, technologies and skills* (pp. 415–442). Springer.

Hasan, A., Ahn, S., Baroudi, B., & Rameezdeen, R. (2021). Structuration model of construction management professionals' use of mobile devices. *Journal of Management in Engineering, 37*(4), 04021026.

Hasan, A., Ahn, S., Rameezdeen, R., & Baroudi, B. (2019). Investigation into post-adoption usage of mobile ICTs in Australian construction projects. *Engineering, Construction and Architectural Management, 28*(1), 351–371.

Hasan, A., Baroudi, B., Elmualim, A., & Rameezdeen, R. (2018). Factors affecting construction productivity: A 30 year systematic review. *Engineering, Construction and Architectural Management, 25*(7), 916–937.

Hemström, K., Mahapatra, K., & Gustavsson, L. (2017). Architects' perception of the innovativeness of the Swedish construction industry. *Construction Innovation, 17*(2), 244–260.

Hussien, A., Waraich, A., & Paes, D. (2020). A review of mixed-reality applications in Construction 4.0. In A. Sawhney, M. Riley, & J. Irizarry (Eds.), *Construction 4.0: An innovation platform for the built environment* (pp. 131–141). CRC Press.

International Organisation for Standardisation. (2018). *ISO 19650–1:2018 – Organization and digitization of information about buildings and civil engineering works, including building information modelling (BIM) – Information management using building information modelling – Part 1: Concepts and principles.*

International Organisation for Standardisation. (2020). *ISO 19650–5:2020 – Organization and digitization of information about buildings and civil engineering works, including building information modelling (BIM) – Information management using building information modelling – Part 5: Security-minded approach to information management.*

Irizarry, J., & Costa, D. B. (2016). Exploratory study of potential applications of unmanned aerial systems for construction management tasks. *Journal of Management in Engineering, 32*(3), 05016001.

Jasperson, J. S., Carter, P. E., & Zmud, R. W. (2005). A comprehensive conceptualization of post-adoptive behaviors associated with information technology enabled work systems. *MIS Quarterly, 29*(3), 525–557.

John Holland. (2024). *Technology & innovation.* https://johnholland.com.au/what-we-do/technology-and-innovation-leaders

Jurison, J. (2003). Productivity. In H. Bidgoli (Ed.), *Encyclopedia of information systems* (pp. 517–528). Elsevier.

Kitsios, F., Kamariotou, M., & Mavromatis, A. (2023). Drivers and outcomes of digital transformation: The case of public sector services. *Information, 14*(1), 43.

Klinc, R., & Turk, Ž. (2019). Construction 4.0-digital transformation of one of the oldest industries. *Economic and Business Review, 21*(3), 393–410.

Koeleman, J., Ribeirinho, M. J., Rockhill, D., Sjödin, E., & Strube, G. (2019). *Decoding digital transformation in construction.* McKinsey & Company. https://www.mckinsey.com/capabilities/operations/our-insights/decoding-digital-transformation-in-construction

Kor, M., Yitmen, I., & Alizadehsalehi, S. (2023). An investigation for integration of deep learning and digital twins towards Construction 4.0. *Smart and Sustainable Built Environment, 12*(3), 461–487.

Lucas, H. C., Jr. (1975). *Why information systems fail.* Columbia University Press.

Mantha, B. R., & de Soto, B. G. (2019 June 29–July 2). *Cyber security challenges and vulnerability assessment in the construction industry* [Paper presentation]. Creative Construction Conference, Budapest, Hungary.

McKinsey. (2017). *Reinventing construction: A route to higher productivity.*

Mell, P., & Grance, T (2011). *The NIST definition of cloud computing.* The National Institute of Standards and Technology, U.S. Department of Commerce.

Mohammed, B. H., Sallehuddin, H., Yadegaridehkordi, E., Safie Mohd Satar, N., Hussain, A. H. B., & Abdelghanymohamed, S. (2022). Nexus between building information modeling and Internet of Things in the construction industries. *Applied Sciences, 12*(20), 10629.

Morgan, B., & Papadonikolaki, E. (2022). Digital leadership for the built environment. In M. Bolpagni, R. Gavina, & D. Ribeiro (Eds.), *Industry 4.0 for the built environment methodologies, technologies and skills* (pp. 591–608). Springer.

Mugge, P., Abbu, H., Michaelis, T. L., Kwiatkowski, A., & Gudergan, G. (2020). Patterns of digitization: A practical guide to digital transformation. *Research-Technology Management, 63*(2), 27–35.

NBS. (2021). *Digital construction report.* http://www.thenbs.com/digital-construction-report-2021/

Oke, A. E., & Arowoiya, V. A. (2021). Evaluation of Internet of Things (IoT) application areas for sustainable construction. *Smart and Sustainable Built Environment, 10*(3), 387–402.

Orlikowski, W. J. (1992). The duality of technology: Rethinking the concept of technology in organizations. *Organization Science, 3*(3), 398–427.

Pärn, E. A., & de Soto, B. G. (2020). Cyber threats and actors confronting the Construction 4.0. In A. Sawhney, M. Riley, & J. Irizarry (Eds.), *Construction 4.0: An innovation platform for the built environment* (pp. 441–459). CRC Press.

The Pennsylvania State University. (2023). *Strategic planning for BIM implementation.* https://psu.pb.unizin.org/bimplanningforowners/chapter/strategic-planning-for-bim-implementation/

Saldanha, T. (2019). *Why digital transformations fail: The surprising disciplines of how to take off and stay ahead.* Berrett-Koehler.

Sawhney, A., Riley, M., & Irizarry, J. (2020). Construction 4.0 – Introduction and overview. In A. Sawhney, M. Riley, & J. Irizarry (Eds.), *Construction 4.0: An innovation platform for the built environment* (pp. 3–22). CRC Press.

Son, M., & Han, K. (2011). Beyond the technology adoption: Technology readiness effects on post-adoption behavior. *Journal of Business Research, 64*(11), 1178–1182.

Succar, B. (2009). Building information modelling framework: A research and delivery foundation for industry stakeholders. *Automation in Construction, 18*(3), 357–375.

Tabrizi, B., Lam, E., Girard, K., & Irvin, V. (2019). Digital transformation is not about technology. *Harvard Business Review.* https://hbr.org/2019/03/digital-transformation-is-not-about-technology

Tezel, A., Papadonikolakiet, E., Yitmen, I., & Bolpagni, M. (2022). Blockchain opportunities and issues in the built environment: Perspectives on trust, transparency and cybersecurity. In M. Bolpagni, R. Gavina, & D. Ribeiro (Eds.), *Industry 4.0 for the built environment methodologies, technologies and skills* (pp. 569–588). Springer.

Thomson Reuters. (2018). *Blockchain for construction/real estate.* https://mena.thomsonreuters.com/en/articles/blockchain-for-construction-and-real-estate.html

van der Vlist, A. J., Vrolijk, M. H., & Dewulf, G. P. (2014). On information and communication technology and production cost in construction industry: Evidence from the Netherlands. *Construction Management and Economics, 32*(6), 641–651.

Verhoef, P. C., Broekhuizen, T., Bart, Y., Bhattacharya, A., Dong, J. Q., Fabian, N., & Haenlein, M. (2021). Digital transformation: A multidisciplinary reflection and research agenda. *Journal of Business Research, 122,* 889–901.

Wernicke, B., Stehn, L., Sezer, A. A., & Thunberg, M. (2023). Introduction of a digital maturity assessment framework for construction site operations. *International Journal of Construction Management, 23*(5), 898–908.

Williams, C., Schallmo, D., Lang, K., & Boardman, L. (2019, June 16–19). *Digital maturity models for small and medium-sized enterprises: A systematic literature review* [Paper presentation]. The International Society for Professional Innovation Management (ISPIM) Conference. Florence, Italy.

Winfield, M. (2022). Legal implications of digitization in the construction industry. In M. Bolpagni, R. Gavina, & D. Ribeiro (Eds.), *Industry 4.0 for the built environment methodologies, technologies and skills* (pp. 391–411). Springer.

Yevu, S. K., Yu, A. T. W., & Darko, A. (2021). Digitalization of construction supply chain and procurement in the built environment: Emerging technologies and opportunities for sustainable processes. *Journal of Cleaner Production, 322,* 129093.

Yevu, S. K., Yu, A. T. W., Darko, A., Nani, G., & Edwards, D. (2023). Modeling the influence patterns of barriers to electronic procurement technology usage in construction projects. *Engineering, Construction and Architectural Management, 30*(10), 5133–5159.

Zhou, T. (2011). An empirical examination of users' post-adoption behaviour of mobile services. *Behaviour & Information Technology, 30*(2), 241–250.

11 Diversity, Equity and Inclusion

Abid Hasan

11.1 Introduction

Gender refers to socially constructed attributes, opportunities and relationships associated with being male and female, learned through socialisation processes. Therefore, the social norms around gender are context and time specific and change over time, as we have seen throughout history. According to the United Nations (2022a), *gender equality* refers to 'the equal rights, responsibilities and opportunities of women and men and girls and boys'. In other words, individuals' rights, responsibilities and opportunities must not depend on gender.

Still, in many organisations, communities and societies, women and men are treated differently regarding responsibilities assigned, activities undertaken and access to and control over resources. For instance, in many patriarchal societies and cultures, women are given domestic and caring duties. They are not allowed to step outside their house or work with men. In comparison, men are responsible for outdoor work and earning wages to support their families financially. Similarly, in organisational contexts, women are primarily assigned office-based and clerical positions in many countries and industries. The involvement of women in the decision-making process depends on broader sociocultural, economic and religious contexts (United Nations, 2022a).

Gender equality is not necessarily a women's issue. Instead, it must be seen as a human rights issue. Unfortunately, the construction industry remains a workplace where women are treated differently because of their gender. Despite the well-known benefits of a diverse and inclusive workplace, construction companies have failed to harness them. Currently, roughly only 1 in 7 people working in the construction industry are women. In contrast, the presence of women in trade roles is less than 1 in 20. While the overall percentage of women in the construction industry mentioned in industry and government reports in many countries offers data on gender distribution at an aggregate level, horizontal and vertical segregation in women's employment in different construction roles and occupations is often overlooked (Hasan et al., 2021).

Women are primarily employed in administrative and office-based positions in the construction industry. They are under-represented in leadership, management and site-based jobs in most construction organisations worldwide. For example, the data compiled by the Workplace Gender Equality Agency (WGEA) in Australia from non–public sector employers with 100 or more employees under the *Workplace Gender Equality Act 2012* show that the percentage of female CEOs in construction organisations is abysmally low – less than one-third of female CEOs in other industries. Similarly, females had lower representation in other managerial and leadership positions (WGEA, 2024).

The gendered pay gap is often found higher in construction for females working in full-time, part-time and casual roles. The gender segregation and inequality in small and medium-sized

DOI: 10.1201/9781003223092-11

enterprises that constitute most construction companies can be expected to be greater as small and medium-sized enterprises face additional structural and cultural barriers and less scrutiny, which may lead to more gender inequality. In short, women in construction often do not receive the same career opportunities and participation in decision-making as men, leading to the widespread issue of gender inequality in the construction industry. Moreover, they are often subjected to various forms of gender-based discrimination (Hasan et al., 2021).

The present chapter will first introduce the concepts of diversity, equity and inclusion (DEI). Next, it will discuss some of the drivers of the lack of gender DEI in the construction industry. Then it will discuss actions various stakeholders can take to achieve DEI in the construction industry. It will also present a few examples of DEI initiatives by governments and construction companies in different countries. Finally, it will outline the benefits of DEI and reiterate the key points discussed in the chapter.

11.2 Diversity, Equity and Inclusion

Examples of *diversity* in workplaces include gender diversity (men, women and non-binary people), age diversity (young, ageing and older workers), ethnic diversity (different cultural traditions and backgrounds), physical ability and neurodiversity. It is important to note that the scope of DEI policies and actions in construction companies must not be limited to only increasing the participation and retention of women but also focus equally on the rights of lesbians, gay, bisexual, transgender and intersex (LGBTI+) individuals whose gender identities are beyond the binary framework. In addition, the DEI initiatives should cover minority groups (for instance, Black, Asian and minority ethnic [BAME] individuals) and individuals with disabilities. There is a lack of research on these minority groups in the construction industry. Still, the limited data available on the employment of diverse individuals in the UK construction industry show that many LGBTI+ employees experience homophobic and derogatory terms at work and hide or disguise their identity at work to avoid discrimination (The Chartered Institute of Building, 2022).

Gender equality was made part of international human rights law by the Universal Declaration of Human Rights, adopted by the United Nations General Assembly in 1948 (United Nations, 2022b). One of the 17 United Nations Sustainable Development Goals necessary for achieving a better and more sustainable future for all is to achieve gender equality and empower all women and girls (i.e., Sustainable Development Goal 5). The United Nations recognises that gender inequalities are still deep-rooted in every society. As a result, women lack access to decent work and are under-represented in political and economic decision-making processes. They also face occupational segregation and gender wage gaps and are victims of violence and discrimination (United Nations, 2023).

Ending all forms of gender violence and securing equal access to education, health, economic resources, political opportunities and leadership and decision-making positions at all levels for all genders are critical elements of gender equality. In addition, it is essential to provide equal opportunities in access to employment and career prospects. Therefore, achieving broadly equal opportunities and outcomes by allowing people to access and enjoy equal rewards, resources and opportunities regardless of gender is fundamental to achieving gender equality. On the other hand, *gender equity* is the process of ensuring fairness in the distribution of resources among men and women, and it is essential for achieving gender equality (Edmond, 2023). Therefore, equity takes into consideration a person's unique circumstances, adjusting treatment accordingly so that the end result is equal (McKinsey & Company, 2022). A strong focus on gender equity is required at workplaces to achieve a sustainable, meaningful and positive change for women and address imbalanced organisational policies and systems.

Finally, an *inclusive workplace* ensures that employees do not experience bias or discrimination because of gender, colour, race or other aspects of their identity. They must have equal access to job and career growth opportunities, decision-making and resources required to perform their job irrespective of whether they belong to the LGBTI+ community, have some form of disabilities or are individuals of colour. In addition, an inclusive workplace allows and encourages employees to express different opinions and ensures that the voices of all employees are heard. Building an equitable and inclusive work culture essentially involves recognising and supporting the fact that a 'one-size-fits-all' approach to work and employment conditions does not work for all employees. For instance, construction companies that promote an equitable and inclusive culture can offer flexible work arrangements to suit individual needs and circumstances. Employees can work remotely or on-site depending on their preference or reduce their work hours if they need more time to fulfil their non-work commitments.

Before we discuss how construction companies can create a diverse, equitable and inclusive workplace, it is crucial to understand why construction has historically remained male-dominated and hostile to minority workers, including women, LGBTI+ and individuals of colour. Without understanding the root causes, addressing them to achieve sustainable progress towards DEI in the construction industry would be challenging. Since most research on DEI in construction covers gender inequality, the discussion in this chapter will focus more on the low participation of women in the construction industry.

11.3 Drivers of Lack of Diversity, Equity and Inclusion

Numerous studies conducted over the last 30 years or so in different parts of the world have identified several causes of the lack of gender diversity in the construction industry, especially in site-based roles and leadership positions. A range of factors, such as social and cultural barriers, poor work–life balance, gender-based discrimination and traditional attitudes and social norms about the status and role of women, discourage women from entering and building a successful career in the construction industry. Some of the main issues and barriers women face in the construction industry are discussed below.

11.3.1 Gender Roles and Societal Norms

The prevalent ideological, religious, ethnic and cultural norms in societies often dictate what women or men can or cannot do. Many cultures and communities do not allow or encourage women to work outside the home, especially for long hours and at male-dominated workplaces. Women have been traditionally seen in the role of caregivers performing domestic duties. Even in many developed societies, women are seen as primary caregivers and, in many instances, expected to sacrifice or compromise their careers to meet family demands. Therefore, gender norms and sociocultural barriers prevent many women from pursuing careers in the male-dominated construction industry.

11.3.2 Long Work Hours and Poor Work–Life Balance

Employment in industries or workplaces that offer flexible work options, such as telecommuting and flexitime, helps women create a synergy between their family and work demands because domestic or family responsibilities allow limited flexibility and mobility. In contrast, site-based jobs in the construction industry are highly mobile and involve long work hours and inflexible work schedules. Most construction sites operate on a 6-day workweek basis, and workers typically spend 10 to 12 hours on site each day. Such a work schedule is quite demanding on workers' personal time and significantly affects their work–life balance. Watts (2009, pp. 53–54) noted that the construction culture 'glorifies employees who work as if they have no personal life'. Consequently, it is very challenging for women, who often have domestic or caring duties, to meet the typical job demands and expectations at construction workplaces, especially in project-based roles. Many women do not consider a career in the construction industry due to concerns about work–life balance. Those working in the industry struggle to balance their work and family demands and often decide to leave their construction career (Hasan et al., 2021). Research shows that single or unmarried women without dependent children remain in the industry more than married women (Morello et al., 2018).

11.3.3 Image Problem and Career Misconceptions

The construction industry is not seen as an exciting or glamorous place to work. Construction work typically involves working in outdoor conditions in untidy, chaotic and unsafe conditions for long hours. Moreover, construction is seen as a dangerous workplace due to a high rate of injuries, fatalities, mental health issues and suicides. The macho culture associated with the construction industry projects an image of the workplace where other genders are not welcomed. Previous studies show that workers are exposed to several physical and psychosocial hazards (e.g., high job demands and poor support). Though the organisational culture and practices are evolving with time and many construction companies are making efforts to create gender-diverse and inclusive workplaces, the industry broadly still has an image problem. The poor image or reputation acts as a deterrent for young women considering a career in the construction industry. Additionally, despite the diverse career options open to women due to the uptake of new technologies and increased focus on areas such as sustainability, stakeholder management and circular economy, misinterpretation and misinformation of career pathways by career counsellors, friends and families reinforce the belief that construction offers weak career prospects for women compared to other industries and construction roles always include working in an outdoor and unsafe environment. As a result, construction is seen as a sector for 'last chancers' or those who are academically weak, damaging the potential skills base (Barnes, 2019). The image problem and misconceptions about construction careers are reflected in the low enrolment of females in construction management or other construction-related programs in tertiary education institutions in many countries.

11.3.4 Biased Recruitment Practices and Gender Stereotypes

Biased recruitment practices make it difficult for women to enter the industry. Women often face sexist attitudes and discrimination during the recruitment process. Women are sometimes not considered suitable for construction trade roles that are physically demanding. Moreover, not all construction jobs are advertised, and many times, future employees are recruited from the personal contacts or professional networks of the current employees. The prevalence of informal recruitment practices in the construction industry limits women's opportunities due to a lack of professional connections and access to job networks. Gender stereotypes around women's roles in the construction industry and their capabilities also affect their participation and job satisfaction. It is often perceived that the outdoor nature of construction work is unsuitable for women. Construction companies sometimes also show reluctance to hire women in site-based roles due to the possibility of sexual harassment, the additional cost of providing separate amenities, maternity leave and other legal and logistical considerations.

11.3.5 Slow Career Progression

Women face discriminatory practices and lack of opportunities, preventing them from advancing their careers in the construction industry. For instance, many construction companies are reluctant to assign site-based roles and responsibilities to women. However, site experience is necessary for on-the-job training and career growth in the construction industry. For many middle and senior managerial positions in construction companies, site experience is considered a mandatory requirement that many women fail to demonstrate due to their primarily office-based positions and limited exposure to site work. The office-based jobs and administrative roles make it challenging for women to move between functional areas and from support areas into site management roles. Additionally, women working in the industry are often excluded from social groups and networking events, limiting their professional network and growth opportunities. In some instances, women make conscious decisions to stay in office-based roles to avoid the stressful, chaotic site environment and long work hours and achieve a better work–life balance at the expense of their careers. Career interruptions due to maternity leave or other domestic responsibilities also affect their career if adequate organisational support is not provided.

11.3.6 High Job Turnover

Retention relates to people staying in their careers, and *progression* relates to people advancing in their careers. The retention of women in construction roles has been low, and they leave the construction profession faster than men. A higher turnover rate means fewer women working in the construction industry than entering the industry. Incidents of discrimination and harassment, inadequate support from colleagues and management, gender pay gap, pay inequity and poor work–life balance affect the job satisfaction and well-being of many women and force them to look for career opportunities in other industries. The realities of the construction industry are very different from the sheltered and protected environment of educational institutions. As a result, women who have not experienced working in the industry through internships or do not have a family member or friend already working in the industry can be ill-prepared. The stressful and chaotic work environment and gendered treatment could make female entrants rethink their career choices (Regis et al., 2019).

11.3.7 Hostile Workplace Culture and OHS Concerns

The workplace culture in the construction industry is often shaped around the majority view or male-dominated work culture. As a result, inappropriate or discriminatory behaviour against women and minority groups is accepted as normal practice at many workplaces. Alcohol and drug consumption and risk-taking behaviour are also commonly found in the construction industry. Several studies report that women in construction roles are subjected to hostile environments like harassment and abuse (Hasan et al., 2021). They experience additional occupational mental health stressors, resulting in a greater number of psychological injuries than men (Kamardeen & Hasan, 2023). Improperly fitting personal protective equipment (PPE) and personal protective clothing not designed for women could compromise their health and safety, especially for construction tradeswomen (United States Department of Labor, 2022). Women have different anthropometrics or body shapes and measurements than male workers. Women are not small men, and therefore smaller sizes of PPEs designed for men are ineffective in protecting them from hazards and injuries (Industrial Accident Prevention Association, 2022). While in some countries such as the United States, the International Safety Equipment Association lists manufacturers offering safety equipment appropriate for the female body, women workers do not have access to gender-specific PPE in many parts of the world or on many construction projects, either due to the unavailability of a full range of stock and sizes of PPE in the local market or lack of willingness of construction companies to purchase separate PPE for women workers.

11.4 Actions to Achieve Diversity, Equity and Inclusion

The reasons for a lack of gender diversity in the construction industry, as discussed in the previous section, are complex and multifaceted. These issues affect women's participation and growth in the construction industry at all stages of their careers. However, despite a wealth of knowledge on the systemic issues that affect the recruitment, retention and career progression of women in the construction industry, the absence of a holistic approach to achieving gender equality has led to a failure to address the key barriers. Inconsistent and flawed approaches to addressing a multi-layered and extremely complex issue fail to achieve significant improvements in participation, retention, career growth and job experiences of women in the construction industry.

Mainstreaming gender perspectives and attention to the goal of gender equality and inclusion require a root and branch solution to equality and must be central to all activities, including government policies and legislation; organisational strategies; policies, procedures and processes; procurement strategies; recruitment; career growth support and retention strategies; employment and pay conditions; decision-making processes; engagement and dialogue; etc. A sustainable change will require intervention and actions at all different levels (i.e., policy, strategic and operational). Moreover, it will require an inclusive approach with coordinated efforts and support from both management and workers. Research shows that gender equity initiatives and activities designed or implemented without proper consultation with all relevant stakeholders and their support often have limited impact, as women-specific or targeted efforts could be easily marginalised in male-dominated construction workplaces. Therefore, embedding gender concerns in all aspects of social and business activities and the involvement of everyone is necessary to achieve gender equity goals and create sustainable changes in workplaces (Padayachie, 2019).

Achieving gender equality demands both top-down and bottom-up efforts. Construction companies would need to work along with other stakeholders, such as the government, clients, professional organisations, universities, trade schools, etc., to promote construction careers among women and encourage them to join the industry. Moreover, making construction careers and workplaces attractive, safe and inclusive will reduce turnover rates and help organisations improve gender DEI in the construction industry.

11.4.1 Government Policies and Quotas

Globally, the United Nations Economic and Social Council is one of the six principal organs of the United Nations and is responsible for coordinating the economic and social fields of the organisation. It serves as the central forum for discussing international economic and social issues and formulating policy recommendations addressed to Member States and the United Nations. The strategy of mainstreaming is defined in the United Nations Economic and Social Council (1997) Agreed Conclusions, 1997/2, as

> the process of assessing the implications for women and men of any planned action, including legislation, policies or programmes, in all areas and at all levels. It is a strategy for making women's as well as men's concerns and experiences an integral dimension of the design, implementation, monitoring and evaluation of policies and programmes in all political, economic and societal spheres so that women and men benefit equally and inequality is not perpetuated. The ultimate goal is to achieve gender equality.

Nationally, governments can formulate and enforce acts and gender equality policies to close the gender gaps in workplaces. Different legislative and procedural requirements concerning gender quotas and reporting exist in different countries. Iceland was the first country to legislate an 'equal pay standard', where workplaces must undergo an audit and receive certification that their employees are paid equally for work of equal value. In Belgium, Austria and France, organisations are required to make their gender pay gaps publicly available or available to their employees on request. Employers are also encouraged to publish an action plan that explains how they intend to tackle the gender pay gap in their organisations (Government Equalities Office, 2022). Similarly, several countries in the European Union have adopted gender equality reporting models where union representatives or employee work councils monitor the gender pay gap. If women's remuneration is less than men's, the organisation is obliged to tackle pay inequality.

In Australia, the *Workplace Gender Equality Act 2012* requires various employers to lodge reports each year containing information relating to various gender equality indicators (e.g., equal remuneration between women and men). The WGEA was established under the Act to advise and assist employers in promoting and improving gender equality in the workplace. Organisations tendering for government contracts may need to satisfy a requirement to comply with the Act (Australian Government, 2023).

The government can also use public project procurement to drive the adoption of gender equality measures. Contractors working on public projects may be required to implement a quota system employing a certain minimum number of women and minority groups across different roles, among other things. For instance, London 2012 developed employment and training programs with partner organisations and contractors to encourage women, BAME individuals, people with disabilities and those who were previously unemployed to improve their skills and apply for jobs in construction and other areas where they have traditionally been under-represented. The London 2012 Women into Construction project recruited 270 women and placed them directly into jobs on the Olympic Park through this project, exceeding its original target of 50. One thousand seven hundred thirteen women worked on the major construction phase of the Olympic Park and Olympic and Paralympic Village, and 5,092 of the Park and Village workers declared that they were from a BAME background (The UK Government, 2023).

Similarly, the Building Equality Policy was developed by the Victoria State Government in Australia in consultation with various stakeholders (Victoria State Government, 2022). It adopts a broader look at the barriers to employment, retention and career growth of women in the construction industry, acknowledging that externally imposed quotas will have limited effects if the fundamental issues are ignored.

Building Equality Policy (Victoria, Australia)

The Building Equality Policy introduced by the Victorian government effective 1 January 2022 applies to all publicly funded construction projects valued at $20 million or more and sets the following minimum on-site gender equality targets:

1. Trade Covered Labour: women are required to perform at least 3% of the contract works' total estimated labour hours for each trade position.
2. Non-trade Construction Award Covered Labour: women are required to perform at least 7% of the contract works' total estimated labour hours for each non-trade Construction Award covered labour position.
3. Management/Supervisory and Specialist Labour (staff): women are required to perform at least 35% of the contract works' total estimated labour hours for each staff position.

Additionally, suppliers are required to engage women who are registered apprentices or trainees to perform building and construction work for at least 4% of the contract works' total estimated labour hours for apprentices and trainees. Finally, suppliers are required to provide both project-specific and organisation-wide gender equality action plans when submitting an expression of interest or tender for government-funded construction work, which will form part of the procurement contract if the tenderer is successful. The six workplace gender equality indicators of gender equality action plans are (a) workplace prevention and responses, (b) inclusive and respectful workplace, (c) flexible and empowering workplace, (d) diverse and representative workforce, (e) improve leadership, representation, and accountability and (f) collect and report data about gender equality and the gender pay gap.

While the quota system and mandatory requirements from the government will have a trickle-down effect throughout the construction industry and can be expected to bring changes to the male-dominated work culture by increasing the participation of women in the construction workforce across all different roles, many researchers argue that practical solutions and a well-developed business case for diversity can only bring long-term changes, rather than legislation or quotas for equity (Dainty et al., 2002; English & Bowen, 2012). Moreover, the scope of the legislation and the compliance vary considerably across states and regions, size and value of projects and type of projects and organisations, limiting their effectiveness in bringing large-scale changes to the industry. Often, small and medium-sized companies that form most of the organisations in the construction industry fall outside the purview of such requirements due to the low headcount of employees. Similarly, many conditions are recommendations or non-mandatory requirements, leaving it to the construction companies to determine whether they want to adopt them.

Generic solutions, cookie-cutter policies and targets could fail to achieve their intended goals if they do not consider the realities of the broader environment and context and therefore must be accompanied by a range of initiatives and actions by other stakeholders. The government or policymakers need to work with construction companies and other stakeholders to devise a holistic approach that addresses various structural, cultural and organisational barriers present at various levels. For instance, the government can work with banks and lending institutions to offer low-interest credit to encourage women entrepreneurs who want to start a construction business. Similarly, the government could work with training institutes and universities to provide necessary training to women to help them join the industry.

11.4.2 Organisational Policies and Workplace Actions

In addition to legislation, organisational policies, cultural change and practical solutions to achieve DEI are crucial. Construction companies have the most significant role to play in ensuring the success of DEI efforts. Sustained efforts within construction companies at the level of project teams and site work environment are necessary to bring sustainable changes to the status and treatment of minority groups of workers such as women and LGBTI+ individuals. They need to create equal employment opportunities, close the gender pay gap and bring structural and cultural changes across the company to remove the barriers to the full and equal participation of women and other under-represented groups in the workforce.

To start with, gender equality considerations should be a key element of a construction company's strategy and its business goals and decisions. Construction companies must implement comprehensive gender equity policies to support gender equality in different areas of human resource management practices such as recruitment, job assignment, flexible work arrangements, training and development, pay and other employment benefits, performance management, promotion, etc. Companies need to openly embrace DEI, set specific targets and pledge to achieve gender equity in all aspects of their operations.

Research shows that workplace culture in the construction industry has remained hostile to women workers, showing little progress towards adopting an inclusive culture that promotes gender diversity and offers equal opportunities irrespective of gender. The *Oxford Advanced Learner's Dictionary* (2024) defines *culture* as 'the beliefs and attitudes about something that people in a particular group or organisation share'. Organisational culture has a significant influence on how accessible and comfortable the workplace environment is perceived by the employees. Poor culture is among the key reasons many women either decide not to choose a construction career or leave after working for some time.

However, changing organisational culture could be a slow process requiring policy, actions and interventions at all levels across the company. Efforts to change the dominant culture are likely to face considerable resistance from workers. Moreover, cultural changes take time and considerable effort to realise and are possible only where the will exists to promote and implement them (Wright, 2013). Policies or strategies to promote an inclusive culture and prevent physical and mental harm, any sort of discrimination, harassment and psychosocial risks at the workplace will only be effective if they are implemented and enforced and receive the support of all employees. Therefore, the first and foremost action by construction companies towards achieving DEI is to nurture, promote and maintain a workplace culture where everyone feels respected, safe, included and supported at work.

Construction companies need the support of all employees to drive cultural changes at their workplaces and bring sustainable changes in the mindsets and attitudes toward women workers. The leaders and top management need to show their commitment to achieving gender equality and providing an inclusive culture to employees. Efforts to strengthen positive relationships with women and minority workers involve active engagement with them at all levels, including the grassroots level, to learn about their experiences and listen to their feedback. Moreover, companies should identify and support gender equality and inclusive culture champions at various operational and functional levels to challenge gender stereotypes and promote DEI. Managers and senior staff in the team need to act as role models to instil the desired behaviour in project teams and other site staff. Leaders and managers who are serious about addressing gender inequality must make gender equity actions a priority for their company. For instance, the CEOs of some large construction companies in Australia are pay equity ambassadors, meaning their companies are committed to undertaking a pay gap analysis of their workforce every 2 years, acting on the results, reporting pay equity metrics to the executive and board and communicating their pay equity initiatives to their employees. These ambassadors must sign the pay equity pledge and commit themselves to work with the agency to promote and improve gender equality. They must also renew their commitment to pay equity every 2 years (WGEA, 2022).

The gender pay gap is another serious concern leading to employee dissatisfaction and turnover. The gender pay gap is not the same as equal pay. Gender pay equity goals would ensure that employees receive equal pay for work of the same or similar value irrespective of their gender. In many countries (e.g., Australia), organisations are required by law to provide equal pay to employees who are performing work of equal or comparable value. In contrast, the gender pay gap measures the difference between the average earnings of women and men in the workforce. Women and men working in different jobs, with female-dominated jobs attracting lower wages, are among the leading causes of the gender pay gap in construction. In many parts of the world, women still earn substantially less than men in the construction industry (Regis et al., 2019). In the United Kingdom, the gender pay gap in construction is higher than the national average (Macdonald & Company, 2021). The average hourly rate for women at the top 40 contractors was still 28% less than that for men (Barnes, 2019). Construction companies should carry out a gender pay gap audit using payroll data. If a gender pay gap exists in the organisation, concrete actions must be taken to close the gap.

As discussed earlier, the rigid work schedules and long work hours prevalent in the construction industry are major deterrents for women. Construction companies should offer flexible employment conditions to employees who need them. Flexible working has been associated with improved organisational productivity, improved employee well-being, increased gender diversity, increased proportion of women in leadership, employee retention and future-proofing the workplace (WGEA, 2021). Inflexible work schedules and long work hours in construction not only affect the health and well-being of employees irrespective of gender but also the career of employees needing flexible working arrangements. Therefore, flexible work arrangements responsive to the needs of employees could help construction companies create a diverse, equitable and inclusive workplace.

It is advised that flexible work arrangements be extended to all employees. For instance, parental leave policy should recognise all genders, encourage gender-neutral parental care and recognise the role of men in caring responsibilities rather than stereotyping it as a gendered role. Acknowledging the changes in gender roles and accepting a broad definition of families is essential for formulating and implementing inclusive workplace policies. An employee who is a parent (biological, step, adoptive or foster), guardian or carer of a child, responsible for someone dependent on them for care or with a disability or without any of these special requirements may need flexible working arrangements encompassing changes to the hours, pattern and location of work to meet their personal life, family and caring demands or on some occasions to better take care of their own health and well-being. Encouraging men to access flexible work arrangements and leave entitlements would also help reduce the gender pay gap.

Regarding job allocation and responsibilities, construction companies must provide equal opportunities to employees irrespective of their gender. Not allowing women to work on-site or in site-based roles or recruiting them for office-based positions only limits their career options and growth opportunities, as project site experience is considered valuable in the construction industry. Moreover, site-based jobs have higher salaries than office-based or administrative jobs.

Consequently, fewer women in site-based positions creates a gender pay gap. Similarly, there is a lower representation of women in senior management and leadership positions, limiting their participation in decision-making. Adequate organisational support, fair promotion policies and mentoring from experienced professionals and leaders could help more women advance their careers in the construction industry. Taking action to increase the number of women in leadership positions will also help construction companies reduce the gender pay gap. Inclusive decision-making, equity in job assignments and career growth opportunities and pay equity require active organisational approaches and leadership and management commitment to a diverse, equitable and inclusive workplace.

Construction organisations must also pay more attention to providing a safe and healthy work environment. Additionally, measures to support employees affected by harassment and domestic violence should be an integral part of workplace policies or strategies. Inclusive initiatives, such as flexible employment conditions, training, better site amenities and health and safety improvement measures, can be implemented more easily as they benefit all employees. In contrast, policies and initiatives favouring a particular gender or minority group or minority firms to meet affirmative action goals and gender quotas are often resisted. When policies and initiatives are not accompanied by fundamental changes at the grassroots level and the support of workers and other stakeholders at all levels, this could further reinforce negative attitudes, gender stereotypes and biases about the capabilities of women and other minority workers or firms (English & Hay, 2015; Kim and Arditi 2010a, 2010b). All employees should be included in conversations concerning DEI at the workplace, and their views should be taken seriously. Therefore, engagement and consultation with various stakeholders are critical for formulating and implementing DEI initiatives successfully.

11.4.3 Support Groups and Not-for-Profit Organisations

Many not-for-profit organisations and professional networking and support groups have emerged worldwide to help women join or advance their careers in the construction industry. They organise site visits, networking events, industry seminars, workshops for personal and professional development, business know-how and more, all of which provide opportunities to network with other construction professionals in a relaxed and friendly environment. For instance, the National Association of Women in Construction (NAWIC), founded in 1953, is an international not-for-profit association dedicated to the advancement of women in the construction industry. The global network of NAWIC organisations includes national and several local branches in countries such as the United States, Canada, the United Kingdom, Ireland, New Zealand and Australia. NAWIC provides a forum for its members to meet and exchange information, ideas and solutions and an opportunity to expand personal and business networks, maintain awareness of industry developments, improve skills and knowledge and mentor or assist other women in the construction industry. In the United Kingdom, Women into Construction is an independent not-for-profit organisation that promotes gender equality in construction. It supports women wishing to work in the construction industry; offers them advice, training, work placements, mentoring and employment opportunities; and assists contractors in recruiting highly motivated, trained women, helping reduce skills gaps and create a more gender-equal workforce.

11.5 Case Studies: Organisational Diversity, Equity and Inclusion Initiatives

In recent years, there has been more focus on building an inclusive culture and providing flexible and family-friendly employment opportunities to support all employees to feel valued and have equal access to opportunities. Many companies in the construction industry are making efforts to prioritise DEI. Following are some examples of the organisational initiatives towards achieving DEI in construction workplaces:

> Aurecon, a large engineering, design, and infrastructure advisory company, focuses on leadership, policies, language and behaviour, and targets in its business strategy to improve gender balance and inclusion in its workforce. Language and behaviour of employees, especially leaders and senior managers, have a significant important influence on organisational culture. AURECON reviewed all organisational policies to ensure they were gender-neutral and explicitly inclusive. It formed the Diversity & Inclusion Council, made up of both leadership and employees, to guide and champion efforts to build diverse and inclusive workplaces for all. Moreover, it introduced policies such as inclusive parental leave and flexible working policies. Some of the outcomes of these policies were achieving the national workforce target of 35% female representation by 2019 and increased target for subsequent years, 88% female employees and 91% male employees saying they have the flexibility they allow them to manage work and other commitments, and the proportion of paid parental leave taken by men increasing from 7% to 40% since the new policy was introduced in late 2017.
>
> (WGEA, 2023)

Diversity and inclusion initiatives at John Holland extend to all employees, including those who do not identify themselves as either male or female. The initiatives promote the organisation as an inclusive workplace that does not discriminate based on a person's gender, colour, ethnicity or background. The following excerpt from their corporate website offers insights into their diversity and inclusion initiatives (John Holland, 2024).

> **Respect for diversity**
> We value people of all genders, ages and cultural backgrounds. We welcome everyone who identifies as LGBTQI+, has a disability and/or is of Aboriginal or Torres Strait Islander background. As an employee, we strongly encourage you to take part in our networks and specialist working groups, including the Pride Network, Celebrate Women Network, GROW Network, Indigenous Inclusion Network, and the Bulabul and Reconciliation Action Plan working group.
>
> **Championing equality**
> Women excel in all types of roles with us, from carpentry to finance, project direction and executive. You can focus on your career development and strengthen your networks with our own women's mentoring program, or in a business-wide mentoring program. We're striving for 40/40/20 gender equality and driving action on gender diversity. Women currently make up 21% of our overall employee base. While this is above the industry average, we want to do more and have a tangible plan to get to 30% by 2025. It goes without saying that people of all genders are paid equally. We monitor salaries in like-for-like positions every year to guarantee this commitment.

Foot in the door

We have priority jobs and training for veterans, refugees and asylum seekers, Indigenous Australians, youths, and people with disabilities, often breaking the cycle of poverty. We partner with social enterprises such as Vets in Construction and Soldier On, which open the door for veterans to work in construction. Refugees and asylum seekers may be interested in our partnership with CareerSeekers, which provides pathways to employment on our projects. The program is open to qualified professionals and graduates, many of whom find employment with us at completion. Young Indigenous Australians who have interacted with the youth justice system can gain work experience on our sites with our ON TRACK initiative.

Healthy work-life balance

At John Holland, we want you to decide how you work best. You are welcome to choose different types of flexible work options. You might consider shifting your start and end times, take extended breaks or work from home. You can also 'buy' additional leave and share your position with a colleague.

Supporting equal parenting

We were the first Australian construction company to implement a paid parental leave scheme. Today, we are encouraging more men to take primary carer's leave. We offer 18 weeks of paid primary carer's leave and 3 weeks of secondary carer's leave. Parents continue to receive superannuation payments while on paid or unpaid leave for 18 weeks. You are welcome to 15 'keep in touch' days while you're on parental leave. When you're ready to come back, we'll support your return with employee and manager toolkits.

AECON is a well-positioned company in the Canadian construction industry, and it is recognised as an industry leader in the development and construction of infrastructure. They have a roster of ongoing major projects here and abroad, a record backlog diversified across multiple sectors and durations and a robust pipeline of future project pursuits. Inclusion is one of the four values of the company in addition to safety, integrity and accountability. Some of the other innovative programs of AECON to improve diversity and inclusion are listed below (AECON, 2024):

AECON Women Inclusion Network (aWIN)

aWIN is a grassroots community for female employees and their allies, committed to building community, driving career development, and championing the advancement of women at Aecon and across our industry. The aWIN community, including many of senior female leaders, are passionate community ambassadors, visiting local schools, sponsoring and participating in community events, and attending speaking engagements that encourage girls and young women to pursue interests in Science, Technology, Engineering, Math (STEM) and trades. Internally, aWIN regularly hosts networking and mentoring sessions, discussion groups, and learning sessions to empower, inspire and support one another.

AECON Women in Trades (AWIT)

Launched in 2019, the AWIT program is dedicated to bringing more women into the trade workforce – helping them thrive and achieve continued success. The AWIT program offers women career-building opportunities in the trades through hands-on training, mentorship and field experience. It is led by Aecon Utilities, a leading utility solutions provider offering innovative, nationwide construction services in the areas of oil and gas, telecommunications infrastructure, and power distribution networks. No prior experience in the trades is required. AWIT covers everything participants need to know on the job. Moreover, contrary to the culture of unpaid training in the industry, participants are paid minimum wage for all the hours of training they complete.

The NexGen Builders Program
Aecon has partnered with the Toronto Community Benefits Network to support The NexGen Builders program, which prepares workers who have been historically underrepresented in the construction industry to successfully enter the workforce. NexGen supports groups including Black, Indigenous and People of Colour (BIPOC), women and newcomers to Canada by providing mentorship by mentors who are experienced construction journeymen and professionals.

Aecon Diversity in the Trades
The Aecon Diversity in the Trades (ADT) program supports construction trades workers who identify as Black, Indigenous and People of Colour (BIPOC). The ADT program operates within Aecon's Civil East division – a diverse, comprehensive market leader focused on expanding and improving Canada's infrastructure and transportation networks.

Aecon-Golden Mile
Aecon-Golden Mile is an innovative joint venture to expand economic opportunities for residents of the Greater Golden Mile area in Toronto who are facing barriers to the labour market. Aecon-Golden Mile (A-GM) is a partnership between Aecon and the Centre for Inclusive Economic Opportunity (CIEO). A-GM is majority-owned by CIEO (51%) and will eventually be operated and staffed by local residents. A-GM strives to be a first-choice employer and contractor for the redevelopment of the Golden Mile area. The area along Eglinton Avenue East between Victoria Park and Birchmount alone will be redeveloped with over 75 towers containing more than 30,000 residential units and hundreds of thousands of square feet of retail and office space over the next few decades. A-GM prioritizes hiring and training residents from Greater Golden Mile postal codes. A-GM hires and trains workers from the Greater Golden Mile, and successful candidates receive tailored job training, competitive pay and opportunities for advancement. We encourage applications from traditionally underrepresented groups in the construction industry, including Women, Black, Indigenous, People of Colour (BIPOC) and newcomers to Canada.

Similarly, there are other programs, such as Aecon Techs in Trades, that offer 3 weeks of paid training, competitive salaries and benefits and a unique opportunity to launch a rewarding career in the trades. Explore Working Warriors is a platform that enables Indigenous job candidates to search for employment opportunities in a more centralised manner.

AECON is also committed to building an LGBTI+ inclusive workplace. It is a proud partner of Pride at Work Canada. Through dialogue, education and thought leadership, Pride at Work Canada empowers employers to build workplaces that celebrate all employees regardless of gender expression, gender identity and sexual orientation.

Another example of how construction companies can help women with their career progression and growth is from Laing O'Rourke.

Laing O'Rourke realised that only recruiting more women would not be enough to create a diverse workforce until attention was paid to their retention and promotion. Few women were promoted to senior and leadership roles compared to men. Especially, the number of women working on construction sites moving into leadership roles was quite low. To address this gap, the company introduced a sponsorship program in partnership with Cultivate Sponsorship. The program matched Executive and Senior Leaders with high-potential women. The program resulted in an increase in the proportion of women promoted to new roles and opportunities across the business. Additionally, more women progressed through project delivery and engineering streams into Construction Manager and Project Leader roles. The program's success also inspired other women in the company by showing them that there are opportunities for women to work in senior leadership roles.

11.6 Benefits of Diversity, Equity and Inclusion

DEI has many benefits for individuals, teams and organisations. Research shows that workplace gender equality is associated with improved overall economic performance, productivity, organisational performance, recruitment and retention and organisational reputation (WGEA, 2018). A diverse workplace is an asset to any industry or organisation. A report by McKinsey & Company (2015) found that gender diversity leads to better performance, and every 10% increase in gender diversity in the senior executive team (and a 1.4% increase for the board) increased earnings before interest and taxes by 3.5%. An inclusive culture supporting all employees to feel valued, have a sense of belonging and have equal access to opportunities could lead to high employee satisfaction, engagement and commercial performance (WGEA, 2023).

Based on a data set encompassing 15 countries and more than 1,000 large companies, McKinsey & Company (2020) found a correlative relationship between business performance and diversity due to greater access to talent and increased employee engagement. The report also found that the greater the representation of gender diversity, the higher the likelihood of outperformance. Companies with women as 30% of the executives were more likely to outperform companies with a lower percentage of women in executive roles. Notably, the likelihood of outperformance continued to be higher for diversity in ethnicity than in gender (McKinsey & Company, 2020).

Other than the above benefits, promoting DEI has become a necessity for construction companies, given the acute shortage of construction workers in recent years due to the aging workforce, low rate of entry and high attrition (Hasan et al., 2021; Morello et al., 2018). The skills shortage in the construction industry demands that construction companies focus on gender diversity and attract talent from traditionally under-represented worker groups, such as women, minority ethnic groups, people with disabilities and LGBTI+ individuals. Moreover, they must retain and support them to sustain their growth, productivity and competitiveness. A lack of a diverse workforce means construction companies will lack innovation and different viewpoints or ideas and experience lower productivity, earning potential and business growth. The need to achieve DEI in construction is not just limited to the notion of 'fairness' or 'the right thing to do' but is linked to the survival and sustainable growth of businesses. Construction companies must recognise the importance of increasing DEI to strengthen their business operations and connect with untapped talent.

11.7 Key Questions

The following list of questions (Table 11.1) is indicative only, not exhaustive. A comprehensive list with specific targets could be developed to assist construction companies in identifying the gaps and taking necessary actions to achieve DEI. The detailed framework/metrics need to undertake a holistic view of strategy, policy and procedures, review and audit and milestones and targets covering recruitment, retention and career advancement of women across different roles. Research shows that more fundamental issues leading to the lack of DEI are related to the organisational and cultural factors and gendered views of construction work, which require both top-down and bottom-up approaches for sustainable change. The targets for metrics would work best when developed in consultation with the relevant stakeholders rather than externally imposed.

Table 11.1 Example questions for achieving DEI

S. No.	Domain	Key questions
1	Organisational strategy and policies	Is DEI an integral part of corporate strategy and business objectives? Does the company have DEI policies, targets and action plans, including recruitment, retention and career support? Does the company have a gender pay equity policy and action plan to reduce pay disparity? Does the company have policies and support available to deal with/prevent harassment and discrimination, sexual harassment and bullying? Does the company have policies and support (e.g., paid leave, etc.) available to deal with domestic and family violence? Does the company monitor the implementation and take-up of policies? Do the leaders and managers demonstrate their commitment internally and externally to achieving DEI?
2	Milestones and targets	Does the company have set DEI targets for (a) executive/leadership/ board positions, (b) management positions, (c) non-management positions, (d) graduate positions, etc.? Does the company regularly monitor gender imbalance in different roles? Do the leaders and managers have specific tasks, responsibilities and accountabilities concerning DEI? Does the company have a clear plan with milestones to achieve the DEI targets? Does the company make the DEI targets and its performance publicly available? Does the company conduct a gender pay gap analysis?
3	Recruitment and employment practices	Does the company remove potential gender biases in the recruitment process, task/role assignment, performance review and promotion? Does the company offer adequate support (caring facilities, caring support, parental leave, etc.) to those with caring responsibilities? Does the company support return to work after a break? Does the company have flexible working arrangements? Are flexible work practices adopted and promoted by managers and leaders? Does the company ensure that employees' flexible work arrangements do not affect their performance reviews? Does the company acknowledge individual circumstances while offering flexible work arrangements rather than a one-size-fits-all policy?

(Continued)

Table 11.1 (Continued)

S. No.	Domain	Key questions
4	Training, mentoring, career growth opportunities	Does the company organise training and professional development programs to create awareness of gender equity, inclusive workplace practices, gender-based harassment and discrimination, sexual harassment and bullying? Does the company have sponsorship programs? Does the company have formal succession planning? Does the company have formal mentoring programs? Does the company support participation in networking events and professional organisation memberships? Does the company consider gender balance for career development, mentoring and sponsorship? Does the company have employee-led gender networks and formal committees informing DEI policies and practices?
5	Role models	Does the company have DEI champions and ambassadors? Does the company use successful employees from different genders and diverse backgrounds as role models? Does the company recognise and reward DEI initiatives and actions?
6	Community awareness	Does the company promote construction careers in schools and the community? Does it work with the community to bust myths about construction careers?

11.8 Conclusion

In many countries, construction is one of the significant contributors to the economy and employment. Moreover, a career in the construction industry offers varied and challenging assignments, a chance to contribute to society by contributing to housing and critical infrastructure development, decent pay and a sense of accomplishment. Still, in the 21st century, when women actively work and contribute across different sectors, the construction industry has failed to attract and retain them. The under-representation of females, non-binary people and other marginalised groups in construction is a broad, complex and long-standing issue and has serious implications for construction companies and society. Many construction companies in different countries face acute skill shortages in trades and professional roles. These skill shortages affect the schedule, cost and productivity of construction projects, contributing to other issues such as unaffordable housing, waste of public money and poor infrastructure.

In construction companies, promoting DEI in the workplace will require removing the long-standing barriers to recruitment, retention and promotion. Retaining women, in addition to attracting more women and supporting their careers, is an important consideration for construction companies aiming to achieve better gender diversity. Women who are joining the industry and want to build a career in construction need to be recognised and better accommodated by making suitable adjustments to workplace policies, practices and the environment. Therefore, construction companies must support individuals with family and caring responsibilities and close the gender pay gap. At the same time, they should extend family-friendly policies such as parental leave and flexible working arrangements to all employees. Moreover, changing employment and work practices to allow people with different abilities and family responsibilities to work in various construction roles and build successful careers is critical. Construction companies must also provide equal employment opportunities and a safe and inclusive work environment to minority ethnic groups, people with disabilities, and LGBTI+ individuals.

References

AECON. (2024). *Training programs*. https://www.aecon.com/join-our-team/diversity-programs/

Australian Government. (2023). *Workplace Gender Equality Act 2012*. https://www.legislation.gov.au/Details/C2016C00895

Barnes, D. (2019). *Response to the APPG for Excellence in the Built Environment inquiry into the recruitment and retention of more women into the construction sector*. The Chartered Institute of Building. https://www.ciob.org/industry/politics-government/consultations/Response-APPG-Excellence-Built-Environment-Inquiry-Recruitment-Retention-More-Women-Construction-Sector

The Chartered Institute of Building. (2022). *Future of construction: Equality, diversity and inclusion*. https://www.ciob.org/industry/policy-research/policy-positions/equality-diversity-inclusion

Dainty, A. R. J., Bagilhole, B. M., Ansari, K. H., & Jackson, J. (2002). Diversification of the UK construction industry: A framework for change. *Leadership and Management in Engineering, 2*(4), 16–18.

Edmond, C. (2023). *International Women's Day: What's the difference between equity and equality?* https://www.weforum.org/agenda/2023/03/equity-equality-women-iwd/

English, J., & Bowen, P. (2012). Overcoming potential risks to females employed in the South African construction industry. *International Journal of Construction Management, 12*(1), 37–49.

English, J., & Hay, P. (2015). Black South African women in construction: Cues for success. *Journal of Engineering, Design and Technology, 13*(1), 144–164.

Government Equalities Office. (2022). *Gender pay gap reporting: Guidance for employers*. https://www.gov.uk/government/collections/gender-pay-gap-reporting

Hasan, A., Ghosh, A., Mahmood, M. N., & Thaheem, M. J. (2021). Scientometric review of the twenty-first century research on women in construction. *Journal of Management in Engineering, 37*(3), 04021004.

Industrial Accident Prevention Association. (2022). *Personal protective equipment for women: Addressing the need*. https://elcosh.org/record/document/1198/d001110.pdf

John Holland. (2024). *Inclusion, diversity & equity*. https://johnholland.com.au/join-us/diversity-and-inclusion

Kamardeen, I., & Hasan, A. (2023). Analysis of work-related psychological injury severity among construction trades workers. *Journal of Management in Engineering, 39*(2). https://doi.org/10.1061/JMENEA.MEENG-5041

Kim, A., & Arditi, D. (2010a). Performance of MBE/DBE/WBE construction firms in transportation projects. *Journal of Construction Engineering and Management, 136*(7), 768–777.

Kim, A., & Arditi, D. (2010b). Performance of minority firms providing construction management services in the U.S. transportation sector. *Construction Management and Economics, 28*(8), 839–851.

Macdonald & Company. (2021). *The Macdonald & Company rewards, attitudes and salary report*. https://report.macdonaldandcompany.com/report/2020-2021/

McKinsey & Company. (2015). *Diversity matters*. https://www.mckinsey.com/~/media/mckinsey/business%20functions/people%20and%20organizational%20performance/our%20insights/why%20diversity%20matters/diversity%20matters.pdf

McKinsey & Company. (2020). *Diversity wins: How inclusion matters*. https://www.mckinsey.com/featured-insights/diversity-and-inclusion/diversity-wins-how-inclusion-matters

McKinsey & Company. (2022). *What is diversity, equity, and inclusion?* https://www.mckinsey.com/featured-insights/mckinsey-explainers/what-is-diversity-equity-and-inclusion

Morello, A., Issa, R. R., & Franz, B. (2018). Exploratory study of recruitment and retention of women in the construction industry. *Journal of Professional Issues in Engineering Education and Practice, 144*(2), 04018001.

Oxford Advanced Learner's Dictionary. (2024). Culture. In *Oxfordlearnersdictionaries.com dictionary*. https://www.oxfordlearnersdictionaries.com/definition/english/culture_1?q=Culture

Padayachie, N. (2019). *Gender equality at work: Some are more equal than others*. Knowledge Resources.

Regis, M. F., Alberte, E. P. V., dos Santos Lima, D., & Freitas, R. L. S. (2019). Women in construction: Shortcomings, difficulties, and good practices. *Engineering, Construction and Architectural Management, 26*(11), 2535–2549.

UK Government. (2023). *Setting new standards for construction*. https://assets.publishing.service.gov.uk/media/5a7a529ce5274a34770e5d7e/ODA_Venue_Factfile_Standards.pdf

United Nations. (2022a). *Concepts and definitions*. https://www.un.org/womenwatch/osagi/conceptsandefinitions.htm

United Nations. (2022b). *Peace, dignity and equality on a healthy planet*. https://www.un.org/en/global-issues/gender-equality

United Nations. (2023). *Gender equality and women's empowerment*. https://sdgs.un.org/topics/gender-equality-and-womens-empowerment

United Nations Economic and Social Council. (1997). *ECOSOC AC 1997.2.DOC: Concepts and principles*. https://www.un.org/womenwatch/osagi/pdf/ECOSOCAC1997.2.PDF

United States Department of Labor. (2022). *Women in construction*. https://www.osha.gov/women-in-construction/ppe

Victoria State Government. (2022). *Building equality policy*. https://www.vic.gov.au/building-equality-policy#download-a-copy-of-the-policy

Watts, J. H. (2009). 'Allowed into a man's world' meanings of work–life balance: Perspectives of women civil engineers as 'minority' workers in construction. *Gender, Work & Organization, 16*(1), 37–57.

Workplace Gender Equality Agency. (2018). *Workplace gender equality: the business case*. https://www.wgea.gov.au/publications/gender-equality-business-case

Workplace Gender Equality Agency. (2021). *Flexible work post-COVID*. https://www.wgea.gov.au/publications/flexible-work-post-covid#_edn2

Workplace Gender Equality Agency. (2022). *Pay equity ambassadors*. https://www.wgea.gov.au/what-we-do/pay-equity-ambassadors

Workplace Gender Equality Agency. (2023). *Case study: Elevating women in STEM*. https://www.wgea.gov.au/sites/default/files/documents/Aurecon_elevating_women_in_STEM.pdf

Workplace Gender Equality Agency. (2024). *WGEA data explorer*. https://www.wgea.gov.au/data-statistics/data-explorer

Wright, T. (2013). Uncovering sexuality and gender: An intersectional examination of women's experience in UK construction. *Construction Management and Economics, 31*(8), 832–844.

12 Knowledge Management

Abid Hasan

12.1 Introduction

Construction companies continuously create knowledge in the organisational context. They operate in an information-rich environment and rely heavily on the knowledge of employees, supply chain partners and other stakeholders to ensure the successful completion of various project activities (Egbu & Robinson, 2005). However, unlike many other organisations, knowledge in project-based construction companies is often dispersed across individuals and teams working with different stakeholders on different projects at various locations.

Each project undertaken by a construction company is temporary and unique by definition. The fragmented and dispersed supply chain makes it challenging for construction companies to institutionalise the retention, transmission and reuse of the new knowledge generated in construction projects. Project team members such as project managers, engineers and technical staff are repositories of knowledge. They possess the knowledge that has been deemed the most critical asset, leading to successful project performance outcomes (Idris & Kolawole, 2016; Shokri-Ghasabeh & Chileshe, 2014). Their extensive construction knowledge, procedural and business acumen and strong relationships with communities and stakeholders are necessary for successfully delivering construction projects. If not managed properly, the knowledge that resides in their minds, created through their interactions and engagement with other stakeholders and performing various organisational tasks, will be lost when they leave the company and therefore cannot be used by others in the company to maintain its competitive advantage.

For example, when a senior project manager who has worked in the construction industry for several years and possesses extensive experience in managing complexities associated with construction projects leaves the company, the company will likely lose that valuable experience and knowledge if it does not elicit and catalogue their expert knowledge. Therefore, employee knowledge that gives a construction company a competitive advantage must be retained. Similarly, knowledge gained on different projects or from stakeholders can be beneficial for construction companies in improving their decisions and organisational processes or practices. Although every construction project is unique, each project team can still obtain helpful information from other project teams' experiences to avoid repeating the same mistake and improve their processes and performance.

Poor knowledge management in a construction company could mean employees making similar costly mistakes in different projects. Similarly, employees will not be up to date with the innovations or improvements made by others in the company and thus spend resources reinventing the wheel. Such mistakes due to deficiencies in knowledge management can be costly, leading to poor project and organisational performance. However, the research shows that many construction companies invest most of their resources in executing projects and prioritising cost

DOI: 10.1201/9781003223092-12

and time, giving little consideration to knowledge management (Yang et al., 2020). Knowledge management practices are still not widespread in companies within the construction industry despite well-known benefits (Yap & Toh, 2020).

The present chapter first introduces the concepts of knowledge and knowledge management. Next, it discusses knowledge management in the context of construction companies. The subsequent sections discuss various knowledge management elements, processes and tools. Next, the chapter presents benefits, challenges and best practices concerning knowledge management in construction companies. Finally, the conclusion summarises the main points discussed in the chapter.

12.2 Knowledge and Knowledge Management

12.2.1 Knowledge

Knowledge can be described as 'a state or fact of knowing', with *knowing* being a condition of 'understanding gained through experience or study; the sum or range of what has been perceived, discovered, or learned' (Alavi & Leidner, 2001, p. 110). It is the awareness of what one has learned through study, reasoning, experience, association or other forms of learning (McInerney, 2002). Knowledge can be explicit (articulated, generalised) or tacit (rooted in actions, experience and involvement in a specific context; Alavi & Leidner, 2001).

Explicit knowledge is easy to identify, record and transmit. For example, explicit knowledge can be gained from reading building and construction codes and project documents. In contrast, tacit knowledge resides in the minds of individuals and is based on their experiences. It evolves from past practices and experiences of individuals and through socialisation with others (Esmi & Ennals, 2009). For example, knowledge of major stakeholders of the company is explicit knowledge. In contrast, the ability and skills to understand their expectations, resolve their concerns and earn their trust and support for the organisational decisions are derived from tacit knowledge. Similarly, employees can read the safety manual and safe work procedures to gain explicit safety knowledge. Still, they would require tacit knowledge for effective hazard identification, risk assessment and prevention in site conditions where multiple contributing factors interact.

Therefore, applying explicit knowledge in real-life construction projects requires a high degree of tacit knowledge due to a dynamic and complex construction site environment. It is one of the reasons construction jobs often need considerable relevant experiences in most roles, as theoretical knowledge cannot always be applied directly in a construction work environment. Similarly, relationship-oriented leadership styles are preferred to task-oriented leadership styles in construction (Esmi & Ennals, 2009). Research shows that innovative products also require a higher degree of tacit knowledge (Esmi & Ennals, 2009). In short, knowledge employees gain from their personal and work experience, or tacit knowledge, is vital to the success of companies in the construction industry.

However, it is essential to note that tacit and explicit knowledge are not segregated but intertwined. Tacit knowledge can lead to the creation of new explicit knowledge (e.g., best practices or lessons learned register) and vice versa (e.g., learning or understanding developed from reading and discussion; Alavi & Leidner, 2001). Construction professionals must use both explicit and tacit knowledge to make the right decisions. Although tacit knowledge is more difficult to formalise, impart, exchange or purchase (Huseman & Goodman, 1998), Marwick (2001) identified activities, such as dialogue among team members, response to questions and storytelling, that can enable the sharing of tacit knowledge. 'Through conceptualisation, elicitation and ulti-mately articulation, typically in collaboration with others, some proportion of a person's tacit knowledge may be captured in explicit form' (Marwick, 2001, p. 815).

In addition to these two main types of knowledge, Jallow et al. (2019) identified four other forms of knowledge, viz. factual knowledge (gained through an observation to verify data and can be measured), conceptual knowledge (based on systems and perspectives), expectational knowledge (mainly derived from expectations or judgements) and methodological knowledge (deals with solving problems and decision-making).

12.2.2 Knowledge Management

There are several definitions and interpretations of *knowledge management*. Alavi and Leidner (2001) defined knowledge management as the systematic process of acquiring, organising and communicating tacit and explicit knowledge of organisational members so that others may use it to be more effective and productive. Alavi and Leidner also presented different knowledge perspectives, describing knowledge as vis-à-vis data and information, a state of mind, object, process, access to information and capability. These different perspectives have important implications for knowledge management practices and systems in companies. For instance, knowledge as a state of mind perspective means that knowledge management should focus on 'enhancing individual's learning and understanding through provision of information' (Alavi & Leidner, 2001, p. 111). In contrast, a process-oriented view of knowledge requires knowledge management to focus 'on knowledge flows and the process of creation, sharing, and distributing knowledge' (Alavi & Leidner, 2001, p. 111).

Becerra-Fernandez and Sabherwal (2014) argued that knowledge management is not lim-ited to knowledge that is recognised and already articulated in some form, such as knowledge about processes, procedures, intellectual property, documented best practices, forecasts, lessons learned, solutions to recurring problems, etc. Instead, it also focuses on managing important knowledge residing solely in the minds of organisations' experts. Similarly, knowledge man-agement should not only focus on the specific knowledge to be created, stored, transferred and applied at an individual level but also address strategic concerns at group and organisational levels (Renukappa et al., 2021). Therefore, knowledge management should be embedded into a company's business vision, strategy and work processes.

In terms of organisational maturity towards knowledge management, McElroy (2002) sug-gested that first-generation knowledge management systems centre around people sharing ideas and knowledge with each other via emails, document management systems, intranet, informa-tion portals, etc. On the other hand, second-generation knowledge management focuses on the social process of employees working together to create knowledge and promote innovation and is more sustainable. Online collaborative workspaces and tools offer informal and unstruc-tured formats and encourage discussions that are ideal for capturing tacit knowledge, provide advanced search capabilities to find and retrieve relevant knowledge quickly and are available anytime/anywhere (McAfee, 2006).

12.3 Benefits of Knowledge Management

Research shows several benefits of knowledge management in organisations. The knowledge present within a company is one of its most vital resources if it can make it available or utilise it when needed. Effective knowledge management could help construction companies remain ahead in the highly competitive and challenging construction business environment. With proper knowledge management processes and technology governance, construction companies can create new resources and processes to develop and maintain their long-term competitive advantage (Aghimien et al., 2023).

The ability to assimilate the knowledge of employees and generate new knowledge provides competitive power to companies (Smith et al., 2005). Yeong and Lim (2010) found that poor knowledge management implementation in project planning and execution led to project failures, as construction companies failed to leverage the knowledge acquired from previous projects. Therefore, proper knowledge management can help construction companies achieve the desired project performance outcomes and provide a competitive advantage.

Knowledge management can facilitate an optimal operating environment as employees can obtain relevant information promptly. Knowledge management in organisations enables individuals to access the knowledge available to expand their knowledge and apply it to meet the organisation's needs. Moreover, it promotes the creation, sharing and leveraging of the collective knowledge and expertise residing in the minds of an organisation's employees, customers and vendors (Becerra-Fernandez & Sabherwal, 2014). Von Krogh et al. (2000) found that knowledge management impacts organisational performance in the areas of risk minimisation, efficiency improvement and innovation.

Knowledge management strategies promote organisational learning, innovation and competitive advantage (Cheung et al., 2013). It allows less experienced team members to learn from the more experienced members, expanding the organisation's knowledge base (Banerjee et al., 2022). Innovation management is essentially about knowledge management, as innovation in organisations depends on knowledge management processes. Knowledge is both an input to innovation and an output of that process (Quintas, 2005). Knowledge management drives innovative ideas across the organisation while improving collaboration (Jallow et al., 2019). Senaratne et al. (2023) also found that knowledge sharing through social networks leads to increased innovation, efficiency and competitiveness for construction companies.

12.4 Knowledge Management in Construction Companies

In the context of construction companies, Renukappa et al. (2021, p. 714) defined knowledge management as 'a systematic and integrative process of coordinating the organisation-wide activities of mapping, capturing, and sharing knowledge by individuals and groups in pursuit of the major organisational sustainability goals and objectives'.

Knowledge management is necessary to transfer relevant information and innovation in project-based construction companies. It enables improvements in organisational processes and helps construction companies maintain a competitive advantage in the dynamic construction environment (Arif et al., 2017). Creating and applying new knowledge to improve processes and innovate is necessary for construction companies to remain competitive locally and internationally. It also allows them to learn from their experiences to make continuous improvements.

For instance, in this era of rapid technological advancement, construction companies undergoing digital transformation must continuously learn from their digital transformation journey and build on experiences through effective knowledge management processes. The knowledge generated during the company-wide implementation of one technology could assist the company in adopting subsequent technology. Suppose that employees were not consulted before adopting a technology that affected their work processes and thus they resisted integrating it into their daily work or routine tasks. Such user resistance could result in limited use of technology and, as a result, lower than expected return on investment in technology. If the company can document and share these valuable learnings across the organisation, it can help improve employee consultation processes for successful technology implementation in the future.

Knowledge management strategies depend on the organisation's motives, as different organisations may want to achieve different objectives through knowledge management (Butler, 2000). For instance, Renukappa et al. (2021) suggested that construction companies could be interested in knowledge management to achieve improvements such as controlled innovation and change, cost reduction, reduced re-work and improved productivity. Therefore, they need to pay greater attention to their knowledge base and use of their existing knowledge.

Duryan et al. (2020) found that knowledge transfer across construction supply chains is necessary to improve occupational health and safety (OHS) management practices and develop a safety culture in construction companies by removing inconsistencies and sharing good practices and lessons learned. Reusing and effectively utilising safety knowledge from previous projects could improve OHS management in construction companies (Deepak & Mahesh, 2023). A knowledge-based culture in construction companies comprising principles and values could foster gathering, storing, sharing and transferring knowledge and learning processes required to constantly improve OHS management practices (Deepak & Mahesh, 2023; Jarvis et al., 2014).

12.5 Knowledge Management Processes

Knowledge management can be considered to consist of four distinct but interdependent basic processes: (a) knowledge creation/construction, (b) knowledge storage/retrieval of knowledge, (c) knowledge transfer, and (d) application of knowledge (Alavi & Leidner, 2001). Table 12.1 shows the basic knowledge management processes, examples and IT systems that can support these processes.

Awareness of knowledge gaps within a company can drive knowledge creation in areas where fresh knowledge is needed. Companies can strategically concentrate their efforts on acquiring and cultivating the expertise necessary to bridge knowledge gaps (Alavi & Leidner, 1999). The process may also involve recruiting individuals with the required skills and expertise, establishing partnerships or collaboration with external entities or upskilling or retraining employees.

The newfound knowledge must be captured and stored in a format that is easy to access and share. Different capture methods, ranging from interviews, case studies, lessons learned and workshops to data mining and artificial intelligence tools, can help companies systematically gather and consolidate knowledge. However, before preserving/transferring knowledge, attention must be paid to the accuracy, completeness, validity and relevance of the captured knowledge. The preserved knowledge becomes an organisational asset that can be transferred to other employees to use in its current form or create new knowledge or innovation in the form of fresh ideas, insights and solutions.

A secure and efficient knowledge repository or platform that suits the volume and type of knowledge must be selected for knowledge storage. Given the rise of cybersecurity threats, data integrity and security measures are crucial to safeguard sensitive information and prevent unauthorised access to knowledge assets. Knowledge stored on cloud-based storage solutions, enterprise content management systems or specialised knowledge management platforms can be vulnerable to data breaches. Data encryption, data backup and recovery procedures, establishing

Table 12.1 Knowledge management processes, examples, and IT systems

Processes	Description	Examples	Support tools, mechanisms and systems
Knowledge creation/ construction	New or replacement of the existing tacit and explicit knowledge through individual, team and collaborative processes	Written documents and reports, work procedures, employee surveys, social interactions, external training	Applications to create new knowledge and record observations, collaborative software, simulation, data mining
Knowledge storage/ retrieval	Developing organisational memory through storage, organisation and retrieval of knowledge	Structured and codified data and information	Searchable electronic databases; cloud-based storage for easy access; online systems
Knowledge transfer	Transfer of knowledge between individuals, from individuals to explicit sources, from individuals to groups, between groups, across groups and from the group to the organisation	Apprenticeships, mentoring, coaching, lessons learned, demonstration and training, knowledge bank	Informal (e.g., informal meetings and discussions) and formal (e.g., training, workshops and seminars) mechanisms, knowledge and expertise mapping, newsletters, blogs, videos and online discussion forums
Application	The source of competitive advantage resides in the application of the knowledge rather than in the knowledge itself	Work methods, checklists, system redesign, problem-solving, productivity and safety improvements in existing systems and processes	Knowledge integration and application tools, workflow systems, automated processes, collaborative platforms

Source: Adapted from Alavi and Leidner (2001)

access permissions, authentication protocols, user awareness and regular cybersecurity audits could protect companies against cybersecurity threats.

Effective mechanisms of knowledge transfer are crucial to disperse knowledge across the company. They enable employees to access, utilise and contribute to the collective knowledge base. However, considering the distributed nature of organisational cognition, transferring knowledge to individuals and locations where it is needed and can be used may not be a simple process in companies with weak knowledge management systems (Huber, 1991). Post-project reviews, discussion forums and face-to-face meetings are the most popular and widely used knowledge transfer mechanisms in construction companies (Carrillo et al., 2006, 2013). However, post-project reviews are often done on management's demand due to time constraints and review results are seldom distributed (Carrillo et al., 2011, 2013).

Finally, the most crucial step is the application of the knowledge to improve organisational processes and performance, drive innovation and gain competitive advantage. Knowledge storage and transfer or sharing alone does not lead to innovation and improvements if it is not applied to improve the existing organisational processes or devise new ones. Stagnant shared knowledge is often ignored or unused (Dave & Koskela, 2009). Therefore, knowledge should be used to drive positive changes in the company.

Since knowledge management consumes valuable organisational resources, companies also need to monitor the impact of knowledge applications on key performance metrics such as productivity, innovation and client or customer satisfaction (Grant, 1996). This ongoing evaluation allows companies to gauge their knowledge management system's effectiveness and identify improvement areas to extract maximum value from their knowledge assets.

It is not necessary that all four knowledge management processes take place in a linear order. For instance, new knowledge can be created and applied in a given context by an individual without storing or sharing it with others. As a result, knowledge is stored only in their memory, and they may decide not to record and share it at a later date. Furthermore, when the same individual applies their new knowledge the next time in a team environment, it may be recorded after application and incorporated into an organisational routine (Alavi & Leidner, 2001).

12.6 An Example of Knowledge Management Process – Lessons Learned

Due to the temporary nature of project teams in the construction industry, team members often move from team to team and/or project to project. As a result, they do not necessarily share experience and knowledge with their colleagues because of the lack of strong working relationships or collaboration or the lack of an organisational system and culture that facilitates and encourages knowledge sharing in temporary teams. Moreover, team composition and size often change during the life of the project, further changing the team dynamics, knowledge networks and relationships. As a result, considerable knowledge is lost as it is forgotten or becomes limited to filing cabinets but never shared or actioned across other projects in the organisation (Orange et al., 1999).

Post-project review or debriefing sessions at the end of each project phase can systematically capture, document and share lessons learned with other employees within and across projects to improve the existing processes (Busby, 1999). During post-project reviews, knowledge about the causes of failures or poor performance, how they were addressed and the best practices could be captured. Lessons learned databases are used in many project-based companies to reposit knowledge and learnings from past projects. The documented lessons learned can be transferred to subsequent projects. However, if this technique is to be effectively utilised, it should take place immediately after a project is completed, and adequate time should be allocated for it (Al-Ghassani et al., 2005).

The Project Management Institute (2004) defined *lessons learned* as the learning gained from the process of performing the project. It allows organisations to retrieve knowledge on past decisions and their implications and thereby help them avoid repeat mistakes (Rasoulkhani et al., 2020). The Construction Industry Institute also recommends lessons learned to organisations that want to promote internal innovation and remain competitive (Banerjee et al., 2022). New knowledge can be created and shared by carefully examining project success, mistakes and unique experiences.

Banerjee et al. (2022) found that lessons learned databases present several opportunities for construction companies to maintain their competitiveness by facilitating enhanced communication among project teams, better knowledge transfer compared to word-of-mouth, problem-solving, avoidance of repeated mistakes and internal innovation among personnel. Construction companies can learn from failures as well as successes of past projects to improve competitiveness and performance (Love et al., 2000). Moreover, it allows employees to expand their professional knowledge and align their individual learning with the broader organisational knowledge management objectives.

Lessons learned between projects can help construction companies prevent repeating the same errors. However, documenting lessons learned is only one of the four knowledge processes and will not be helpful if not followed by other processes. It is also important to note that while lessons learned constitute useful knowledge, it is a reactive form of learning as it is based on completed activities or projects. Therefore, to be effective, the lessons learned must be formally recorded, shared with others in the company and implemented (Orange et al., 1999). For instance, an online searchable database or library of knowledge learned on individual projects and lessons gathered from project participants can be created and made available to other employees. While some construction companies record lessons learned in a project at project closeout, incomplete recording and ad hoc dissemination of findings reduce its effectiveness (Sun et al., 2019).

12.7 Knowledge Management Tools

Knowledge management tools are not necessarily information management tools, as knowledge management processes can be performed without IT tools. Al-Ghassani et al. (2005) categorised knowledge management tools into 'knowledge management techniques' (non-IT tools) and 'knowledge management technologies' (IT tools).

12.7.1 Knowledge Management Techniques

Knowledge management techniques include face-to-face meetings, brainstorming and group discussions, recruitment, training, seminars, post-project reviews, etc. These techniques are widely used in construction companies of different types and sizes as they are generally easy to implement and maintain and can help companies manage tacit knowledge. Abuezhayeh et al. (2022) found that face-to-face meetings, training and seminars are commonly used within companies to build up their employees' knowledge as they are relatively cheap and uncomplicated methods.

Knowledge management techniques, such as brainstorming, can help create new knowledge. Face-to-face interaction can be a powerful knowledge management technique for sharing tacit knowledge. Similarly, construction companies can infuse new knowledge and expertise from outside by recruiting experts and experienced professionals. Apprenticeship allows apprentices to work with their supervisors and learn through observation, imitation and practice. In contrast, senior members of the company can mentor or offer advice to junior members to facilitate their professional development. Construction companies can also use job observation and rotation systems to bring fresh perspectives and encourage knowledge management and innovation.

12.7.2 Knowledge Management Technologies

Examples of knowledge management technologies include IT hardware (computers and smartphones) and software (e.g., online applications) that help facilitate the implementation of knowledge management. Effective knowledge management in project-based companies requires technology to deliver its primary objective; that is, the right knowledge to the right person at the right time. On the other hand, knowledge sharing or transfer will be limited to immediate co-workers or teams without the right knowledge management system and IT support. Therefore, IT can increase the breadth and reach of knowledge transfer by extending the individual's reach to others in the organisation. Right technologies can help construction companies not just better manage knowledge residing within the organisation but also acquire knowledge from outside and other supply chain partners. Moreover, it can help them share knowledge with other stakeholders.

Construction companies often use corporate intranets or employee portals for collaborative knowledge sharing, and information such as contact lists, standard forms and databases, corporate news, and other relevant information are readily available to employees anytime from anywhere on cloud-based platforms. Similarly, companies share information with supply chain partners and other stakeholders through various means such as quarterly and annual reports, business updates and company websites. On the other hand, social networking applications, online blogs, web-based discussion forums and other collaborative platforms are widely used for sharing tacit knowledge (Dave & Koskela, 2009).

The use of technology could support organisational knowledge management activities if it is appropriate for the nature and type of organisational knowledge. For instance, for accessing relevant information, the role of IT would be to facilitate effective search and retrieval mechanisms. In contrast, if knowledge management is about capability and capacity building, IT would need to support the development of individual and organisational competencies to enhance intellectual capital in the organisation (Alavi & Leidner, 2001).

Sometimes, IT can hinder knowledge management processes if it is expensive to implement and maintain, difficult to use and not fit for purpose. Moreover, IT systems can sometimes only help capture explicit knowledge. Emails, intranet, extranet and document management systems can also negatively impact knowledge management capabilities due to information overload resulting from unorganised, frequent and ad hoc information exchange (Dave & Koskela, 2009). Moreover, it is essential to remember that IT can support and facilitate knowledge management processes to a limited extent if it is used incorrectly or is not supported by employees, internal policies and organisational processes.

Therefore, one must not take an entirely positive or negative view of the role of IT in knowledge management. Instead, the appropriateness of IT, its ease of use and user-friendliness, drawbacks and implementation process to support knowledge management must be carefully examined to maximise the benefits. Moreover, while technology could help streamline knowledge management processes, knowledge management can help select the most appropriate technology for the project or assist construction companies in successfully implementing technology. Poor knowledge management can lead to poor decisions concerning the selection and use of technology, which, in turn, could further affect the knowledge creation, retention and sharing processes. Consequently, both knowledge management and technology influence each other.

12.8 Challenges in Knowledge Management

Knowledge management is a challenge for all types of organisations, especially in project-based organisations, due to the complexity, uniqueness and temporary nature of project work, cost and time pressures and high staff turnover (Butters & Duryan, 2019). Construction companies may face several challenges while implementing knowledge management. For instance, lack of trust and fear of sharing sensitive information or losing competitive advantage affect knowledge sharing among companies at the industry level. Within a company, lack of leadership and management support, lack of motivation, constantly changing technology and lack of clarity on the right way to measure knowledge and how to derive anticipated results are some knowledge management challenges. Chinowsky et al. (2007) found six major barriers to successful knowledge management: (a) lack of support from senior management, (b) lack of support from employees, (c) lack of time, (d) lack of money, (e) lack of value measurement and (f) lack of knowledge-sharing infrastructure.

One major challenge to knowledge management implementation in construction companies is the lack of robust processes to support the development of knowledge management and its systematic application in organisational contexts (Zuofa et al., 2015). Banerjee et al. (2022) found that while it is easier for organisations to document explicit knowledge, tacit knowledge is difficult to record. Although methods such as case studies and interviews can help capture tacit knowledge, they are often time-consuming and resource intensive. Banerjee et al. also found that if software applications to record and retrieve knowledge are not user-friendly or straightforward to use, some employees may avoid using them, reducing the effectiveness of the knowledge management system. Similarly, extra effort and time required to enter the data and information could discourage some employees from actively contributing to knowledge management.

Furthermore, organisational culture can hinder knowledge management. A lack of a supportive culture would create mistrust and demotivate employees to share information with other people. Employees with special knowledge and skills may show reluctance to share their expertise if they fear losing their competitive edge over others by divulging such information (Banerjee et al., 2022). Low profit margins, lack of resources, lack of standard processes and infrastructure and the conservative nature of construction companies also affect the implementation of knowledge management (Carrillo & Chinowsky, 2006; Kivrak et al., 2008). Establishing effective and robust knowledge management systems to capture, save and make knowledge accessible could require significant effort and organisational resources (Abuezhayeh et al., 2022), which could be a bottleneck for construction companies working in a slim-margin industry.

Finally, multi-disciplinary and diverse project teams and the presence of specialists and teams from different backgrounds, cultures and disciplines could make it challenging for construction companies to integrate noncommon heterogeneous knowledge (Grant, 1996). However, Fong (2005) found that team members from differing knowledge domains in multi-disciplinary project teams were more likely to discuss their uniquely held information and knowledge than those who had common information. A main challenge, however, is how construction companies capture masses of experience from different individuals and teams and convert it into actionable knowledge to benefit others (Jallow et al., 2019). In the absence of a proper and easily accessible knowledge management system, the potential inconvenience or effort while accessing knowledge from people with different technical or professional expertise (cognitive effort) or who are located in different locations and time zones (physical effort) may reduce people's desire to approach other people for knowledge (Poleacovschi et al., 2019; Woudstra et al., 2012).

12.9 Best Practices for Knowledge Management

The management of information within the organisation is characterised by culture, structure, strategy and systems, as well as the capabilities of individuals (Beijerse, 2000). Therefore, previous studies have identified several factors, such as organisational culture, the commitment of top management, the willingness of employees to share knowledge, individual competencies, incentives, information technology support and friendly systems, which are crucial for successful knowledge management in construction companies (Tan et al., 2012). Essentially, construction companies should focus on three major elements of knowledge management: people, technology and processes (Anumba et al., 2005).

Employees should be encouraged to share their valuable knowledge, especially tacit knowledge. Knowledge management systems and tools should be able to capture the knowledge that lies within their employees' minds and comprises their skills, abilities and practical experience. Gunasekera and Chong (2018b), in their review of critical success factors for knowledge management in construction companies, found support from senior management and leadership for knowledge management implementation as one of the most important critical success factors to bring about the desired project performance outcomes. Gunasekera and Chong (2018a) found that transformational leadership style, shared leadership, camaraderie and trust amongst team members support knowledge management implementation in construction companies.

Senaratne et al. (2023, p. 1224) stated that 'it is the "people" who convert the knowledge to its advantages, therefore a knowledge-sharing culture is essential for an organisation in construction'. A culture that encourages knowledge sharing and collaboration between individuals and teams is crucial for sustainable knowledge management in construction companies. It is necessary to foster an environment that motivates employees to share their expertise and actively engage in knowledge management processes (Marqués & Simón, 2006). A knowledge-based culture would motivate employees to build, share and use knowledge with others in the organisation (Abuezhayeh et al., 2022). The culture should focus on continuous learning and innovation, creating opportunities for experimentation and acknowledging and rewarding innovative ideas.

Active contributors to knowledge management processes and innovation should be recognised and rewarded to motivate others and reinforce the value placed on knowledge management. Moreover, this should break down the departmental and project silos, creating opportunities for cross-functional collaboration and knowledge exchange. Employees are more willing to share information in cohesive working environments (Cheung et al., 2013). Abuezhayeh et al. (2022) found that factors such as the feeling of being competent, job meaningfulness, participation in the decision-making process and decision-making power, improving employees' qualifications and the sense of being efficient, confident and productive at work improve employees' capacity for creating knowledge.

As discussed earlier, technology can significantly affect knowledge management in construction companies with dispersed employees and projects. IT can systematise, enhance and expedite large-scale intra- and inter-firm knowledge management (Alavi & Leidner, 2001). It not only optimises the knowledge management processes but has the potential to automate it. A centralised and user-friendly knowledge repository facilitates effective knowledge management implementation. It ensures efficient access and use of relevant information, promotes collaboration and innovation and fosters a culture of transparency and trust (Andreeva & Kianto, 2012). Tools and methods such as knowledge maps, semantic searches and ontology developments can be used to simplify the process of recording tacit knowledge (Banerjee et al., 2022). Additionally, artificial intelligence tools can help organisations easily extract useful knowledge from historical contract documents, manuals, lessons learned registers and project specifications.

Abuezhayeh et al. (2022) argued that knowledge management should not be seen as a separate process but needs to be aligned and integrated with the organisational strategy, business processes and culture to inform business decisions and achieve innovation and efficiency. Knowledge management should be aligned with the organisational strategy to contribute to organisational success. Knowledge management efforts should be purposeful and directly support the broader strategic vision and corporate objectives (Gupta et al., 2000). For instance, knowledge management can inform organisational decisions concerning the selection of projects and supply chain partners, technology adoption, assigning roles and responsibilities and understanding market requirements to gain a competitive advantage over other construction companies.

Construction companies must pay attention to knowledge integration at the organisational level by synthesising different types of specialised knowledge from individuals dispersed across different teams and locations. The common integrated knowledge can improve teamwork and decision-making among multi-disciplinary project members as they can easily acquire, process and combine specialised knowledge or expertise from multiple sources (Tsai et al., 2015). Knowledge management integration is particularly important at the design and planning stages of a project to build a common understanding of the project objectives and processes and avoid mistakes and confusion in the later stages of the project (Yang et al., 2020). Similarly, knowledge governance should be established to provide formal and informal governance mechanisms, such as rules and policies, organisational structure, resources, incentives and routine, to support and improve knowledge-based processes in a coordinated and systematic manner (Yang et al., 2020).

It is also crucial to ensure that knowledge is accessible to employees with diverse learning styles and preferences. Utilising a combination of different formats, such as reports, videos, infographics and interactive modules, could make knowledge engaging and accessible (Alavi et al., 2005; Gupta et al., 2000). Knowledge management training programmes in many organisations include communication and leadership skill development training, on-the-job training and mentoring, etc., that need to be augmented with more holistic training on knowledge management focusing on other aspects such as capturing knowledge, sharing knowledge, knowledge mapping, creating a culture for knowledge management and efficient use of a blend of IT- and non-IT-based knowledge management tools (Renukappa et al., 2021).

Regular reviews and adaptations of knowledge management are essential to ensure that knowledge management is aligned with changing organisational needs, technological advancements and industry trends responsive to evolving requirements (Carrillo et al., 2004). Moreover, construction companies should conduct benchmarking exercises to identify superior practices in terms of successful knowledge management implementation and ensure that they are not wasting resources by reinventing the wheel (Gunasekera & Chong, 2018a).

12.10 Conclusion

In today's rapidly changing business ecosystem, knowledge management in project-based construction companies could help them develop their capacity to enable sustainable growth. Construction companies can efficiently store and share knowledge to transform it into new products and services or improve their existing processes and gain a competitive advantage. However, they must bridge the knowledge flow gaps between operational and strategic management levels with the help of knowledge management processes, tools and practices. Specific strategies, tools and technologies and a supportive knowledge management culture involving the participation of management and employees can encourage formal and informal information and knowledge sharing across teams, projects, business units and geographies.

Construction companies must be able to capture the constant production of new knowledge at dispersed locations by using efficient knowledge management processes and tools. Consequently, knowledge management in construction companies must be a deliberate and conscious effort based on a structured and coherent approach supported by a knowledge-based culture and top management to achieve and maintain competitive advantage and other strategic goals. Knowledge existing in construction companies must be organised systematically to facilitate its easy access, retrieval and application as and when needed. In the long run, knowledge management processes should be seamlessly integrated into existing organisational workflows and processes, making them a natural part of work rather than a separate initiative.

References

Abuezhayeh, S. W., Ruddock, L., & Shehabat, I. (2022). Integration between knowledge management and business process management and its impact on the decision making process in the construction sector: A case study of Jordan. *Construction Innovation, 22*(4), 987–1010.

Aghimien, D., Aigbavboa, C., & Matabane, K. (2023). Dynamic capabilities for construction organizations in the fourth industrial revolution era. *International Journal of Construction Management, 23*(5), 855–864.

Alavi, M., Kayworth, T. R., & Leidner, D. E. (2005). An empirical examination of the influence of organizational culture on knowledge management practices. *Journal of Management Information Systems, 22*(3), 191–224.

Alavi, M., & Leidner, D. (1999). Knowledge management systems: Issues, challenges, and benefits. *Communications of the Association for Information Systems, 1*(7), 2–36.

Alavi, M., & Leidner, D. E. (2001). Review: Knowledge management and knowledge management systems: Conceptual foundations and research issues. *MIS Quarterly, 25*(1), 107–136.

Al-Ghassani, A. M., Anumba, C. J., Carrillo, P. M., & Robinson, H. S. (2005). Tools and techniques for knowledge management. In C. J. Anumba, C. O. Egbu, & P. M. Carrillo (Eds.), *Knowledge management in construction* (pp. 83–102). Blackwell.

Andreeva, T., & Kianto, A. (2012). Does knowledge management really matter? Linking knowledge management practices, competitiveness and economic performance. *Journal of Knowledge Management, 16*(4), 617–636.

Anumba, C. J., Egbu, C., & Carrillo, P. (Eds.). (2005). *Knowledge management in construction*. Blackwell.

Arif, M., Al Zubi, M., Gupta, A. D., Egbu, C., Walton, R. O., & Islam, R. (2017). Knowledge sharing maturity model for Jordanian construction sector. *Engineering, Construction and Architectural Management, 24*(1), 170–188.

Banerjee, S., Alsharef, A., Jaselskis, E. J., & Piratla, K. R. (2022). Review of current practices for implementing organization-wide knowledge repositories. In F. Jazizadeh, T. Shealy, & M. J. Garvin (Eds.), *Construction research congress 2022* (pp. 1045–1054). American Society of Civil Engineers.

Becerra-Fernandez, I., & Sabherwal, R. (2014). *Knowledge management: Systems and processes*. Routledge.

Beijerse, R. P. (2000). Knowledge management in small and medium-sized companies: Knowledge management for entrepreneurs. *Journal of Knowledge Management, 4*(2), 162–179.

Busby, J. S. (1999). An assessment of post-project reviews. *Project Management Journal, 30*(3), 23–29.

Butler, Y. (2000). Knowledge management – If only you knew what you knew. *The Australian Library Journal, 49*(1), 31–43.

Butters, R., & Duryan, M. (2019). *Boundary objects as facilitators of knowledge transfer in project based organisations*. Academic Conferences International.

Carrillo, P., & Chinowsky, P. (2006). Exploiting knowledge management: The engineering and construction perspective. *Journal of Management in Engineering, 22*(1), 2–10.

Carrillo, P., Harding, J., & Choudhary, A. (2011). Knowledge discovery from post-project reviews. *Construction Management and Economics, 29*(7), 713–723.

Carrillo, P., Robinson, H., Al-Ghassani, A., & Anumba, C. (2004). Knowledge management in UK construction: Strategies, resources and barriers. *Project Management Journal, 35*(1), 46–56.

Carrillo, P. M., Robinson, H. S., Anumba, C. J., & Bouchlaghem, N. M. (2006). A knowledge transfer framework: The PFI context. *Construction Management and Economics, 24*(10), 1045–1056.

Carrillo, P., Ruikar, K., & Fuller, P. (2013). When will we learn? Improving lessons learned practice in construction. *International Journal of Project Management, 31*(4), 567–578.

Cheung, S. O., Yiu, T. W., & Lam, M. C. (2013). Interweaving trust and communication with project performance. *Journal of Construction Engineering and Management, 139*(8), 941–950.

Chinowsky, P., Molenaar, K., & Realph, A. (2007). Learning organizations in construction. *Journal of Management in Engineering, 23*(1), 27–34.

Dave, B., & Koskela, L. (2009). Collaborative knowledge management – A construction case study. *Automation in Construction, 18*(7), 894–902.

Deepak, M. D., & Mahesh, G. (2023). A framework for enhancing construction safety through knowledge-based safety culture indicators. *International Journal of Construction Management, 23*(12), 2039–2047.

Duryan, M., Smyth, H., Roberts, A., Rowlinson, S., & Sherratt, F. (2020). Knowledge transfer for occupational health and safety: Cultivating health and safety learning culture in construction firms. *Accident Analysis & Prevention, 139*, 105496.

Egbu, C. O., & Robinson, H. S. (2005). Construction as a knowledge-based industry. In C. J. Anumba, C. O. Egbu, & P. M. Carrillo (Eds.), *Knowledge management in construction* (pp. 31–49). Blackwell.

Esmi, R., & Ennals, R. (2009). Knowledge management in construction companies in the UK. *AI & Society, 24*, 197–203.

Fong, P. S. W. (2005). Building a knowledge-sharing culture in construction project teams. In C. J. Anumba, C. O. Egbu, & P. M. Carrillo (Eds.), *Knowledge management in construction* (pp. 195–212). Blackwell.

Grant, R. M. (1996). Prospering in dynamically-competitive environments: Organizational capability as knowledge integration. *Organization Science, 7*(4), 375–387.

Gunasekera, V. S., & Chong, S. C. (2018a). Knowledge management critical success factors and project management performance outcomes in major construction organisations in Sri Lanka: A case study. *VINE Journal of Information and Knowledge Management Systems, 48*(4), 537–558.

Gunasekera, V. S., & Chong, S. C. (2018b). Knowledge management for construction organisations: A research agenda. *Kybernetes, 47*(9), 1778–1800.

Gupta, B., Iyer, L. S., & Aronson, J. E. (2000). Knowledge management: Practices and challenges. *Industrial Management & Data Systems, 100*(1), 17–21.

Huber, G. P. (1991). Organizational learning: The contributing processes and the literatures. *Organization Science, 2*(1), 88–115.

Huseman, R. C., & Goodman, J. P. (1998). *Leading with knowledge: The nature of competition in the 21st century.* SAGE.

Idris, K. M., & Kolawole, A. R. (2016). Influence of knowledge management critical success factors on organizational performance in Nigeria construction industry. *Ethiopian Journal of Environmental Studies & Management, 9*(3), 315–325.

Jallow, H., Renukappa, S., Suresh, S., & Alneyadi, A. (2019). *Strategies for knowledge management in the UK construction industry: Benefits and challenges.* Academic Conferences International.

Jarvis, M., Virovere, A., & Tint, P. (2014). Knowledge management – A neglected dimension in discourse on safety management and safety culture – Evidence from Estonia. *Safety of Technogenic Environment, 5*, 5–17.

Kivrak, S., Arslan, G., Dikmen, I., & Birgonul, M. T. (2008). Capturing knowledge in construction projects: Knowledge platform for contractors. *Journal of Management in Engineering, 24*(2), 87–95.

Love, P. E., Li, H., Irani, Z., & Faniran, O. (2000). Total quality management and the learning organization: A dialogue for change in construction. *Construction Management & Economics, 18*(3), 321–331.

Marqués, D. P., & Simón, F. J. G. (2006). The effect of knowledge management practices on firm performance. *Journal of knowledge management, 10*(3), 143–156.

Marwick, A. D. (2001) Knowledge management technology. *IBM Systems Journal, 40*(4), 814–830.

McAfee, A. P. (2006). Enterprise 2.0: The dawn of emergent collaboration. *Enterprise, 2*, 15–26.

McElroy, M. W. (2002). *The new knowledge management.* Routledge.

McInerney, C. (2002). Knowledge management and the dynamic nature of knowledge. *Journal of the American Society for Information Science and Technology, 53*(12), 1009–1018.

Orange, G., Cushman, M., & Burke, A. (1999, March 18–19). *COLA: A cross organisational learning approach within UK industry* [Paper presentation]. Fourth International Conference on Networking Entities, Krems, Austria.

Poleacovschi, C., Javernick-Will, A., Tong, T., & Wanberg, J. (2019). Engineers seeking knowledge: Effect of control systems on accessibility of tacit and codified knowledge. *Journal of Construction Engineering and Management, 145*(2), 04018128.

Project Management Institute. (2004). *A guide to the project management body of knowledge (PMBOK®)* (3rd ed.).

Quintas, P. (2005). The nature and dimensions of knowledge management. In C. J. Anumba, C. O. Egbu, & P. M. Carrillo (Eds.), *Knowledge management in construction* (pp. 10–30). Blackwell.

Rasoulkhani, K., Brannen, L., Zhu, J., Mostafavi, A., Jaselskis, E., Stoa, R., Li, Q., Alsharef, A., Banerjee, S., & Chowdhury, S. (2020). Establishing a future-proofing framework for infrastructure projects to proactively adapt to complex regulatory landscapes. *Journal of Management in Engineering, 36*(4), 04020032.

Renukappa, S., Suresh, S., & Alosaimi, H. (2021). Knowledge management–related training strategies in Kingdom of Saudi Arabia construction industry: An empirical study. *International Journal of Construction Management, 21*(7), 713–723.

Senaratne, S., Jin, X., & Denham, K. (2023). Knowledge sharing through social networks within construction organisations: Case studies in Australia. *International Journal of Construction Management, 23*(7), 1223–1232.

Shokri-Ghasabeh, M., & Chileshe, N. (2014). Knowledge management: Barriers to capturing lessons learned from Australian construction contractors perspective. *Construction Innovation, 14*(1), 108–134.

Smith, K. G., Collins, C. J., & Clark, K. D. (2005). Existing knowledge, knowledge creation capability, and the rate of new product introduction in high-technology firms. *Academy of Management Journal, 48*(2), 346–357.

Sun, J., Ren, X., & Anumba, C. J. (2019). Analysis of knowledge-transfer mechanisms in construction project cooperation networks. *Journal of Management in Engineering, 35*(2), 04018061.

Tan, H. C., Carrillo, P. M., & Anumba, C. J. (2012). Case study of knowledge management implementation in a medium-sized construction sector firm. *Journal of Management in Engineering, 28*(3), 338–347.

Tsai, K. H., Liao, Y. C., & Hsu, T. T. (2015). Does the use of knowledge integration mechanisms enhance product innovativeness? *Industrial Marketing Management, 46*, 214–223.

Von Krogh, G., Ichijo, K., & Nonaka, I. (2000). *Enabling knowledge creation: How to unlock the mystery of tacit knowledge and release the power of innovation.* Oxford University Press.

Woudstra, L., van den Hooff, B., & Schouten, A. P. (2012). Dimensions of quality and accessibility: Selection of human information sources from a social capital perspective. *Information Processing & Management, 48*(4), 618–630.

Yang, X., Yu, M., & Zhu, F. (2020). Impact of project planning on knowledge integration in construction projects. *Journal of Construction Engineering and Management, 146*(7), 04020066.

Yap, J. B. H., & Toh, H. M. (2020). Investigating the principal factors impacting knowledge management implementation in construction organisations. *Journal of Engineering, Design and Technology, 18*(1), 55–69.

Yeong, A., & Lim, T. T. (2010). Integrating knowledge management with project management for project success. *Journal of Project, Programme & Portfolio Management, 9*(2), 8–19.

Zuofa, T., Ochieng, E., & Burns, A. (2015). Appraising knowledge management perceptions among construction practitioners. *Proceedings of the Institution of Civil Engineers - Management, Procurement and Law, 168*, 89–98.

13 Organisational Resilience

Abid Hasan

13.1 Introduction

Modern construction companies face several challenges in a high-risk and volatile business environment. While any one major event, change or disruption, such as natural disasters, complex and disruptive technologies, globalised supply chains, acute skill shortages and an accelerating rate of change in the economy, society and the environment, can be enough to exert significant adverse effects on the functioning and growth of an organisation, many construction companies these days face more than one major disruption concurrently. For instance, in the post-pandemic world, construction companies are dealing with multiple challenges, such as inflation pressure, high interest rates, global economic slowdown, skills and resource shortages, war and unrest in some countries, etc.

Construction companies have little or no control over these external factors resulting from local and global events and increasingly uncertain macroeconomic and geopolitical environments. However, they can plan and prepare for slow and fast, known and unknown disruptions to minimise the negative impact on their business continuity and turn some disruptive events into opportunities. In other words, construction companies can develop resilience to survive and thrive during times of adversity. Organisational resilience is fundamental to running a successful construction company.

In a highly competitive and slim-margin industry like construction, a lack of resilience could quickly lead to insolvencies or business failures. However, when a construction company fails, it has far-reaching consequences due to the vital role of construction in supporting the economy and our communities. Construction companies provide essential and critical infrastructure, employment and economic growth. Therefore, resilient construction companies are needed for the proper functioning of the economy and the community. Similarly, resilient construction companies play a crucial role by continuing their services to communities during and after disasters and crises. They help communities better respond and recover faster from disruptions. Therefore, improving organisational resilience in the construction industry is essential for creating a more resilient economy and society (Sapeciay et al., 2017; Seville et al., 2006).

Research shows a lack of awareness of the benefits of organisational resilience practice and the implications of a failure for resilience in the construction industry, especially among small and medium-sized enterprises (SMEs) and subcontractors (Sapeciay et al., 2017). SMEs may have a low level of resilience due to the lack of knowledge, experience and exposure to crisis management practice. The reluctance to invest time, effort and resources in developing organisational resilience because of slim profit margins and work pressure resulting from tight project deadlines is another challenge leading to low resilience capabilities of many construction companies, particularly SMEs.

DOI: 10.1201/9781003223092-13

This chapter provides guidance on how construction companies can develop resilience to maintain a competitive edge. It discusses the attributes and indicators of resilience, drawing recommendations from broader research and practice. Moreover, it discusses some pressing issues that demand the attention of construction companies and other stakeholders to make the sector more resilient. Finally, it offers insights into actions and opportunities for improving the resilience of construction companies.

13.2 Organisational Resilience

Organisations deal with uncertainties and unexpected events all the time, and managing them presents opportunities and risks for the organisation (Seville et al., 2006). Organisations that embrace change are more likely to prosper. According to the Resilient Organisations (2023), organisational resilience determines an organisation's ability to survive a crisis and thrive in an environment of change and uncertainty. An organisational crisis is 'a significant and unexpected disruptive event that threatens to harm the organisation or its stakeholders and can threaten an organisation's goals, performance and even the viability of the organisation' (Australian Government, 2018). Organisations that are resilient can survive and perform in the face of expected or unexpected disturbances, disruptive surprises and crisis situations using situation-specific responses and ultimately engage in transformative activities to capitalise on these events (Lengnick-Hall et al., 2011; Williams et al., 2017).

While sustainability focuses on the long-term performance of an organisation, *resilience* is the ability to achieve immediate organisational objectives in uncertain and non-routine times (Australian Government, 2011). Similarly, 'resilience is not just sound risk management, effective emergency/crisis management or business continuity management. It is an organisational approach that embraces asset and resource protection, performance and strategic leadership, organisational development, and a responsive and adaptive culture' (Australian Government, 2011). Being resilient is more than creating business continuity plans; it is about building adaptive capacity and flexibility into organisational planning and processes. Resilient organisations prepare themselves to prevent potential or emerging crises using foresight and situation awareness and can turn crises into a source of strategic opportunity (Resilient Organisations, 2023).

Different organisations tend to approach adverse events and situations differently, as shown in Table 13.1. Some organisations may cease to exist or shrink in size when an adverse situation or crisis strikes, while others in the same situation may bounce back quickly or grow due

Table 13.1 Organisational resilience approaches and objectives in adverse events and situations

S. No.	Approach	Description
1	Decline	An organisation accepts that adversity may cause it to cease operating.
2	Survive	An organisation's resilience objective is to exist in a reduced form after adversity.
3	Bounce back	An organisation's resilience objective is to regain its pre-adversity position quickly and effectively.
4	Bounce forward	An organisation's resilience objective is to improve aspects of its functioning so that during times of adversity, it not only survives but possibly gains from the situation.

Source: Australian Government (2020b)

to their higher resilience. Resilient organisations therefore use times of adversity to achieve positive change – so they do not just focus on bouncing back but also take opportunities to bounce forward.

For instance, construction companies that solely rely on imported materials from a particular region could be severely affected by decisions and events such as geopolitical tensions, import restrictions or natural disasters affecting their supply sources. In contrast, construction companies with a more diversified and resilient material supply chain would not only continue operating but also grow their market share while their competitors struggle. Similarly, digital disruptions could negatively affect construction companies that are not prepared to adopt new technologies and change their processes and practices. However, resilient companies will successfully implement emerging and novel technologies to improve their productivity and gain efficiency in business processes. Therefore, construction companies can maintain function and structure in the face of major disruptions and grow if they are resilient.

13.3 Attributes and Indicators of Organisational Resilience

Cultivating organisational resilience is not an easy or straightforward process, especially in the current turbulent economic, social, political and technological environment that presents multiple disruptions (the global pandemic, Great Attrition in talent, economic slowdown, increasing number of natural disasters and extreme climate events and so on) at once (Maor et al., 2022). Table 13.2 shows three interdependent attributes and 13 indicators of organisational resilience identified by Resilient Organisations (2023).

Table 13.2 Attributes and indicators of resilience

S. No.	Attribute	Indicator	Description
1	Leadership and culture	Leadership	Strong leadership to provide good management and decision-making during times of crisis, as well as continuous evaluation of strategies and work programs against organisational goals
		Decision-making	Staff have the appropriate authority to make decisions related to their work, and authority is clearly delegated to enable a crisis response. Highly skilled staff are involved or are able to make decisions where their specific knowledge adds significant value or where their involvement will aid implementation
		Employee engagement	The engagement and involvement of employees who understand the link between their own work, the organisation's resilience and its long-term success. Employees are empowered and use their skills to solve problems
		Situation awareness	Staff are encouraged to be vigilant about the organisation, its performance and potential problems. Staff are rewarded for sharing good and bad news about the organisation, including early warning signals, and these are quickly reported to organisational leaders
		Innovation and creativity	Staff are encouraged and rewarded for using their knowledge in novel ways to solve new and existing problems and for utilising innovative and creative approaches to developing solutions

(Continued)

Table 13.2 (Continued)

S. No.	Attribute	Indicator	Description
2	Networks and relationships	Effective partnerships	An understanding of the relationships and resources the organisation might need to access from other organisations during a crisis and planning and management to ensure this access
		Internal resources	The management and mobilisation of the organisation's resources to ensure its ability to operate during business as usual, as well as being able to provide the extra capacity required during a crisis
		Leveraging knowledge	Critical information is stored in a number of formats and locations, and staff have access to expert opinions when needed. Roles are shared, and staff are trained so that someone will always be able to fill key roles
		Breaking silos	Minimisation of divisive social, cultural and behavioural barriers, which are most often manifested as communication barriers creating disjointed, disconnected and detrimental ways of working
3	Change-ready	Unity of purpose	An organisation-wide awareness of what the organisation's priorities would be following a crisis, clearly defined at the organisation level, as well as an understanding of the organisation's minimum operating requirements
		Planning strategies	The development and evaluation of plans and strategies to manage vulnerabilities in relation to the business environment and its stakeholders
		Stress-testing plans	The participation of staff in simulations or scenarios designed to practise response arrangements and validate plans
		Proactive posture	A strategic and behavioural readiness to respond to early warning signals of change in the organisation's internal and external environment before they escalate into a crisis

Source: Resilient Organisations (2023)

The indicators shown in Table 13.2 can be used to develop resilience health checks and benchmarking tools. For instance, an Organisational Resilience HealthCheck developed by the Resilience Expert Advisory Group of the Trusted Information Sharing Network for Critical Infrastructure Resilience is freely available on the Department of Home Affairs, Australian government website to evaluate resilience attributes and identify opportunities to improve resilience capability in organisations (Australian Government, 2020a). The Health-Check measures 64 resilience indicators (leadership and culture – 30 resilience indicators; networks and relationships – 18 resilience indicators; and change-ready – 16 resilience indicators) on a scale of 1 (*low resilience*) to 4 (*high resilience*). The health check exercise can be performed by an individual (e.g., a risk manager or senior executive) or as a group. Table 13.3 shows some of the examples of high resilience indicators (maximum marks; i.e., 4).

The attribute leadership and culture contributes maximum resilience indicators or points (i.e., 30 out of 64 indicators or 120 out of 256 points) towards the Organisational Resilience HealthCheck score (Australian Government, 2020a). Sapeciay et al. (2017) also found leadership to be the most important organisational resilience indicator, followed by staff engagement, decision-making and situation awareness.

Leaders play a crucial role in developing an organisation's resilience capacity. They must anticipate or notice change around them even when their organisations function in a stable

Table 13.3 Examples of high organisational resilience indicators

Indicator	High resilience indicator
Leadership	Leaders display decisive leadership and innovation and seek opportunity, including in times of adversity.
	Leaders 'walk the talk' and demonstrate behaviours aligned with the values of the organisation.
	Leaders are balanced and strategically focused to ensure that the organisation is acting with control and foresight.
	Leaders are outcome driven/results focused.
	Leaders care for the well-being of their people and their ability to thrive in times of adversity.
	Leaders are empowered to make decisions and are supported in doing so by senior management.
	Leaders act as resilience champions and advocates of the resilience agenda.
Decision-making	The organisation has clear and communicative protocols for mobilisation during adverse events.
	Solutions to problems are encouraged at all levels in the organisation, displaying rapid adaptive behaviour.
	The organisation has clear and transparent processes for escalation.
	Employees are encouraged to use their authority to make decisions in an adverse event.
	Decision-making follows a clear and transparent process.
	Key decisions are recorded and well documented.
	Decision-making is congruent with the organisation's purpose and values to meet expectations.
Employee engagement	The organisation recognises the importance of high employee morale and considers this in planning and response.
	The organisation demonstrates authentic 'care' for employees.
	Employees have a high sense of 'teaming' and collaboration, pulling together in adversity – 'one in, all in'.
	Employees are very clear about their decision-making ability and feel empowered and supported to take action.
	Employees feel strongly connected to the organisation and are likely to go out of their way to support it in times of adversity.
Effective partnerships	The organisation actively collaborates and works with others in partnership.
	The organisation has strong links with its industry peers.
	The organisation is active in the community in which it operates.
	The organisation works hard to develop trusted relationships with suppliers and key customers.
	The organisation has constructive relationships with regulators/authorities.
Stress-testing plans	Plans are rigorously tested to confirm capability with adequate resources available to implement plans and make continuous improvements in line with organisational changes.
	Exercises are designed to identify weaknesses and opportunities for improvement as part of quality assurance and continuous improvement.
	Plans are regularly stress-tested against various scenarios relevant to changing contexts and environments.
	Plans are exercised and tested regularly with other business areas and organisations.

Source: Australian Government (2020a)

environment and ensure that they do not fall into the trap of confirmation bias; that is, only looking for information to confirm their position while ignoring information that contradicts their interpretation of an event or suggests that a solution they developed may not work (Parson, n.d.). Moreover, during times of crisis, leaders are required to make complex decisions amidst uncertainties and identify new directions and opportunities to bounce forward. Effective leaders build hope, optimism and resilience amongst employees and other key stakeholders (Parson, 2010). Therefore, developing and implementing effective adversity leadership practice is vital to strengthening an organisation's resilience capacity to manage uncertainty, complexity and novel situations.

Wilkinson et al. (2016) argued that construction companies need strong leadership, an awareness and understanding of their operating environment, an ability to manage vulnerabilities and an ability to adapt in response to change to be resilient. However, most managers in construction companies are so occupied with daily or routine activities that they do not have time to collect and analyse information that can predict emerging or developing risks in the broader business environment. Construction companies need people from diverse backgrounds who have an inquiring mind and the ability to think systematically in complex situations, think outside their operating sphere, are networked and accept uncertainty to effectively perform risk management tasks (i.e., identify sources of risk, assess risk, control risk and review risk control measures).

Moreover, factors such as the lack of clearly defined roles and responsibilities, executive and management support, resources, employee engagement along with silos within and across organisations, poor corporate culture and complacency in construction companies can inhibit their resilience preparations and efforts. Structural rigidity within companies could stop them from being able to adapt to changing environments, no matter how historically successful they may have been. Organisational boundaries, hierarchical leadership and silo-based planning often impede growth. Companies must ensure effective information exchange to solve complex problems in a rapidly developing context.

Effective decision-making is another important aspect of organisational resilience. Decentralisation of decision-making power is important in construction companies that operate in a highly fragmented and dynamic business environment to make favourable decisions quickly in response to market conditions. *Decision Making During a Crisis: A Practical Guide* (Australian Government, 2018) provides six characteristics of decision-making that companies can implement during a crisis, as listed in Table 13.4.

Table 13.4 Characteristics of decision-making during a crisis

Characteristic	Description
1. Confirm authority and direction	Due to the uncertainty, complexity and dynamic nature of a crisis that usually occurs in the initial phase, there will be a requirement for someone to lead decision-making. In a crisis, changing circumstances may prompt a change in management style from consultative to a 'command and control' approach.
2. Establish psychological safety	Members of the crisis management team need to have a 'safe environment' where they can voice any concerns. However, creating a psychologically safe environment does not mean there has to be a consensus among the crisis management team to make quality decisions. Absolute consensus in a team or groupthink can lead to irrational and dysfunctional outcomes.
3. Use explicit and tacit knowledge	Crisis management teams can use explicit and tacit knowledge to identify indicators and recognise patterns that allow them to choose a course of action that they consider will achieve the best outcome. Accomplished crisis management teams develop and consider multiple options to address the immediate situation and implications for the future.

(*Continued*)

Table 13.4 (Continued)

Characteristic	Description
4. Manage expectations	In crisis management, stakeholders tend to demand action very quickly and, by doing so, create an 'obligation to act' by the organisation. Internal expectations include upholding the organisation's values, complying with internal governance structures and recognising the board's and shareholders' positions. External expectations can involve assumptions made by the public, the media and other organisations currently involved in or impacted by a crisis. It is important to have a well-prepared communications strategy and a dedicated team responsible for managing communication with customers, staff and stakeholders.
5. Manage bias	Cognitive bias or systematic error caused by simplified information processing strategies employed by our subconscious mind may be amplified under dynamic, complex and uncertain conditions. Three common types of cognitive bias that feature in decision-making in a crisis are the following: • Anchoring bias: the common human tendency to rely too heavily, or 'anchor', on one trait or piece of information and then adjust to that value to account for other elements of the circumstance when making decisions. Often, the anchoring occurs early in the decision-making process with the first viable intelligence received. • Confirmation bias: the tendency to process information by looking for or interpreting information that is consistent with one's existing beliefs. This biased approach to decision-making often results in ignoring inconsistent information. • Overconfidence bias: when someone believes subjectively that his or her judgement is better or more reliable than it objectively is. This bias is the tendency for people to be more confident in their own abilities. Leaders and teams can overcome these biases by objectively considering all possibilities, collecting information from different sources and actively listening to dissenting views.
6. Record decisions	Decisions made by crisis management teams potentially affect the whole organisation, stakeholders and the impacted community. Therefore, crisis-driven decisions must be accurately documented. Moreover, decision-makers may later need to defend their decisions and provide justification for decisions. Visualising decisions and decision options also enables better thinking and analysis and allows the crisis management team to share their progress with newcomers.

Source: Australian Government (2018)

It is also important to remember that in an industry environment, resilience cannot be achieved in isolation. Construction companies do not function in silos and are dependent on a large number of stakeholders, such as suppliers, skilled workers and clients, for continued survival and growth. Organisational resilience strategy and efforts must recognise these interdependencies and aim to achieve collective rather than individual ownership wherever possible.

Construction companies must work with other stakeholders to address sector-level disruptions and challenges emerging from skill shortages, material and equipment shortages, climate change and regulatory changes. Advocacy and efforts from several organisations could yield better and faster outcomes while dealing with broader policy and supply chain issues. However, achieving the collective commitment to actively manage resilience issues is a significant challenge due to factors such as a lack of shared goals and purpose and different motivations, difficulty in ensuring ongoing commitment and the influence of different organisational cultures (Seville et al., 2006). Moreover, organisations fear that close collaboration to improve resilience may require sharing sensitive business information with other supply chain partners.

13.4 Key Resilience Challenges for Construction Companies

Nowadays, construction companies deal with several challenges affecting both short- and long-term resilience. Recently, we witnessed how price escalations, skill shortages, border closures and other supply chain challenges caused by COVID-19 pandemic disruptions impacted construction companies worldwide, leading to a large number of insolvencies in the sector. Construction companies that relied on a migrant workforce and imported construction materials and equipment struggled financially and operationally from border closures. The government-imposed restrictions reduced migration and disrupted the flow of skilled workers and other resources. Many construction companies suffered significant losses in fixed-price or lump sum contracts due to project delays and cost overruns. Similarly, a rise in geopolitical instability and political conflicts in different countries continue to impact resource supplies in different parts of the world. Moreover, adverse and unpredictable weather events due to climate change severely affect construction projects in some regions.

Various supply chain disruptions and uncertainties are causing very high inflation in material, tool and other resource prices. Price uncertainties have disrupted assumptions used to estimate costs of materials, labour and other resources. It is becoming extremely difficult for businesses to accurately budget for and forecast contingencies for new construction projects. At the same time, an increase in the number of insolvencies, contract defaults and disputes has affected clients' confidence in the construction industry. In short, both slow changes (e.g., an ageing workforce, skill shortage, etc.) and rapid or unexpected events (e.g., the COVID-19 pandemic, natural hazards such as earthquakes, geopolitical tensions, etc.) can create significant disruptions to construction companies if they are not prepared to respond and recover from them and attain growth amidst all challenges. Let us look at some of these long-term resilience challenges in more detail.

13.4.1 *Financial Vulnerabilities*

Poor financial management is not uncommon in the construction industry, as reflected in the high number of insolvencies. Coggins et al. (2016) discussed several characteristics of the construction marketplace that explain the high vulnerability of construction companies to insolvency. First, due to the prevalence of subcontracting and the existence of pyramidal subcontracting chains on construction projects, insolvency, payment default or poor payment practices of a party higher up in the contracting chain may create a financial strain on lower-level parties due to the chain effect. The failure of timely payment by one organisation to another could affect several other organisations in the supply chain. The snowball effect of a failure of large construction companies forces many SMEs and suppliers into insolvency, as often witnessed in practice.

Second, there is a predominance of trade credit in the construction industry; that is, a gap or delay of weeks or months between delivering the product and services and receiving payments. While advance payment arrangements could improve liquidity, the way the construction industry works is the opposite, as contractors and suppliers often have to wait for payment after completion of the job. Moreover, disputes are common in variation claims processes that further delay the payment process, causing a severe financial burden on construction companies. Construction companies are often undercapitalised and rely heavily on credit and borrowed capital. Insufficient working capital and the resultant dependency on timely cash flow mean that many construction companies are not financially resilient in the event of any financial problems.

Third, to win a contract on the basis of the lowest cost to the client in a highly competitive industry, many construction companies underbid at a price lower than cost, adopt a marginal cost pricing strategy or operate at a very low profit margin. As a result, in the event of a crisis leading to price increases, it becomes impossible for them to deliver fixed-price contracts. For instance, a 2022 report by the Australian Constructors Association and Arcadis found that construction material input prices for materials such as processed bar reinforcement, structural steel, ductwork and cables rose by up to 70% or more in the 12 months leading up to the publication of the report. The report also found that changing dynamics, from rising material costs and supply challenges to labour and skill shortages, have created an extremely volatile market.

Fourth, inequitable risk allocation in construction contracts exposes construction companies to many intertwined and complex financial risks. Additionally, regular use of retention monies, performance bonds, prequalification of tendering contractors and contractual 'proof of payment' clauses create a financial burden. Finally, the lack of financial management skills and expertise in some companies further aggravates cash flow and financial mismanagement issues.

13.4.2 Supply Chain Vulnerabilities

Supply chain vulnerability refers to 'the existence of random disruptions that lead to deviations in the supply chain operations from normal or planned activities, all of which cause negative consequences for the involved construction parties' (Svensson, 2000, p. 732). Vulnerabilities could arise from internal and external factors related to strategy, management, personnel, process, technology, market pressure, the regulatory environment, etc. For example, the knock-on effect of COVID-19 transportation restrictions and manufacturing unit closures resulted in material and equipment shortages and a steep rise in the cost of imported materials and equipment prices in many countries, causing financial pressure on construction companies, especially in fixed-price contracts. Similarly, technology and information system failures or industrial actions and policy changes can disrupt the supply chain network and functions.

The transient project-based supply chain with complex interdependencies between many organisations presents unique challenges in the construction industry. Due to a lack of information sharing and the fragmented nature of the industry, construction companies are exposed to overlapping risks that are wider than their immediate contractual responsibilities (Loosemore, 2000). The limited visibility of the supply chain and lack of information and collaborative data sharing among project team members make it difficult for construction companies to detect any disruptions arising from their supply chain (Zainal Abidin & Ingirige, 2018). The poor performance of the supply chain in the construction industry has been identified as one of the leading causes of low productivity, project delays and cost overruns (Hasan et al., 2018; Zainal Abidin & Ingirige, 2018).

13.4.3 Skill Shortages

The shortage of construction workers is a global problem (Farmer, 2016) and has become more acute in the post-pandemic world. Building and construction businesses list worker shortages as their biggest issue. This includes attracting and recruiting staff as well as the availability of subcontractors and other service providers (Master Builders Australia, 2023). The labour shortage is forecasted to worsen, with many ageing trade workers retiring in the coming years.

Skilled workers and professionals are a critical resource for construction companies, as construction work execution primarily depends on human resources. In many countries such as Hong Kong, Singapore, Australia, Gulf Cooperation Council countries and Western countries, the industry depends on migrant workers and professionals as skill demand far exceeds skill availability in the local market. For instance, Master Builders Australia (2023) estimated that 486,000 workers will need to enter the Australian building and construction industry by the end of 2026. However, the dependence on foreign supply puts many construction companies at risk, especially during disasters and geopolitical crises. Moreover, the process of attracting overseas human resources is heavily regulated and impacted by systemic issues in many countries.

Construction companies need new skilled workers to maintain and achieve capacity growth and to replace those retiring or leaving the company. However, the high attrition rate of skilled workers is common in the construction industry due to the fluctuating availability of work, burnout from the physical nature of construction and relocation requirements. Unfortunately, despite being a people-intensive sector, construction companies are known for poor human resource management (HRM; Dainty et al., 2007). Many SMEs generally do not have a separate, adequately resourced HRM department. Moreover, the high rate of subcontracting that has existed in the industry for a long time adversely affects HRM functions (Debrah & Ofori, 1997). When the demand for skilled workers exceeds the supply, the turnover rate is high in companies with poor HRM practices. Moreover, inadequate employment practices, hazardous work conditions, unclear career paths for construction workers, lack of training and education and poor work culture are common in many construction companies, which reduces the attractiveness of construction workplaces among young people.

Worker shortages could jeopardise the future of construction companies, affecting their capability to deliver the current projects and limiting their capacity to expand or grow their business. Skills shortages and turnover risk for critical talent persist in the construction industry in many countries. LinkedIn Learning's 2023 Workplace Learning Report found that many organisations are grappling with unprecedented employee turnover in the wake of the pandemic (LinkedIn Learning, 2023). Skills shortages in the available worker supply create inflationary pressures on salaries and site operating costs (Smith, 2019). The consequences of skills shortages include project delays and cost overruns, lack of housing and infrastructure, greater reliance upon foreign workers, quality and safety issues and other compromises within the built environment due to a shortage of people who possess the necessary knowledge, skill and expertise to perform critical tasks properly and safely (Bilau et al., 2015; Karimi et al., 2018). Moreover, skill shortages and resulting higher employee turnover could encourage construction companies to poach talent rather than create talent.

In a skill-short market, some companies offer higher salaries and other incentives to attract talent from competitors that cannot match those offers, leading to an increased turnover rate. As a result, companies may stop or reduce investment in employee training and skills development as they fear employees could join another organisation after upskilling and the money and resources spent will be lost. However, if many companies stop investing in developing talent and refuse to upskill their current workforce, the talent pool will dry up, and productivity will decline in construction companies and across the sector. Moreover, companies will not be prepared to take on emerging or unknown challenges or avail themselves of new opportunities due to their reduced resilience capacity.

13.4.4 Poor Safety, Health and Well-being

A safe and healthy work environment is extremely important for a productive and resilient workforce and companies. However, the construction industry worldwide remains a high-risk workplace due to the hazardous nature of work and work environment, resulting in a very high number of incidents, injuries, diseases and fatalities. It is one of the riskiest industries for work-place accidents. Construction fatality rates in developing countries are much worse – three to six times higher than those of developed nations (Umar & Egbu, 2018). Government data and empirical research show higher rates of work-related musculoskeletal diseases and other health-related problems, such as hypertension and obesity, among construction workers than among workers in many other sectors (Chung et al., 2018; Fuller et al., 2022).

Construction workers have been found to experience high stress and long working hours and adopt unhealthy lifestyles such as harmful alcohol and drug consumption and smoking (Duckworth et al., 2024). Mental health problems are prevalent in the construction industry due to poor work–life balance, high job demand, poor cultural norms and mental health stigma, chronic bodily pain, lack of social support, workplace injustice and job insecurity. Statistics on suicides reflect the severity of mental health problems in the construction industry. In countries such as Australia and the United Kingdom, the number of deaths among construction workers from suicide is more than the national average (Duckworth et al., 2024).

Poor safety, health and well-being in the construction industry have significant direct and indirect costs to individuals, their families, construction companies, society and the government and lead to an increase in the total labour cost, compensation and insurance premiums, sick leaves and absenteeism, employee turnover and reduced employee performance, productivity and morale (Loudoun & Townsend, 2017). Concerns related to poor safety, health and well-being are also significant deterrents to attracting young people and females to the construction industry, thus leading to skill shortages.

13.4.5 Unsustainable Construction Practices

Construction companies consume excessive material resources in construction projects, exerting pressure on natural resources and materials reserves worldwide. Many construction material manufacturing processes and construction building methods are energy intensive and contribute significantly to greenhouse gas emissions (Lim et al., 2015). Moreover, the construction industry produces significant liquid and solid waste during new construction as well as renovation and demolition of existing facilities. The average annual construction and demolition waste is more than 2 billion tons, which poses significant environmental threats (Silva et al., 2019). Gounder et al. (2023) argued that the future of the environment significantly relies on the decisions made by construction companies and other stakeholders concerning the selection and use of construction materials in building projects.

The focus on sustainability and sustainable construction practices has increased in recent years due to climate change and greater awareness of the severe adverse impacts of the construction industry on the environment. The pressures posed by global warming and climate change have led construction stakeholders to question the viability of the widely adopted economic linear production and consumption model of making, using and disposing of the used material or product as waste and demand concrete actions from construction companies to reduce their carbon footprint. Furthermore, changes in building and infrastructure regulations in different countries driven by net zero targets will likely force companies to eliminate or reduce their unsustainable practices. Construction companies need to disturb the current culture of design, build and waste by replacing linear models of material consumption with circularity or circular construction with the help of new business models and processes.

13.4.6 Technological Disruptions

The increased adoption and use of construction technologies and digital tools (e.g., robotics, drones, building information modelling [BIM], etc.) present great opportunities for improving efficiency and productivity and the development of the construction industry, as discussed elsewhere in the book in the digital transformation chapter. However, digitalisation also poses threats to construction companies that continue to use traditional methods and have low digital maturity. Many construction companies are struggling to understand how to manage the change from current construction practices to digital construction (Lavikka et al., 2018). The limited financial capacity of SMEs to invest in technologies, uncertainty about benefits, low motivation, lack of the right skills, etc., are other challenges to the digital transformation of construction companies.

As digitisation and technologies such as artificial intelligence and BIM continue to penetrate further into the industry, driven by the push from public and private sector clients and organisations with higher digital maturity, the vast majority of construction companies will need to adjust and adapt their practices to survive and grow. Similarly, construction companies with better integration of digital supply chains and industrialised construction (IC) in their business process and delivery models are likely to be more competitive and productive than companies solely relying on traditional methods of construction and supply chain management. As a result, many construction companies that are slow to respond to technological disruption will struggle with the increased adoption of IC and digital technologies in the construction industry.

13.5 Opportunities and Actions for Improving Resilience of Construction Companies

Construction companies are exposed to regular changes in the regulatory environment, market dynamics, new competitors, disruptive technologies and other short- and long-term shocks such as natural disasters or pandemics like COVID-19. The constantly changing business environment and unexpected changes demand that construction companies be able to adapt and change their strategy and practices to suit the needs of the new conditions. Organisational resilience is critical for the survival of construction companies in an increasingly uncertain and dynamic business environment.

It is in the interests of the government and other stakeholders to work together on solutions to make construction companies more resilient. Legislative reforms, along with capacity building, innovations, payment security, proactive risk management and the development of a robust supply chain, could help construction companies mitigate various risks and build resilience at a time when it is becoming difficult to predict external factors in an increasingly dynamic business environment.

13.5.1 Financial Resilience

Improving financial resilience through better management of cash flow and security of payment regulations throughout the construction industry is critical for a sustainable, resilient construction sector. Construction companies must maintain high working capital requirements and upfront cash outlays in construction projects to support mobilisation costs, wages and other direct and indirect costs (Crookes, 2022).

Many cost items are susceptible to market conditions, and prices of materials and equipment can fluctuate considerably over the project's life. The reasons for the price increases are largely beyond the control of construction companies. Pricing for unforeseen escalation could make the bid price less competitive and thus force the contractor to take financial risks. In some cases, it may turn out to be a positive risk if the prices decline. Construction companies should carefully review the price adjustment clauses (rise and fall) in contracts to ensure that unaccounted fluctuations are dealt with fairly.

Special attention should be given to construction projects where project scope and design are poorly developed and risks cannot be quantified because lump sum price contracts in those projects can put construction companies at significant financial risk. Early ordering and storage of price-sensitive materials could protect construction companies against hyper price escalation in cases where the contract does not allow for a cost adjustment (e.g., lump sum contract or fixed-price contracts negotiated and locked in years prior) and suppliers and subcontractors are unable or unwilling to offer fixed prices for the duration of their services.

13.5.2 Supply Chain Resilience

Supply chain resilience can be defined as 'the supply chain's ability to react to the negative effects caused by disruptions that occur at a given moment to maintain the supply chain's objectives or recover to a better state' (Zainal Abidin & Ingirige, 2018, p. 416). Supply chain resilience can be measured based on supply chain vulnerability and capabilities. Vulnerabilities were discussed earlier in the chapter. Capabilities are the 'attributes that enable an enterprise to anticipate and overcome disruptions' (Pettit et al., 2010, p. 6). Supply chain capabilities could enable supply chains to prepare and respond effectively to adverse uncertainties and disruptions (Ponomarov & Holcomb, 2009).

Supply chain resilience capabilities consist of (a) proactive capabilities (provide withstanding abilities to the supply chains) and (b) reactive capabilities (provide abilities of supply chains to respond to change by adapting their initial stable configurations; Wieland & Wallenburg, 2013). Examples of capabilities are flexibility, capacity, market position, financial strength, resourcefulness, etc. Organisations can use capabilities to address the vulnerability and prevent or mitigate disruptive events or enable adaptation following a disruption (Pettit et al., 2010; Zainal Abidin & Ingirige, 2018).

Construction companies with digital technology capabilities could utilise BIM and Internet of Things platforms to collaborate from manufacturing to delivery, facilitate error-free design and resolve product quality issues. Similarly, enterprise resource planning software could provide better supply chain visibility and timely warnings on resource shortages (Ekanayake et al., 2022). The skill shortage and associated challenges require innovations in supply chain management. For instance, to avoid skilled worker shortages and very high labour costs in the Hong Kong construction industry, companies can procure prefabricated components from suppliers located in Mainland China as they are more cost-effective due to lower labour costs in China (Ekanayake et al., 2021).

Construction companies should ensure the availability of resources for continuous operation even if their most used or preferred supply chain is disrupted. Having backup supply chain partners that can provide materials, equipment and other resources is of paramount importance to avoid supply–demand mismatch or shortages over time. Developing redundancies in the supply chain as a capacity-building measure can help construction companies maintain continuity and facilitate quick recovery by overcoming disruptions, such as supply chain breakdowns due to quality issues, supply shortages and labour disputes (Ekanayake et al., 2022; Sheffi & Rice, 2005). Therefore, construction companies can improve their supply chain resilience by developing and maintaining relationships with multiple suppliers located in different regions and using digital technologies and innovative measures.

13.5.3 Meeting the Skill Demands

Construction companies can combat skill shortages to some extent with the right measures and strategies. For example, they can invest in their employees and HRM practices to improve job satisfaction, productivity and retention. Lawani et al. (2022) recommended that health and safety concerns, unpaid sick leaves, lack of holiday pay, lack of innovation and lack of job security be addressed within the industry to attract and retain workers. Employees will likely stay with the employer that offers financial and non-financial benefits and opportunities such as attractive remuneration, learning opportunities, work–life balance, a supportive work environment and career growth opportunities. Moreover, building and maintaining a positive corporate image and reputation and a positive perception of the company's culture and values are essential in attracting and retaining new talents.

It is important to remember that skills shortage is a complex, multi-layered issue that demands actions at different levels and coordinated efforts from different stakeholders. There is no one 'silver bullet' for the skills shortages, and a series of measures need to be adopted. A few suggestions from the Australian Construction Industry Forum (2022) are as follows:

- A properly funded and resourced vocational education and training system and adequate funding for university courses for construction-related disciplines.
- Attracting and retaining more women in construction.
- Attracting more skilled migrants.
- Promoting greater digitalisation in construction.
- Promoting greater prefabrication in construction.
- Greater alignment of future workforce and construction planning.
- Encouraging greater interstate and intrastate migration and ensuring that the Automatic Mutual Recognition of licences system works effectively.

Construction companies must focus on expanding the worker pool rather than competing for the same pool of workers (e.g., one company poaching workers from other companies). They must work closely with other stakeholders, such as the government, vocational training providers and higher education institutions, to create opportunities for bridging the skill gaps and shortages. Governments can help educational institutions offer construction courses and training at discounted fees or provide scholarships to encourage people to choose construction careers. Scholarship opportunities for retraining could enable people to change their professions and join the construction industry. Educational institutions should recognise and provide work-integrated learning to prepare industry-ready graduates.

Governments can implement measures to attract more foreign employees to the construction industry by increasing annual quotas and providing new migrant visa benefits (Ceric & Ivic, 2020). Similarly, government departments or professional organisations authorised by the government could help construction companies assess migrant workers' qualifications, skills and experience. For example, a migration skills assessment outcome letter is needed from Engineers Australia, the peak body for the engineering profession in Australia, if a person wants to migrate to Australia as an engineer.

However, it must be noted that many construction companies are reluctant to hire foreign workers and professionals even after successful skill assessments by an independent body. If the concern is a lack of experience in the local construction industry, construction companies can collaborate with relevant stakeholders to provide additional training and exposure to migrants to successfully absorb them in the local industry to meet the skill demands. With adequate learning, development and training, employees are better positioned to accomplish the job they were trained for and prepare themselves for the future. Diverse and practical training offered to workers can reduce turnover and attract new talent. Empowering employees with career development tools expands workforce skills. As a result, construction companies can act quickly on valuable opportunities.

Construction companies should also try to utilise ageing and older workers' strengths, such as experience, transferable skills and trustworthiness, by keeping them in the workforce longer by moving them into supervisory roles or offering them roles that are less physically demanding. Similarly, construction companies should seek workers from other sectors possessing transferable skills and those looking to return to work by providing additional training. Some construction companies in the United Kingdom and Australia have provided employment opportunities to ex-offenders, whereas others have increased their apprenticeship intakes. Similarly, many construction companies have adopted measures to attract and retain more females, as discussed elsewhere in the book. The construction sector remains predominantly male and able-bodied, which needs to be changed by implementing top-down and bottom-up measures (Hasan et al., 2021; Powell & Sang, 2013).

Construction companies also need to initiate employee expansion with modern recruitment tactics to reach skilled workers. Some useful recruitment tactics to effectively expand the construction workforce are job fairs, referrals and word of mouth by getting employees involved, attracting the next generation of construction workers and professionals by offering work placements and internships and using social media platforms such as LinkedIn. Construction companies can also hire recruitment agencies or recruiters specialising in finding and recruiting workers and professionals.

Essentially, construction companies must engage the entire talent pool to address the skills shortage by hiring a diverse workforce and attracting prospective employees from new sources of talent, including veterans, persons with disabilities and returning workers, in addition to women and underrepresented ethnic groups (Deloitte Research Center for Energy & Industrials, 2023). Moreover, construction companies should embrace new technologies and automation and use manufacturing methods to move from manual to industrialised production. Prefabrication or IC can reduce workforce requirements on project sites and improve construction productivity. Similarly, more applications of machines and robotics should be explored for performing site-based tasks to reduce the dependence on manual labour. Modern methods of construction, including off-site manufacturing, are expected to become the new normal in the construction industry in the coming years (Oxford Economics, 2021).

13.5.4 Resilient Workforce

A safe and healthy work environment supported by a positive, respectful and supportive workplace culture is important for the success and well-being of employees, who play a key role in building organisational resilience. A resilient workforce will be more adept and capable of handling business changes during uncertain and disruptive times. On the other hand, a non-resilient workforce will resist changes, show mistrust in organisational leadership and decisions and feel vulnerable and de-motivated.

Construction companies must focus on improving the physical and mental health aspects of workforce resilience, given the high prevalence of physical and mental health disorders among the construction workers, as discussed earlier. Safety could be promoted by implementing safe design and prevention through design principles to eliminate or minimise risks in the design and early stages of the project. Construction companies should develop a positive safety culture by placing a high level of importance on safety beliefs, values and attitudes within the company. Moreover, leaders must show their commitment to providing safe workplaces by prioritising health and safety in their organisational work practices and decisions and allocating sufficient resources for proper safety management (Hasan & Kamardeen, 2023).

Kamardeen and Hasan (2023) suggested regular psychosocial hazard risk assessment and mental health audits to identify and reduce or eliminate psychosocial risk factors. Campbell and Gunning (2020) recommended several strategies for construction companies, such as establishing manageable workloads; ensuring good work–life balance; maintaining a strong, supportive and inclusive culture; and zero tolerance for bullying and harassment to improve workers' mental health.

13.5.5 Sustainable Construction

Sustainability has become a business imperative, and construction companies must adapt to evolving market trends and environmental regulations and meet customer demands for sustainable construction and buildings. They need to shift their design and construction principles and practices, focusing on reducing the embodied carbon of key construction materials, procuring lower-carbon construction materials, adopting passive design principles, and using energy-efficient equipment. Moreover, they can incorporate circular economy principles early in the design and construction process, reusing materials, designing structures for disassembly and reducing waste. Construction companies also need to pay attention to ethical sourcing of resources such as material and labour.

Both sustainable design and focus on sustainability during the design stage are critical for achieving sustainability in the construction industry. As decisions made during the design phase of the project influence design parameters, the choice of construction materials and energy consumption in completed buildings, stakeholders need to work together to ensure that those decisions are based on sustainability considerations. Factors such as government regulations and incentives, market demand, cost-efficiency of sustainable materials and availability of expertise and information are critical for promoting sustainability in the construction industry (Gounder et al., 2023).

Technology and techniques such as BIM, Internet of Things, data analytics, energy optimisation techniques and simulation software can allow construction companies to model energy performance and reduce their carbon footprint. Moreover, IC, 3D printing and advanced manufacturing could help companies reduce construction waste and optimise material use. Cost-effectively delivering sustainable projects can help construction companies position themselves as ethical organisations that care about the environment. It can also help them differentiate themselves in a saturated and competitive market to gain a competitive advantage (Deloitte Research Center for Energy & Industrials, 2023). Therefore, a long-term strategy for achieving sustainability must be developed to achieve a resilient future.

13.6 Conclusion

While globalisation has offered several opportunities to various sectors, including construction, it has exposed organisations to global economic and geopolitical volatilities. Events such as the COVID-19 pandemic or wars in some countries have far-reaching consequences due to interdependent supply chain networks. The risks associated with economic, technological and geopolitical landscapes are putting increased pressure on construction companies. As a result, the viability and sustainability of construction companies continue to be tested in a world that is constantly changing.

The level of resilience can determine the failure, survival or growth of construction companies operating in an ever-changing business environment pressured by the economic slowdown, skill shortages, geopolitical tensions, supply chain disruptions and regulatory and policy reforms. Construction companies must invest resources in increasing their resilience capacity and capabilities to not only bounce back but also bounce forward after a major disruption. A regular resilience health check and scenario planning could assist companies in checking their resilience level and identifying areas for improvement.

The chapter discussed some of the critical issues that will continue to test the resilience of construction companies, such as skill shortages, financial vulnerabilities, supply chain vulnerabilities, technological disruptions, sustainability and health, safety and well-being concerns. It also offered suggestions to build resilience in those areas. In addition to resilience capacity building within the company, construction companies must work with other stakeholders to address sector-level disruptions and challenges. Cultivating organisational resilience is not a quick or straightforward process and may require significant time, effort and resources, but it is critical for the survival and growth of construction companies.

References

Australian Construction Industry Forum. (2022). *Comprehensive strategy needed for skills shortages.* https://www.acif.com.au/newsletters/id/689/idString/yfntj13323

Australian Constructors Association & Arcadis. (2022). *Market sentiment survey: February–March 2022.* https://www.constructors.com.au/wp-content/uploads/2022/06/Market-Sentiment-Survey-Results-2022_FINAL-V3-3.pdf

Australian Government. (2011). *Organisational resilience: A position paper for critical infrastructure.*

Australian Government. (2018). *Decision making during a crisis: A practical guide.* https://www.organisationalresilience.gov.au/Documents/decision-making-during-a-crisis-a-practical-guide.pdf

Australian Government. (2020a). *HealthCheck.* https://www.organisationalresilience.gov.au/HealthCheck/indicators#

Australian Government. (2020b). *Organisational resilience.* https://www.organisationalresilience.gov.au/

Bilau, A. A., Ajagbe, M. A., Kigbu, H. H., & Sholanke, A. B. (2015). Review of shortage of skilled craftsmen in small and medium construction in Nigeria. *Journal of Environment and Earth Science, 5*(15), 98–110.

Campbell, M. A., & Gunning, J. G. (2020). Strategies to improve mental health and well-being within the UK construction industry. *Proceedings of Institution of Civil Engineers: Management, Procurement and Law, 173,* 64–74.

Ceric, A., & Ivic, I. (2020). Construction labor and skill shortages in Croatia: Causes and response strategies. *Organization, Technology & Management in Construction: An International Journal, 12*(1), 2232–2244.

Chung, J. W. Y., Wong, B. Y. M., Yan, V. C. M., Chung, L. M. Y., So, H. C. F., & Chan, A. (2018). Cardiovascular health of construction workers in Hong Kong: A cross-sectional study. *International Journal of Environmental Research and Public Health, 15*(6), 1251.

Coggins, J., Teng, B., & Rameezdeen, R. (2016). Construction insolvency in Australia: Reining in the beast. *Construction Economics and Building, 16*(3), 38–56.

Crookes, A. (2022). *Building resilience into the Australian construction industry.* https://www.laingorourke.com/thinking/building-resilience-into-the-australian-construction-industry/

Dainty, A., Green, S., & Bagilhole, B. (2007). People and culture in construction – Contexts and challenges. In A. Dainty, S. Green, & B. Bagilhole (Eds.), *People and culture in construction – A reader* (pp. 3–25). Taylor & Francis.

Debrah, Y. A., & Ofori, G. (1997). Flexibility, labour subcontracting and HRM in the construction industry in Singapore: Can the system be refined? *International Journal of Human Resource Management, 8*(5), 690–709.

Deloitte Research Center for Energy & Industrials. (2023). *2024 Engineering and construction industry outlook.* https://www2.deloitte.com/us/en/insights/industry/engineering-and-construction/engineering-and-construction-industry-outlook.html

Duckworth, J., Hasan, A., & Kamardeen, I. (2024). Mental health challenges of manual and trade workers in the construction industry: A systematic review of causes, effects and interventions. *Engineering, Construction and Architectural Management, 31*(4), 1497–1516.

Ekanayake, E. M. A. C., Shen, G. Q., Kumaraswamy, M., & Owusu, E. K. (2021). Critical supply chain vulnerabilities affecting supply chain resilience of industrialized construction in Hong Kong. *Engineering, Construction and Architectural Management, 28*(10), 3041–3059.

Ekanayake, E. M. A. C., Shen, G., Kumaraswamy, M., Owusu, E. K., & Xue, J. (2022). Capabilities to withstand vulnerabilities and boost resilience in industrialized construction supply chains: A Hong Kong study. *Engineering, Construction and Architectural Management, 29*(10), 3809–3829.

Farmer, M. (2016). *The Farmer review of the UK construction labour model: Modernise or die.* Construction Leadership Council.

Fuller, T., Hasan, A., & Kamardeen, I. (2022). A systematic review of factors influencing the implementation of health promotion programs in the construction industry. *Engineering, Construction and Architectural Management, 29*(6), 2554–2573.

Gounder, S., Hasan, A., Shrestha, A., & Elmualim, A. (2023). Barriers to the use of sustainable materials in Australian building projects. *Engineering, Construction and Architectural Management, 30*(1), 189–209.

Hasan, A., Baroudi, B., Elmualim, A., & Rameezdeen, R. (2018). Factors affecting construction productivity: A 30 year systematic review. *Engineering, Construction and Architectural Management, 25*(7), 916–937.

Hasan, A., Ghosh, A., Mahmood, M. N., & Thaheem, M. J. (2021). Scientometric review of the twenty-first century research on women in construction. *Journal of Management in Engineering, 37*(3), 04021004.

Hasan, A., & Kamardeen, I. (2023). Construction project risk management in developing countries. In G. Ofori (Ed.), *Building a body of knowledge in project management in developing countries* (pp. 320–354). World Scientific.

Kamardeen, I., & Hasan, A. (2023). Analysis of work-related psychological injury severity among construction trades workers. *Journal of Management in Engineering, 39*(2), 04023001.

Karimi, H., Taylor, T. R. B., Dadi, G. B., Goodrum, P. M., & Srinivasan, C. (2018). Impact of skilled labor availability on construction project cost performance. *Journal of Construction Engineering and Management, 144*(17), 04018057.

Lavikka, R., Kallio, J., Casey, T., & Airaksinen, M. (2018). Digital disruption of the AEC industry: Technology-oriented scenarios for possible future development paths. *Construction Management and Economics, 36*(11), 635–650.

Lawani, K., McKenzie-Govan, S., Hare, B., Sherratt, F., & Cameron, I. (2022). Skill shortage of bricklayers in Scotland. *Journal of Engineering, Design and Technology, 20*(1), 321–338.

Lengnick-Hall, C. A., Beck, T. E., & Lengnick-Hall, M. L. (2011). Developing a capacity for organizational resilience through strategic human resource management. *Human Resource Management Review, 21*(3), 243–255.

Lim, Y. S., Xia, B., Skitmore, M., Gray, J., & Bridge, A. (2015). Education for sustainability in construction management curricula. *International Journal of Construction Management, 15*(4), 321–331.

LinkedIn Learning. (2023). *Building the agile future.* https://learning.linkedin.com/resources/workplace-learning-report-2023

Loosemore, M. (2000). *Crisis management in construction projects.* ASCE Press.

Loudoun, R., & Townsend, K. (2017). Implementing health promotion programs in the Australian construction industry: Levers and agents for change. *Engineering, Construction and Architectural Management, 24*(2), 260–274.

Maor, D., Park, M., & Weddle, B. (2022). *Raising the resilience of your organization.* McKinsey & Company. https://www.mckinsey.com/capabilities/people-and-organizational-performance/our-insights/raising-the-resilience-of-your-organization

Master Builders Australia. (2023). *Future-proofing construction: A workforce blueprint*. https://master builders.com.au/wp-content/uploads/2023/05/Master-Builders-Australia-Future-proofing-construction-April-2023.pdf

Oxford Economics. (2021). *Future of construction: A global forecast for construction to 2030*. https://resources.oxfordeconomics.com/hubfs/Future%20of%20Construction_Full%20Report_FINAL.pdf

Parson, D. (n.d.). *Adversity leadership*. Department of Home Affairs, Australian Government. https://www.organisationalresilience.gov.au/Documents/AdversityLeadership.pdf

Parson, D. (2010). Organisational resilience. *Australian Journal of Emergency Management, 25*(2), 18–20.

Pettit, T. J., Fiksel, J., & Croxton, K. L. (2010). Ensuring supply chain resilience: Development of a conceptual framework. *Journal of Business Logistics, 31*(1), 1–21.

Ponomarov, S. Y., & Holcomb, M. C. (2009). Understanding the concept of supply chain resilience. *The International Journal of Logistics Management, 20*(1), 124–143.

Powell, A., & Sang, K. J. C. (2013). Equality, diversity and inclusion in the construction industry: Editorial. *Construction Management and Economics, 31*(8), 795–801.

Resilient Organisations. (2023). *Resilient organisations resilience benchmark tools*. https://www.resorgs.org.nz/about-resorgs/what-is-organisational-resilience-2/

Sapeciay, Z., Wilkinson, S., & Costello, S. B. (2017). Building organisational resilience for the construction industry: New Zealand practitioners' perspective. *International Journal of Disaster Resilience in the Built Environment, 8*(1), 98–108.

Seville, E., Brunsdon, D., Dantas, A., Masurier, J. L., Wilkinson, S., & Vargo, J. (2006). *Building organisational resilience: A New Zealand approach*. University of Canterbury. http://ir.canterbury.ac.nz/handle/10092/649

Sheffi, Y., & Rice, J. B., Jr. (2005). A supply chain view of the resilient enterprise. *MIT Sloan Management Review, 47*, 41.

Silva, R. V., De Brito, J., & Dhir, R. K. (2019). Use of recycled aggregates arising from construction and demolition waste in new construction applications. *Journal of Cleaner Production, 236*, 117629.

Smith, S. (2019). *New housing & future construction skills – Adapting and modernising for growth*. Scottish Government. https://www.gov.scot/binaries/content/documents/govscot/publications/independent-report/2019/05/new-housing-future-construction-skills-adapting-modernising-growth/documents/new-housing-future-construction-skills/new-housing-future-construction-skills/govscot%3Adocument/new-housing-future-construction-skills.pdf

Svensson, G. (2000). A conceptual framework for the analysis of vulnerability in supply chains. *International Journal of Physical Distribution& Logistics Management, 30*(9), 731–750.

Umar, T., & Egbu, C. (2018). Causes of construction accidents in Oman. *Middle East Journal of Management, 5*(1), 21–33.

Wieland, A., & Wallenburg, C. M. (2013). The influence of relational competencies on supply chain resilience: A relational view. *International Journal of Physical Distribution and Logistics Management, 43*, 300–320.

Wilkinson, S., Chang-Richards, A. Y., Sapeciay, Z., & Costello, S. B. (2016). Improving construction sector resilience. *International Journal of Disaster Resilience in the Built Environment, 7*(2), 173–185.

Williams, T. A., Gruber, D. A., Sutcliffe, K. M., Shepherd, D. A., & Zhao, E. Y. (2017). Organizational response to adversity: Fusing crisis management and resilience research streams. *The Academy of Management Annals, 11*(2), 733–769.

Zainal Abidin, N. A., & Ingirige, B. (2018). The dynamics of vulnerabilities and capabilities in improving resilience within Malaysian construction supply chain. *Construction Innovation, 18*(4), 412–432.

Index

Note: For figure citations, page numbers appear in *italics*. For table citations, page numbers appear in **bold**.

Printed in the United States
by Baker & Taylor Publisher Services